Computational Biology

The *Computational Biology* series publishes the very latest, high-quality research devoted to specific issues in computer-assisted analysis of biological data. The main emphasis is on current scientific developments and innovative techniques in computational biology (bioinformatics), bringing to light methods from mathematics, statistics and computer science that directly address biological problems currently under investigation.

The series offers publications that present the state-of-the-art regarding the problems in question; show computational biology/bioinformatics methods at work; and finally discuss anticipated demands regarding developments in future methodology. Titles can range from focused monographs, to undergraduate and graduate textbooks, and professional text/reference works.

Series web page: springer.com > Computer Science > Book Series > Computational Biology

Also in this series

Albert Burger • Duncan Davidson • Richard Baldock

Editors

Anatomy Ontologies for Bioinformatics

Principles and Practice

 Springer

Albert Burger, BSc, MSc, PhD
Duncan Davidson, BSc, PhD
Richard Baldock, BSc, PhD

Computational Biology Series ISSN 1568-2684
ISBN: 978-1-84628-884-5   e-ISBN: 978-1-84628-885-2

British Library Cataloguing in Publication Data
A catalogue record for this book is available from the British Library

Library of Congress Control Number: 2007940841

Printed on acid-free paper

9 8 7 6 5 4 3 2 1

Springer Science+Business Media
springer.com

Foreword

This book is about the ontology of anatomy. With respect to the individual fields of ontology and anatomy, the ontology of anatomy has aspects of both an old and a new topic area. A new aspect for anatomy is that the ontology of anatomy brings medicine together with molecular biology and its related subjects. Similarly, for the field of ontology, biomedical informatics has seen an explosion in the use of ontologies and ontology-like resources. There has been a particular interest in ontologies for human anatomy and also the anatomy of other types of organism. This explosion has pushed the field of ontology into the limelight, with new practical applications of ontology being developed and new formalisms to accommodate the things that biologists need to say.

The ontology of anatomy covers a broad spectrum of life sciences, but why should medics and geneticists, molecular biologists, etc. really be so interested in anatomy? For medics, the reason for this interest is seemingly self evident—medical things happen to bodies and bits of the body. Surgical procedures are carried out on body parts; illnesses and injuries happen to the body and parts of the body. So, if we are to describe medicine, we need to start with anatomy.

For molecular biologists, it is often not immediately obvious that biology and medicine join at the level of anatomy, especially in the study of disease processes and the treatment of disease, particularly through drug action. Genes (mostly) encode proteins that are used within and without cells; cells aggregate in tissues and tissues aggregate in organs, which in turn can be assembled into a body. As the focus of study moves away from the single sequence or gene, anatomy from tissue to whole body, becomes more important. Indeed, this wider focus also brings in systems biology; a field in which biologists begin to model systems that, after all, work within anatomical regions.

For many years however, the orientation of much work in molecular biology has been toward a single sequence of some particular biological interest. Single proteins and their gene etc. were studied, often in the context of "the cell". Certainly the contexts recorded in many standard databases such as Uniprot have a subcellular focus. The advent of high-throughput experimental techniques, especially those in transcriptomics, has radically changed this focus. It is now possible to measure levels of transcription for huge numbers of genes at a given time. Given this ability, there are new generic questions that the molecular biologist can now ask:

- What is the difference in transcription between cell in state x and cell in state y? This state will often be disease, which of course, happens in a part of the body.
- What genes are expressed at different stages of development? Again, what is it that develops? It is tissues, organs and parts of the body.
- What genes are expressed in different tissues etc. etc.

Having established that anatomy is relevant to a broad spectrum of the medical and life sciences, the next question is; why the interest in ontology for representing anatomies? Ontology is very much becoming a mainstream activity in bioinformatics and has been important in medical informatics for many years. It is, however, useful to remind ourselves of why this is so. Biology is unlike physics and much of chemistry in that, although it contains many laws and models, few of these are reduced to a mathematical form. It is not possible to take a sequence of amino acids that represents a protein, apply some formula, and derive a set of characteristics such as accurate three-dimensional shape, functionality, forms of modification, etc. While it is easy to compare, for instance, nucleic acid or polypeptide sequences between bioinformatics resources, the additional knowledge component of these resources is very difficult to use computationally. A similar argument can be applied to medicine—we are not yet diagnosed by mathematics!

Instead of mathematical laws, bioinformaticians use what they understand about characterised entities to make inferences about uncharacterised entities. What we are doing is making these inferences based on a transfer of our knowledge between similar entities.

Biological and medical knowledge, or the symbols used to represent this knowledge, is highly heterogeneous and not computationally amenable, both for humans and computers, because the knowledge is represented in a wide variety of lexical forms. In computer science, ontologies are a technique or technology used to represent and share knowledge about a domain by modelling the entities in that domain and the relationships between those entities. These relationships describe the properties of those things; in essence, what it is to be one of those things in the domain being modelled. An ontology is a set of axioms, often in a logical formalism, that is used to provide a vocabulary that represents a conceptualisation of reality or simply reality (after Guarino[1]). The labels used for the things and their properties in an ontological model can provide a language for a community to talk about their domain.

---

[1] Guarino, N., Formal Ontology in Information Systems, in N. Guarino (ed.) Formal Ontology in Information Systems. Proceedings of FOIS'98, Trento, Italy, June 6-8, 1998. IOS Press, Amsterdam, pp. 3-15.

By agreeing on a particular ontological representation, a common vocabulary or unified understanding can be used to describe and ultimately analyse data. The knowledge about the domain becomes much easier to handle as the same things are referred to in the same manner across the resources in which that knowledge is stored. This provides a common way to talk about what we understand about gene product function, disease, experiments, ways of dying, and so forth. By talking about the same things in the same way across resources, we improve query, analysis and billing.

This need for a vocabulary is where ontology, in the computer science form, has started. Yet, as ontology becomes more formal, both in its distinctions and its representation, it is possible to manipulate the knowledge captured in the ontology in order to make inferences. A formal, logical representation of knowledge means that knowledge can be handled "mathematically", in a form akin to how we are used to manipulate symbols algebraically. Formality, both at the knowledge representation language level and the philosophical-ontological level is a stern friend.

In computer science, ontology is not the same as ontology in philosophy. It should be informed by philosophy, but not necessarily bound by it. It would be a mistake for any community at either level to let dogma, particularly philosophical viewpoints, get in the way of practical usefulness. It is salutary to remember that we are attempting to facilitate science through our efforts to share knowledge and we should not be distracted by pre-scientific arguments. Nevertheless, philosophy of science, regardless of which flavour is chosen, can bring highly useful rigour to the modelling process.

Why do we need a whole book on the simple task of cataloging the parts of the body? Many disciplines are interested in anatomy, but each has its own perspective and will re-use the same words in different ways. To a surgeon the pericardium is part of the heart, but this is not how the developmental anatomist views the antomy of the heart. Regardless of philosophical stance, such differing viewpoints as normal/abnormal, developmental and structural need to be accommodated if ontology of anatomy is to be useful.

We can already see the full range of ontologies or ontology-like resources in anatomy; from simple controlled vocabularies, through structured controlled vocabularies to logical formal descriptions of the world. All can be useful in the appropriate context, yet the goals are not that different—to share what we understand of a domain, especially in a way computers can use. In this book the whole spectrum will be seen, but with an emphasis on enabling science.

Manchester,
November 2007

Robert Stevens
Department of Computer Science
The University of Manchester

Preface

Bioinformatics as a discipline has come of age and there are now numerous databases and tools that are widely used by researchers in the biomedical field. New developments in both areas, computing and biology/medicine, however, pose new challenges. In particular, the next generation world wide web is expected to be one of the key information technology advancements of this decade, while Translational Science and Systems Biology are recognised as key fields in today's medical and life sciences.

The successful development of future bioinformatics and medical informatics applications will depend heavily on an appropriately formalised representation of domain knowledge, and as described by Robert Stevens in the foreword, one such key domain knowledge is that of anatomy.

There exists a substantial body of work on anatomy ontologies, ranging from the more philosophical considerations of mereology to state of the art 3D visualisations and descriptions of anatomy for human and model organisms. The corresponding literature, however, is scattered across a large number of scientific conference proceedings, journals and books for various target audiences, such as computer scientists, bioinformaticians, biologists, medics and philosophers. This book provides a timely, unique and first of its kind collection of papers about anatomy ontologies. It is interdisciplinary in its approach, bringing together the relevant expertise from computing as well as biomedical studies, and covers the more theoretic as well as the applied aspects of the field. Whilst taking account of important work in the past, it also covers the latest developments in the field of anatomy ontologies.

The book is primarily aimed at readers who will be involved in developing the next generation of IT applications in the areas of Life Sciences, Bio-Medical Sciences and/or Health Care. Specifically, the book is relevant to: 1) those who will further develop anatomy ontologies, 2) those who will use them (annotators and scientists wanting to query the relevant databases), and 3) informatics staff involved in the actual development of relevant software applications. The goal is to provide the reader with a comprehensive understanding of the foundations of anatomical ontologies and the state of the art in terms of existing tools and applications that are using or planning to use these ontologies.

There are four major parts to the book. The first focuses on existing anatomy ontologies for human, model organisms and plants, complemented by a chapter on disease ontologies. Part II describes systems and tools dealing with linking anatomy ontologies with each other and other on-line resources, such as the biomedical literature. Anatomy in the context of spatio-temporal biomedical atlases is discussed in part III. A number of modelling principles are presented in part IV, which also concludes the book with a chapter on recent efforts to develop a common anatomy reference ontology (CARO).

We would like to thank first and foremost all authors for their efforts. This book, of course, would not have been possible without their kind contributions. Thanks also go to Karen Sutherland, who has been helping locally with the editing process. Finally, we are grateful for the support from the publishers, Springer Verlag, particularly Helen Callaghan, Catherine Brett, Joanne Cooling, Jeffrey Taub and Wayne Wheeler, who have been incredibly patient with us.

Edinburgh,   *Albert Burger*
November 2007   *Duncan Davidson*
                                                                                             Richard Baldock

Contents

List of Contributors

Dr. Stuart Aitken
Artificial Intelligence Applications
Institute,
University of Edinburgh, UK
stuart@inf.ed.ac.uk

Dimitra, Alexopoulou
Biotechnological Centre,
Dresden University of Technology,
Germany
dimitra.alexopoulou
@biotec.tu-dresden.de

Prof. Richard Baldock
Human Genetics Unit,
Medical Research Council, UK
Richard.Baldock
@hgu.mrc.ac.uk

Prof. Jonathan Bard
Department of Biomedical Sciences,
University of Edinburgh, UK
j.bard@ed.ac.uk

Prof. Thomas Bittner
Department of Philosophy,
Department of Geography,
New York State Center of Excellence in
Bioinformatics and Life Sciences,
National Center for Geographic
Information and Analysis,
University at Buffalo, USA
bittner3@buffalo.edu

Dr. Albert Burger
Human Genetics Unit,
Medical Research Council, UK
and
Department of Computer Science,
Heriot-Watt University, UK
ab@macs.hw.ac.uk

Yin Chen
National e-Science Centre,
University of Edinburgh, UK
yin@nesc.ac.uk

Dr. Duncan Davidson
Human Genetics Unit,
Medical Research Council, UK
Duncan.Davidson@hgu.mrc.ac.uk

Heiko Dietze
Biotechnological Centre,
Dresden University of Technology,
Germany
heiko.dietze@biotec.tu-dresden.de

Dr. Maureen Donnelly
Department of Philosophy,
New York State Center of Excellence in
Bioinformatics and Life Sciences,
University at Buffalo, USA
md63@buffalo.edu

Prof. Mark H. Ellisman
National Center for Microscopy and
Imaging Research,
Center for Research in Biological Struc-
ture and Development of Neurosciences,
University of California, San Diego,
USA
mark@ncmir.ucsd.edu

Lisa Fong
National Center for Microscopy and
Imaging Research,
Center for Research in Biological Struc-
ture and Development of Neurosciences,
University of California, San Diego,
USA
lfong@ncmir.ucsd.edu

Dr. Georgios V. Gkoutos
Department of Genetics,
Cambridge University, UK
g.gkoutos@gen.cam.ac.uk

Prof. Louis J. Goldberg
Departments of Oral Biology and Oral
Diagnostic Sciences,
School of Dental Medicine,
New York State Center of Excellence in
Bioinformatics and Life Sciences,
University of Buffalo, USA
goldberg@buffalo.edu

Dr. Amarnath Gupta
San Diego Supercomputer Center,
University of California, San Diego,
USA
gupta@sdsc.edu

Dr. Melissa A. Haendel
Zebrafish Information Network
University of Oregon, USA
mhaendel@uoneuro.uoregon.edu

Joerg Hakenberg
Biotechnological Centre,
Dresden University of Technology,
Germany
joerg.hakenberg
@biotec.tu-dresden.de

Dr. Katica Ilic
The Arabidopsis Information Resource,
Department of Plant Biology,
Carnegie Institution, Stanford, USA
kat23ica@gmail.com

Prof. Elizabeth A. Kellogg
Department of Biology
University of Missouri - St. Louis, USA
tkellogg@umsl.edu

Prof. Patrick Lambrix
Department of Copmputer and
Information Science,
Linköpings Universitet, Sweden
patla@ida.liu.se

Stephen D. Larson
National Center for Microscopy and
Imaging Research,
Center for Research in Biological Struc-
ture and Development of Neurosciences,
University of California, San Diego,
USA
slarson@ncmir.ucsd.edu

Prof. Paula M. Mabee
Department of Biology,
University of South Dakota, USA
pmabee@usd.edu

Prof. Maryann E. Martone
National Center for Microscopy and
Imaging Research,
Center for Research in Biological Struc-
ture and Development of Neurosciences,
University of California, San Diego,
USA
maryann@ncmir.ucsd.edu

Dr. Jose L.V. Mejino Jr.
Structural Informatics Group,
Department of Biological Structure,
University of Washington School of
Medicine, USA
mejino@u.washington.edu

Asif Memon
National Center for Microscopy and
Imaging Research,
Center for Research in Biological Structure and Development of Neurosciences,
University of California, San Diego,
USA
amemon@ncmir.ucsd.edu

Chris J. Mungall
Lawrence Berkeley National Laboratory, USA
cjm@fruitfly.org

Dr. Fabian Neuhaus
Department of Philosophy,
University at Buffalo, USA
fneuhaus@web.de

Dr. David Osumi-Sutherland
FlyBase,
University of Cambridge, UK
djs93@gen.cam.ac.uk

Prof. Alan Rector
Department of Computer Science,
University of Manchester, UK
rector@cs.man.ac.uk

Dr. Seung Yon Rhee
Department of Plant Biology,
Carnegie Institution, Stanford, USA
rhee@acoma.stanford.edu

Prof. Cornelius Rosse
Structural Informatics Group,
Departments of Biological Structure and
Medical Education and Biomedical and
Health Informatics,
School of Medicine,
University of Washington,
Seattle, USA
rosse@u.washington.edu

Dr. Björn Rozell
Department of Laboratory Medicine,
Karolinska Institute,
Stockholm, Sweden
bjorn.rozell@ki.se

Dr. Paul Schofield
Department of Physiology, Development and Neuroscience,
Cambridge University, UK
ps@mole.bio.cam.ac.uk

Prof. Michael Schröder
Biotechnological Centre,
Dresden University of Technology,
Germany
ms@biotec.tu-dresden.de

Prof. Barry Smith
Department of Philosophy,
University at Buffalo, USA
phismith@buffalo.edu

Prof. Peter F. Stevens
Missouri Botanical Garden, St. Louis,
USA
and
Department of Biology
University of Missouri - St. Louis, USA
peter.stevens@mobot.org

Dr. Robert Stevens
Department of Computer Science,
University of Manchester, UK
robert.stevens
@manchester.ac.uk

He Tan
Department of Copmputer and
Information Science,
Linköpings Universitet, Sweden
hetan@ida.liu.se

Joshua Tran
National Center for Microscopy and
Imaging Research,
Center for Research in Biological Struc-
ture and Development of Neurosciences,
University of California, San Diego,
USA
jtran@ncmir.ucsd.edu

Thomas Wächter
Biotechnological Centre,
Dresden University of Technology,
Germany
thomas.waechter
@biotec.tu-dresden.de

Willy Wong
National Center for Microscopy and
Imaging Research,
Center for Research in Biological Struc-
ture and Development of Neurosciences,
Univ. of California, San Diego, USA
wwong@ncmir.ucsd.edu

Dr. Ilya Zaslavsky
San Diego Supercomputer Center,
University of California, San Diego,
USA
zaslavsk@sdsc.edu

Part I

Existing Anatomy Ontologies for
Human, Model Organisms and Plants

1

Anatomical Ontologies for Model Organisms: The Fungi and Animals

Jonathan Bard

Summary. This chapter reviews how the richness of animal and fungal anatomy can be incorporated into formal ontologies so that knowledge of tissue organisation can be made accessible both to biologists and to other computational resources. The first part of the chapter focuses on the anatomical and bioinformatics principles behind making these ontologies and the problems that have to be solved before they can be made. The next section reviews the anatomical ontologies currently available for the main model animal and fungal organisms. The final section focuses on the current and future uses of these ontologies, together with the problems of curating them and publishing up-to-date versions. The chapter ends with a plea for more and better software to make anatomical ontologies more accessible to the general biological community and so more useful to it.

1.1 Introduction

Ontologies are domains of knowledge formalised so that they can be "understood" by computers and the first section of this chapter discusses the anatomical and informatics problems that have to be solved if anatomy is to be formalised in this way. The second section summarises the major fungal and animal ontologies that are currently available (for plants, see Chapter 2). These sections are aimed at biologists who are not involved in the business of making anatomy ontologies but are interested in these resources and in their intellectual underpinnings. The last section of the chapter considers the uses to which these ontologies are being and could be put, and this is directed as much to those working in the area as to those who just wish to find out more about anatomical ontologies. Perhaps the key point made there is that, while these ontologies were originally intended for organism databases, they can also be used as knowledge resources and for annotating other data with appropriate tissue IDs and so expand computational access (e.g., for data mining). Such annotation is not currently easy to do as the field lacks the appropriate tools, and if these are not provided soon, a great deal of hard work will be inadequately used.

And it does require a lot of hard work to undertake the construction of the full anatomical ontology of an adult animal, let alone its developing embryo: building such an ontology is just too complicated and difficult for it to be enjoyable, and I sometimes think that no one in their right mind would even start making the anatomical ontology of an animal if they knew what they were getting themselves into! Early on in the development of model organism databases, however, it became clear that, if gene expression (G-E) and other tissue-associated data were to be stored and phenotypes adequately described, the field would need access to the tissue names of the main model organisms organised within formal ontologies. Once this was done, these anatomical ontologies could also be used both for accessing their intrinsic knowledge and for annotating (via their unique IDs) other data sets and theoretical models.

It soon however became apparent that formalising anatomy in ontologies was not a trivial exercise for both anatomical and computational reasons. The anatomical reasons centred around the complexities of formalising tissue organisation, while the bioinformatics reasons came from linking this information to the relational databases: these usually hold material in tabular form while anatomy, based as it is on parts having parts having parts, is naturally hierarchical. While it is formally possible to store hierarchies in tabular formats, it is very clumsy and looking for the parts of parts of parts involves a great deal of recursion, something that is not natural to searching algorithms. This difficulty meant that another way of handling anatomy had to be found and here, as in much else in the production of databases for model organisms, the *Drosophila* field led the way, and the way led to the area of ontologies (e.g. [2]), a subject well-established in computer science where it had for some time a strong formal role as it provides a natural way of both formalising and using complex knowledge that can be represented in hierarchical and graphical ways. This is because ontologies are collections of integrated and linked "fact" triads of the general form *<term><relationship><term>* and an example might be *<femur><is_part_of><leg>* (such triads form the basis of the natural language of the semantic web [7]). Bioinformaticians have, for excellent reasons, chosen to handle their ontologies in a way that is more informal but also more intuitive than their colleagues from informatics. The main reason for this is that the database customers who are mainly experimentalists want to be able to access data intuitively and do not take kindly to learning curves when they are not in their laboratories.

1.2 Formalising Anatomy into Ontologies

Although anatomy ontologies have a range of uses, most were initially designed to handle G-E and other tissue-associated data (e.g. phenotypic) and so be linked to relational databases. This in turn meant that the knowledge within an anatomy ontology would be used to interpret information from those databases. Thus, for example, an inquiry about the genes associated with a complex tissue (e.g. the limb) would use the knowledge of its parts that was held in the ontology to collate the data for the answer. Collating and integrating the anatomical information in a format that is easy to combine with database schemas and search algorithms poses problems that the average biologist does not want to know about - he or she just wants easy access to data associated with tissues and is more than happy to ignore any technical complexities involved in formalising the anatomy.

In fact, even the biological side of things is difficult and involves thinking quite seriously about the nature of anatomy, tissues and what is involved in formalising their organisation so that one is quite clear about the nature of the anatomical knowledge that one is handling. The key problem here lies in deciding the nature of the relationships that underpin our knowledge about tissues. Indeed, and although "normal" anatomists would not like to admit it, the business of preparing anatomical ontologies has required us to analyse the structure of anatomical knowledge at a deeper level than has hitherto been necessary. The implicit result of this analysis has been, as we will see, that most people involved in making anatomy ontologies have shied away from handling the full richness on offer. Nevertheless, the explicit result has been that the organisational problems of anatomy have been solved at a level adequate for practical purposes, for the moment at least.

Handling all this computationally has been more straightforward: once the required anatomical knowledge has been made explicit and the required tissues and relationships listed, anatomists have needed only to store them appropriately in a computer-readable file (a standard editor is available for this) and use the code for annotation, linkage and searching. This is conceptually more intriguing than biologists would like to think and it is an important function of this chapter to explain why the informatics side of things is interesting as well as useful.

1.2.1 The Anatomical Problems

A classical anatomist naturally thinks of any organism as a large number of tissues each organised into a hierarchy of smaller parts. Computational approaches to anatomy push this idea a little further. In principle at least, they start by listing every single tissue in the organism (this list comprises the *anatome*[1]; [1]). An additional concept that is sometimes useful is the *abstract organism*; this is the collection of the anatomes for the adult and each developmental stage and thus includes embryological age ([4]). The next step is to decide on the relationships that will be used to organise all these tissues within a hierarchical formalism to meet a set of needs (such as archiving and retrieving G-E data for mouse development).

The basic problems in formalising anatomy can be seen by asking the question: how should we organise the tissues of the early (say Theiler stage 22 or embryonic day 13.5) mouse forelimb so that this anatomical knowledge can be used to annotate G-E data for archiving? This tissue, about four days after it first appeared, is bounded by a jacket of ectoderm with a specialised apical ectodermal ridge. This jacket contains the differentiating mesenchyme that will form bones, joints, muscles, tendons

---

[1] Note that the *anatome* is more than space-filling as it includes, for example, the forearm and all of its parts, the hand and all of its parts (digits, nails, metacarpals etc.) with all tissues being given equal status.

and associated mesenchyme, together with the domain of mesenchyme underlying the ridge known as the "progress zone" as it provides new mesenchyme through active proliferation. Close examination of all this mesenchyme (Figure 1.1) shows that it contains many condensations that are probably pre-muscle masses but whose future identity cannot be recognised, even by experts. Worse, few of these condensations have sharp boundaries. In practice, it is impossible to include in the anatome (or list of parts) any tissues that cannot be uniquely identified and this includes many presumptive tissues that are visible together with the early nerves and minor blood vessels that are not highlighted by standard histology.

Associated with the E13.5 forelimb is a great deal of G-E data (data for ∼9000 genes is stored in GXD, the mouse G-E database[2], some highly detailed (e.g. humerus-associated data from *in situ* work) and some less so (e.g. limb-associated data from northern blots) that has had to be archived. The wrong approach would be to make one table of all these tissues, annotating each with its genes, and another table of genes each annotated with the tissues in which it is expressed. Such an approach hides key knowledge about the anatomical organisation and makes it impossible to answer questions along the lines of "what genes are expressed in bones?" - the list of tissues at no point mentions that the humerus, radius and ulna are bones (actually, cartilage condensations at this early stage). It is also hard to answer the questions "what genes are expressed in the developing lower limb?" because the table does not include any organisational knowledge about which tissues are localised to that region.

So the first obvious problem is how to include organisational information about tissues and the second is how much of it should be included. Indeed, it sometimes feels to the author that there is an infinite amount of anatomical information that might be useful for archiving and searching.

The Boundary Problem

Actually, there is an even more basic problem to solve and that is in defining a tissue, and perhaps we should start with this. It is usual to define standard tissue morphology on the basis of haematoxylin and eosin or other simple, non-specific staining and it is usually easy to point to the middle of a tissue in a histological section and label it. For adult tissues, and for those developmental tissues bounded by an epithelium, it is also easy to draw the boundary. The situations becomes much harder during development when tissue rudiments are still becoming defined and boundaries are ill-formed (Figure 1.1). A similar problem arises with the labelling of high-level structures: there are no biologically meaningful or visually apparent boundaries between the limb and the region of body from which it extended (Figure 1.1), although it is sometimes required (e.g. for a graphical model). This is actually quite a common problem: there

---

[2] www.informatics.jax.org/menus/expression_menu.shtml

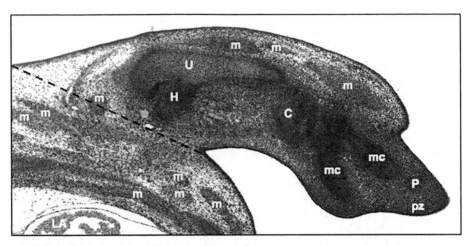

Fig. 1.1. Limb-bud micrograph: Section of the forelimb of an E13.5 mouse stained with haemotoxylin and eosin, with the major mesodermally-derived tissues marked. The dotted line represents a possible division between body wall and forelimb, but it is not convincing! The many mesenchymal condensations (m) cannot be identified nor their fate recognised; furthermore neither they nor many other of the labelled have sharply delimited boundaries; indeed, the domain known as the progress zone (pz) where proliferation and patterning takes place has the superficial ectoderm as its only boundary. It is also worth noting that it is not possible to distinguish nerves or any except the major vessels. (C: carpal mesenchymal condensation; H: cartilage primordium of the distal region of the left humerus; H: cartilage condensation of the left ulna; LA: wall of left atrium of the heart; mc: metacarpal mesenchymal condensations; P: phalange mesenchymal condensation. Mag: 30x). I thank Matt Kaufman for the section.

is no defined boundary between, for example, the ventricles of the heart and the interventricular septum, and even the boundary between the handplate and forearm is not really well-defined as the digit ligaments extend into the forearm. A possible biological solution in these cases would be to identify a sharp and relevant G-E border (this could mark the progress zone of the handplate) and use that as a boundary, but the choice of gene would be arbitrary and another more relevant gene might later be discovered and no one is going to build a database that is intrinsically unstable.[3]

There is in fact no good biological solution to the boundary problem, even in adults, and there will always be archiving problems when a tissue such as a muscle, vessel or a nerve bridges two regions (e.g. trunk and limb) and so has a part in each. This problem will similarly arise where tissues of similar cell type and morphology

---

[3] There is a deeper point here: in making ontologies it is usually best to make the factual base as simple as possible and to exclude information that is essentially orthogonal. Such information is best handled through annotation. In my view, anatomy ontologies should be restricted to simple morphological information, and associated information (e.g. cell types and molecular data) handled through linkage.

abut (e.g. the septa of the heart merges with the myocardium) and there is no well-defined anatomical feature to mark the boundary. The only practical solution for these difficult problems is to include the names in the ontology, trust the biological judgement of the person making the annotation and accept that there is going to be a bit of imprecision in how boundaries are read (as in all searching of archived data, the user needs to confirm from the literature the exact status of any data that seem important).

The Organisational Problem

As for the organisational problem, one soon realises that, if one only tries to include tissue names in the database, one is not only imposing difficult grouping problems on the system but also excluding important anatomical knowledge. In the limb case, for example, one might want to group the bone rudiments under skeleton and this information is not implicit in the tissue names but has to be added by the curator. In general, it makes a great deal of practical sense to add organisational knowledge to lists of tissues, much as has been done in constructing traditional Linnaean taxonomy where species are grouped into a hierarchy of families, phyla, kingdoms etc., even though there can sometimes be an artificial degree of organisation here.

In practice, all ontologies of animal anatomy are primarily based around organ systems (nervous system, urogenital system, circulatory system, etc.) that are incorporated in hierarchical ontologies and curators have spent a considerable amount of effort in adding a great deal of organisational knowledge to the bare list of tissue names. It should be noted that, although the tissues associated with a system are not usually space-filling (e.g. it is not practical to include all the associated vessels and nerves limb) and may not even be connected (the glandular system), they are usually described by a slightly artificial *part_of* relationship (see below for details).

The Complexity Problem

Complexity here does not reflect the numbers of tissues that the ontology will include, but the types and numbers of relationships connecting those tissues. The simplest case is when each tissue is seen as *part_of* a single higher level grouping (the humerus is *part_of* the skeleton) but this approach excludes some obvious knowledge (the humerus is *part_of* the arm); ideally one should include both relationships as well as the fact that the humerus *is_a* bone. One might also wish to include developmental knowledge (bone *derives_from* mesenchyme) and geometric knowledge (the humerus *links_to* the radius and the ulna), but one is immediately aware that the more knowledge that is included, the harder will the ontology be to make - and it is an interesting question as to whether or not one should try to include, for the first pass at least, knowledge that is unlikely to be needed in the context for which the ontology is to be used.

In its simplest form (e.g. an *is_a* classification), terms in an ontology have a single parent and are connected by a single relationship, so that the fact triads combine to make a standard hierarchy. Such a hierarchy will also result from allowing a tissue to be *part_of* only one larger tissue. If, however, we only include the mouse femur as *part_of* the mouse leg and not as *part_of* the mouse skeleton, we are obviously oversimplifying our knowledge of mouse anatomy, and it would be better if the femur was part of two triadic relationships (see below for the informatic implications). More information could be included if other relationships were used so that in a developmental ontology, for example, we could include the relationship that the *Drosophila* antenna *derives_from* the appropriate imaginal disc. The extent to which it matters that an ontology is kept simple depends on the use to which the ontology would be put — and curators have to provide what they think that users will need (actually, they should be a bit more ambitious to allow the ontology to be used for new roles!).

The *Part_of* Problem

It should not be thought that this, the core anatomical relationship is easy to define in any intuitive way. Indeed, there is whole branch of formal logic associated with this relationship that is known as mereology ([11, 12, 13]). Table 1.1 gives just some of the possible (and sensible) uses of the relationship, and there are other, more subtle meanings. In practice, and in the context of G-E work, the meaning of *part_of* that is most useful [4] has two components: first, that any property (e.g. an expressed gene) associated with a child can also be associated with the parent (the relationship implies upwards propagation), and, second, that the relationship is transitive so that, if A is *part_of _B* and B is *part_of _C*, then A is *part_of _C*. This meaning clearly covers cases 1-3, but the final case is more subtle and it may well be better to view the bone marrow as part of the haematopoietic system. There is no right way for the curator to make this choice, but, once made, the user has to know what that choice is.

1. The humerus is *part_of* the forearm	contained, in the natural, biological sense
2. The adrenal is *part_of* the glandular system	part of a distributed group
3. Lymphocytes are *part_of* the blood	part of an inseparable mixture
4. The bone marrow is *part_of* the bone	contained, but in an anatomically ambiguous way

Table 1.1. Uses of *part_of* (Note: It is not actually clear whether the use of *part_of* in item 3 is appropriate for an ontology of anatomical tissues as lymphocytes are cell types and do not form a tissue. It is probably better to annotate blood with its constituent cell types (see below) using the relationship *has_cell_type* .

The other point that should be noted comes from example 2, that the adrenal is *part_of* the glandular system. This relationship looks suspiciously like the *is_a* or class relationship in that the adrenal is a gland, but it differs in one important way: the *is_a* relationship carries downward propagation of properties (more associated knowledge is added as one goes down the hierarchy), whereas the *part_of* relationship carries upwards propagation (if a gene is expressed in the adrenal, it is also expressed in the glandular system). It is thus true, if a bit bizarre, that the *part_of* relationship that is appropriate for handling G-E information implies no sense of geometric containment.

The Granularity Problem

The immediate problem facing anyone setting out to make an anatomy of a complex animal is how many tissues to include. The curators of the *C elegans* anatomical ontology are to be envied as theirs is the only model animal where this is not a problem: as the worm only has about 1000 cells there is a natural upper limit to the number of tissues. Curators of complex organism databases have had to decide how many tissues need to be incorporated into their anatomical ontologies - and this is not a trivial problem given the number of named muscles, blood vessels, nerves and obscure bits and pieces eligible to be included (my own favourite here is the zonule of Zinn). In addition, there are the many, minor un-named tissues that may need to be included, particularly for vertebrates (e.g. the mesenchyme associated with the various parts of the early gut and the early mesenchymal condensations illustrated in Figure 1.1). In short, it is usually impossible for the anatomical ontology of an animal to be complete and the extent to which small, obscure tissues are included is at the discretion of the curator who should aim to judge the needs of the users of the ontology. Under all circumstances, however, curators of even low-granularity ontologies should at least aim to make them spacefilling (at a lowish resolution) so that all domains of gene expression can be included in a way that will be useful.

Even then, there is a further problem which implicitly derives from the nature of data collection: in the *bilateria*, there is a strong degree of mirror symmetry and if, for example, data is to be collected from tissue sections, it is usually impossible to tell whether a tissue is from the left or the right side of the embryo as the polarity of the section is rarely obvious. The usual practice here is to assign left and right to tissues where there is visually recognisable L/R asymmetry (e.g. ventricles) but to include symmetric tissues only once (e.g. somites), with the implicit understanding that the one term (and ID) covers both left and right items. While this is not entirely satisfactory, it does not lead to annotation errors provided, of course, that any associated properties (e.g. gene expression) are also symmetric.

The Variability Problem

Ontologies incorporate idealised knowledge and cannot handle general variability in any easy way. The exception is gender dimorphism: this is straightforward to deal with by saying that the anatome contains male, female and common tissues within the reproductive system. In practice, this leads to no ambiguity in viewing male and female reproductive systems as equivalent but distinct in the reproductive system, with a given animal having the appropriate subset of tissues. Much harder to represent is the real variability that can arise through mutation or developmental abnormality and that could lead to extra or missing tissue items in the anatome or to geometric variability (e.g. in the course of a nerve) that would lead to ambiguity in, say, *next_to* or *linked_to* relationships. Here, the curator can only handle what is viewed as the norm and the user has to accept this limitation - organising knowledge about phenotypic variation is handled by PATO, the phenotype and traits ontology ([6]; bioontology.org/wiki/index.php/PATO:Main_Page, obo.sourceforge.net).

The Incompleteness Problem

Ontologies handle knowledge, and that knowledge is not always complete. One reason reduces to the granularity problem: minor tissues might (accidentally or deliberately) have been excluded from the ontology and users need to realise this and inform a curator if an important tissue is missing from the ontology. A harder case occurs where there is ambiguity: a curator who plans to incorporate developmental lineage into an ontology but realises that the origin of a particular is not known has to decide on how to handle this. The problem is that a user may assume that the knowledge within an ontology is complete so that any relationship not included implicitly means that this relationship does not exist. In practice, users need to be aware that fact triads within the ontology are (should be!) true, but that the set of included facts may not be complete and no inferences should therefore be made from omissions. If a user does discover or know of a missing or incorrect relationship, he or she should immediately email that information to the curator who should regularly issue updated versions of the ontology – and hope that anyone who integrates that ontology into other computational resources also updates their version. Here, communication is all.

In short, before making an anatomical ontology, and independent of whether it is to cover developmental stages as well as the adult, a curator has to answer the following questions:

1. Which tissues should be included so that the full volume of the organism should be covered, even if the granularity is to be coarse (as a simple example: the eyeball could be included but not its constituent parts)?
2. How best to organise the tissues?
3. How to handle boundaries when they are not obvious?
4. What relationships (and hence what sorts of knowledge) to include?
5. How much complexity should be incorporated?

There is no unique answer to these questions and what is possible for a simple organism may be unrealistic to one more complex. The best working rule in making an ontology is to work to the purpose for which the ontology is being designed: a curator must ensure that these needs are met, and preferably in a way that users will find intuitive to handle. Curators should however always be aware that, with increasing knowledge or with more sophisticated databases being made, their ontology may well need to be expanded.

1.2.2 Ontology Informatics

The details of ontology bioinformatics are handled elsewhere in this book, but a few points need to be mentioned here to make the rest of this chapter comprehensible. Ontologies represent areas of knowledge formalised as mathematical graphs. The knowledge is built up by integrating and linking "fact" triads, each of which is a pair of terms (synonyms are nodes, leaves, concepts) linked by a relationship (or edge). The simplest case occurs when the relationship is directed (one-way, e.g. *part_of*) and each child has a single parent: here the graph reduces to a simple hierarchy, and the classic example is Linnaean taxonomy which is actually based on the *is_a* (or "set") relationship. A more complex type of graph occurs where a term is linked to two or more parents by a directed relationship (e.g. the humerus is *part_of* the skeleton and *part_of* the arm) and here the result is what is known as a Directed Acyclic Graph (or DAG) which has the property of having no pathway through it that is circular. The general case occurs when the relationship is reciprocal and not directed (e.g. *next_to*) and the resulting graph may well have circular paths (consider the stations on the circle line of the London Underground).

A key property of biological ontologies is that each term has a unique identifier (ID) that can be used for annotation. This ID has the form <prefix><integer> where the prefix indicates the ontology name and the integer is a unique number associated with the term. This ID is used to link ontologies with databases and to search databases either directly or interoperatively. It is worth noting that many non-biological ontologies do not have such IDs, but assign a unique name to each concept and it is this name that is used for archiving etc. The reason for bio-ontologies using IDs is really because the full name of a tissue will often be cumbersome (e.g. *the mesenchyme associated with the anterior part of the future duodenum at Theiller stage 19*) while an abbreviated term name will be far easier to handle (e.g. associated mesenchyme), with its context defined by its path(s) within the graph. It is thus sensible for ontology concepts and relationships to be uniquely and carefully defined, partly for clarity but mainly to ensure that, when the logic of the ontological relationships is used computationally (e.g. if a request is made for the genes expressed in the developing mouse forelimb at Theiller stage 20, the search is made only over all those tissues that are part of the forelimb at that stage), correct results are obtained (see below). In practice, however, it has to be said that complex anatomies have too many tissues for anyone to have time to write the definitions that would, in any case, be abstracted from standard texts.

A full anatomical ontology contains all triads of the tissues (with their definitions and IDs) and relationships, together with a small amount of meta-data (curator, date, version number etc.); it may also include synonyms and annotated links to other resources (e.g. the ID of the equivalent term in another ontology) and can include non-anatomical data (e.g. cell types associated with tissues) although this may not be wise — my preference is to handle such orthogonal material through look-up-tables as these are easier to curate. One difference between informal biological ontologies and fully formalised ontologies is that, for the latter, each concept in defined within the ontology by an *is_a* relationship and this can help with using the internal logic of the ontology in handling queries. Most anatomy ontologies do not include this set information, partly because of the work involved and partly because of the difficulty (it is hard to formalise these relationships for all the tissues of the adult mouse how would one handle the lens of the eye here?), but mainly because it is not obviously useful for many tissues.

There is a particular problem with developmental ontologies that should be mentioned here and that concerns how tissues should be accessed when they extend across several stages of development. The difficulties particularly occur in rich ontologies where a tissue occurs in several stages and a viewer may wish to obtain data associated with all the stages over which a tissue exists. One solution here is to check each stage for associated data, but a better one is to use the *abstract organism* (the list of every tissue at every stage see earlier) which can readily be annotated with the start and end stages of each tissues. Using appropriate software (and COBrA is good for this), it becomes possible to represent every term in a single hierarchy with its start and end stage (Figure 1.2).

Ontologies can be stored in many flat-file formats (GO, OWL, OBO, etc.) as well as Protégé which is frame-based and currently only required for the Foundation Model of Human Anatomy (FMA). OWL (Web Ontology Language) is the current standard but only OBO is readable by eye (and even then, the triad relationships are hidden). Ontologies are not really meant to exist on paper, but to be viewed in an editor or browser and here the two standards are OBOedit and COBrA. The former handles a single ontology and includes a graphical viewer which shows all the hierarchies associated with a concept. The latter will not only handle up to three ontologies and make links across them which are stored in an ontology format but also translate one format into another (for further details on formats and editors, see Chapter 7). It is also worth noting that there is a general web browser at the European Bioinformatics Institute (www.ebi.ac.uk/ontology-lookup/) that holds all the OBO ontologies and has a search facility that allows a user to identify ID and other properties.

All anatomical ontologies should be available from the OBO site (obo. sourceforge.net/), but it has to be emphasised that the version posted there may well not be up-to-date: curators improve their ontologies and communication among curators is not always as good as it should be. Anyone who wishes to use an anatomy ontology for curation purposes must therefore contact the curator directly to check what is available and to request the most up-to-date version; they should also ask to be notified of any future changes. A version will usually be available in OBO and, while this has the advantage of being easy to read, it has the disadvantage that this format differs from and is less comprehensive than OWL, the web standard. Translation across many of the standard formats can be done using the COBrA tool (see Figure 1.2).

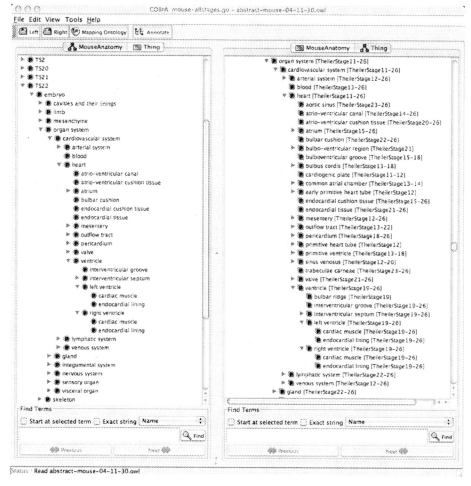

Fig. 1.2. The mouse developmental anatomy ontology in the COBrA environment. The left panel shows staged anatomy and focuses on the hierarchy for the heart of the Theiller stage 22 mouse embryo. The right panel shows the abstract mouse and again shows the hierarchy for the heart; here, each tissue is annotated with its stage limits (TS 26 implies that the tissue extends to the end of embryogenesis and, by implication, to the adult). In COBrA, the alternate left and right panels (labelled "thing"!) show relationships and IDs.

1.3 The Current Anatomy Ontologies

If one set out to capture all of the core anatomical knowledge about an organism and its development within an ontology, it would include *part_of*, *develops_from* , and *is_a* relationships together with some geometric information (*continuous_with* , *next_to* , etc.) to connect a space-filling set of tissues at a reasonable level of granularity. There are further aspects of anatomical knowledge that could, in principle be made, the most obvious being the cell types associated with each tissue, although this is probably better done through a look-up table (see below). Given the realities of time, all this information could probably only be incorporated in the anatomy ontology for a simple organism, and, even then, making all this knowledge available either computationally or via a GUI (graphical user interface) would pose substantial problems.

It is therefore not surprising that no curator has yet attempted to produce such a complex ontology for their organism (see Table 1.2), although that for *Dictyostelium discoideum*, in particular, holds full anatomical and lineage data within a DAG format (Figure 1.3), while that for adult human anatomy (FMA) integrates the largest number of relationships and is the only one that sets out to include geometric information. This is the main reason that the full FMA is the only ontology not publicly available in OBO or OWL format but uses the Protégé environment with the anatomical data being held in a linked database.

If an anatomy ontology is to be used for handling linked data, the key property of its *part_of* relationship has to be upwards propagation so that information associated with a high-level tissue includes all the data associated with its children. As can be seen from Table 1.2 which summarises the properties of the best known fungal and animal ontologies, it seems that all incorporate this property. As can also be seen from Table 1.2, where the properties of the current anatomical ontologies are summarised, a very large amount of careful work has gone into developing them.

1.3.1 Fungal Ontologies

The ontologies for fungi and for *Dictyostelium* are both pretty complete and rich. The fungus ontology, which includes *is_a* and *part_of* relationships, is intended to cover the anatomy of all fungi but is still in a draft state, awaiting community responses. The *Dictyostelium* ontology includes both *part_of* and *develops_from* relationships, together with full definitions (Figure 1.3) and is probably the most complete such ontology currently available. It is to be hoped that both of their respective communities will avail themselves of these rich resources for annotation, curation and modelling.

1.3.2 Animal Ontologies

Each of the anatomical ontologies for the main animal model organisms currently available or that are known to be under construction (see Table 1.2 for details) has

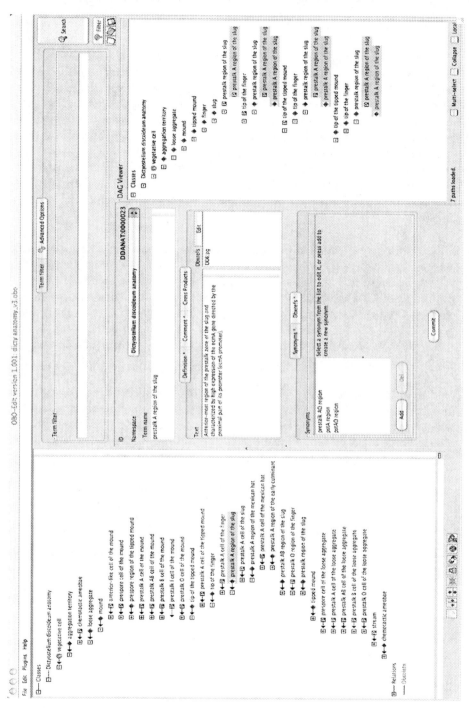

Fig. 1.3. The Dictystelium discoideum ontology of developmental ontology in OBOedit. The left panel shows the essential hierarchy. The central panel includes the definition, synonyms and ID. The right panel shows a graphical representation the developmental and part-of relationships for the prestalk A region of the slug.

its own style with respect to the granularity, relationships and structure, although all include the major organ systems at a very high level. What remains unclear is the extent to which they are actually being used for curation, annotation and the provision of knowledge. A few are linked to G-E databases (e.g. mouse development), some are being used to annotate genetic data (e.g. *Drosophila*) and some have just been produced as information resources but do not seem to be linked to any databases or other computational resources. What follows is a few comments on a selection of these ontologies, but the core information is held in Table 1.2.

Mouse developmental anatomy is available in a staged version where a simple single-parent hierarchy for each Theiller stage is held separately, and this is useful for standard archiving and searching (e.g. in GXD and in EMAP, the graphical database of mouse gene expression, genex.hgu.mrc.ac.uk [5]). It is also held in an abstract form where all the tissues, with their stage data, can be represented in a single hierarchy, and the two versions can be compared using the COBrA environment (Figure 1.2). A similar abstract format is available for the zebrafish anatomy. The ontology of mouse adult anatomy includes multi-parent links and is thus a DAG.

Human developmental anatomy for the Carnegie stages (up to E50 or so) is available as an ontology for annotation, albeit at a coarser granularity than the ontology for mouse developmental anatomy on which it is based. It is also available as a website (www.ana.ed.ac.uk/anatomy/humat/) for consultation and this version gives the basic hierarchies with links to formal notes about the provenance of the data as well as further links to the literature [8]. Adult human anatomy is available through the Foundation Model of Anatomy (FMA, [10]). The full version does, as mentioned above, require downloading a database and the Protégé environment (and currently also requires a license, which is free to academics). This ontology has the most relationships and is a formidable resource, even if it requires some computational skills to handle it. Fortunately, there is an abbreviated version of the anatomy known as FM Explorer (Figure 1.4) that is available online (sig.biostr.washington.edu/projects/fm/FME/index.html) which includes a great deal of data and which many will find useful. In the future, it would be helpful to the community if the curators were to release a *part_of* only version in a flatfile format (OWL or OBO) that would also make the FMA IDs easily available for external annotation purposes.

Invertebrate anatomy is also available. The ontology of *Drosophila* anatomy includes the egg, embryo, pupa and adult stages as separate hierarchies, with that for the embryo integrating the tissues from all the Bownes stages [3] in a single hierarchy. The granularity of the adult anatomy in particular is very fine and it is to be hoped that it together with the PATO ontology will soon be used to annotate *Drosophila* mutations. Thus far, the ontology does not yet appear to be fully integrated with Flybase, the core *Drosophila* database. The ontology for *C. elegans* anatomy is under construction at the time of going to press, but a working version is available from www.xpan.org.

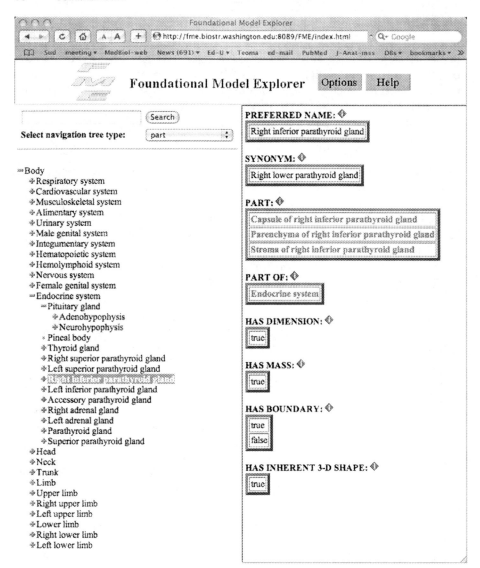

Fig. 1.4. A screenshot of FM Explore (the online version of the FMA) showing the information associated with the right superior parathyroid gland. Note that, in addition to the hierarchy and parts, there is additional information about the tissue itself (right panel).

1.3.3 Associated Resources

There are three associated resources that are worth mentioning: SAEL and EVOC have been produced for low-level anatomical annotation, the former ([9]) is species-independent and intended for microarray annotation, the latter was designed for human annotation but is more generally applicable. Both are closer to controlled vocabularies than substantial ontologies and neither can be used in any formal way as they only use an *is_a* relationship for tissue and their anatomies are not spacefilling. The other associated ontology is that for all cell types: this rich DAG is intended for annotating the tissues in anatomical ontologies for any species (see www.xspan.org). It is in principle quite possible to do this using the relationship *has_cell_type* and assigning to each tissue its constituent cell types with their associated CL IDs. The FMA includes such information (although it was produced before the cell-type ontology was published so lacks the linking IDs). These days, it is probably better to keep the ontologies separate and maintain a look-up-table (LUT) of links and leave the assignments to the software; such an external LUT is far easier to curate than internal annotations.

This was the solution chosen for the XSPAN project, where it is possible to identify tissues in the mouse, human, *Drosophila* and *C. elegans*, on, among other properties, their common cell types and the XSPAN resource has LUTs linking the cell-type ontologies to the tissues of the developing mouse and human, the embryonic, pupal and adult stages of *Drosophila* and the adult *C. elegans*.

1.4 Discussion

There are three main roles for which anatomical ontologies are needed. The first is to provide a means of coding and accessing tissue-associated data (gene expression, micrographs, literature, etc.) in databases that have been annotated with the relevant tissue IDs. The second is as a sort of online textbook to provide knowledge about tissues, and this may or may not be associated with external sources (e.g. www.xspan.org). If an ontology does hold *part_of, develops_from* and other relationships, it is possible, in principle at least, to make this information available to a user (the FMA does this for adult human anatomy). The third, and probably least exploited, is the use of ontology IDs to annotate tissue names in other computational resources (e.g. Pathbase — www.pathbase.net — uses the mouse anatomical ontologies to code information about mouse pathology); once the IDs are in place, it becomes straightforward to link them to other databases for interoperable querying. Obvious uses here include mutation and phenotype data and systems modelling (e.g. of development) that involves tissues.

These important uses require, first, that anatomical ontologies are properly curated to incorporate corrections and knowledge updates, second, that these updates are easily accessible and, third, that user-friendly and adequate visualisation and annotation tools are available for ontology curation, access and use. The new National Center for Biomedical Ontology (www.bioontology.org) will be one natural home for providing these ontologies and tools, much as the Gene Ontology (www.geneontology.org) does for its users, and their recent meeting report on the development of the Phenote Annotation Tool is to be welcomed.

1.4.1 Curation

When anatomical ontologies were first being constructed, perhaps a decade ago, it was easy to believe that true and complete knowledge was being organised and that, once made, any ontology would be stable. Such wishful thinking turned out to be wrong and this was partly because the many problems detailed in the first section were not properly appreciated and handled, partly because large ontologies usually include errors and over-simplifications that need correcting, partly because new knowledge becomes available, partly because new relationships may need to be incorporated, and partly because different users require the anatomical ontologies in different formats.

This last aspect was particularly unexpected but it turns out that not everyone needs the full richness of a complex ontology for their annotations: for example, one person may require stage-dependent tissue IDs (e.g. somite 11 at Theiller stage 16 is different for that at stage 17), while for another a time-independent annotation for somite 11 may be more appropriate. In practice, this has meant that ontologies may need to be available in different formats and in restricted forms. The deeper truth here is that ontologies are made for specific purposes and it may be best to view that which holds the most tissues and richest relationships as the standard but that sub-ontolgoies should be available as required.

The net result is that anatomical ontologies may need to be released in several formats and to be upgraded as new knowledge is incorporated and corrections made. The onus here is shared between users who need to inform the curator of problems and the curator who needs to incorporate changes, release them and publicise the existence of the new version.

1.4.2 Publication

The requirement for anatomy ontology upgrades and format variants brings with it a need for their access. At the moment, the access point is obo.sourceforge.net, but this site is, at the time of writing, particularly spartan: there is no versioning for anatomical ontologies and no associated text explaining the underpinnings and uses of particular ontologies. It is to be hoped that the new NCBO will improve things here. An alternative approach is being developed in Edinburgh through the COBrA-CT group (see Chapter 7). This ontology curation tool aims to produce versions and alternates of ontologies and also to facilitate discussion between developers and users. A further initiative is being provided by the OBO Foundry (obofoundry.org):

this is setting out to provide fuller facilities (with a CVS archive) than OBO for a subset of strongly curated ontologies, but the only anatomical ontology currently included is the FMA. No matter how it is done, the field requires easy and up-to-date access to all ontology information if these resources are to be properly and fully used.

1.4.3 Tools

Anatomy ontologies were originally set up for database curation and the only tool really needed then was something for writing and editing them - and DAGedit (now OBOedit) was the default. The XSPAN project, which aimed to make mappings among the anatomies of the main model animal organisms, required a tool that could handle two or three ontologies and make links across them that could be stored in an ontology format. This requirement led to the production of the COBrA tool which had these functionalities and could also translate among the various ontology formats (OWL, OBO, RDFS, GO, etc.). This tool is now being expanded for general curation www.aiai.ed.ac.uk/project/cobra-ct/COBrA_home.htm).

Such tools are not actually very helpful to people who are not professional computer scientists and curators. They are poor at illustrating knowledge (only OBOedit currently provides any graphical visualisation beyond the basic hierarchy), partly because there can be so much material that a screenshot gives too little of the whole, and partly because the ontologies can be so rich that their visualisation is too complex to allow a user to look at the implications of a single relationship in a multi-relationship ontology. They are thus not necessarily helpful to someone who wishes to use the ontology as a knowledge resource. The FMA is an exception here in that it provides quite a lot of text information, but it uses the Protégé environment, and this is not always as easy to use as one might like.

Worse, none of these tools are of any use to someone who wishes to annotate their own data with ontology IDs. At the time of writing, the only way to do this is manually: neither OBOedit or COBrA allow the copying of IDs to a file. The least that should be provided by the next generation of ontology editors is the facility for a user to select required tissues in turn and have their name and ID added to a file that can later be downloaded.

In addition, and perhaps more important, far better visual support than is currently available is needed for viewing subdomains of ontologies so that that anatomical knowledge can be made more accessible to the wider community. Thus far, OBOedit provides a graphical DAG which shows all the relationships associated with a concept. The Protégé environment has a graphical plug-in. COBrA just provide parent and children data as text, but a graphical viewer is under construction. As ontologies become richer, so the information becomes more complex and harder for the user to appreciate. The best way forward here would be graphical viewers that allow the user to select the subset of information to be shown. If ontologies do not become easy and obvious to handle, working with the intuitive knowledge of biologists, their use will be restricted to the bioinformatics community - and this would be against the whole spirit of the enterprise.

1.5 Conclusion

Anatomy matters: it lies at the basis of all biology and, in the old days, it underpinned medicine on the one hand and taxonomy on the other; these days its use has gone beyond these areas to provide the grounding for gene expression, mutation, pathology, and evolution. This resurgence in the need for anatomy, both adult and developmental, has provided new challenges for the subject and, perhaps surprisingly, it is the demands of molecular investigations that have forced anatomy to become computational.

The organism informatics community has risen to this challenge by providing comprehensive set of anatomical ontologies and it seems a shame to conclude this chapter with a series of complaints about the inadequate public resources available for handling these resources. Nevertheless, if the general curation, provision and availability of these ontologies together with tools for handling them are not much improved, the wider community is not going to get the benefit of all that the curatorial community has done.

Acknowledgements

I thank Stuart Aitken for commenting on the draft manuscript and those curators of anatomical databases who have provided me with up-to-date information.

Name	ID	Size	Granu-larity	Space-filling	Relationships, DAG/hierarchy	Definitions	Upward propagation[1]	Access	Comment
Fungus	FAO	+	Fine	yes	IS_A + PART_OF DAG	Yes	Unclear	Obo.sourceforge.net	See yeastgenome.org/fungi /fungal_anatomy_ontology/
Dictyostelium (development)	DDANAT	+	Fine	yes	DEVELOPS_FROM + PART-OF DAG	Yes	Yes	Under construction	www.dictybase.org does not handle anatomy, but does have a phenotype annotation ontology
C elegans	WBDAG	+	Fine	yes	PART-OF hierarchy	No	Yes	xspan.org	There is no public version but a working version is available from www.xspan.org - current wormbase details available from raymond@caltech.edu
Drosophila (development + adult)	FBbt	+++	Fine	Yes	IS_A, PART_OF & DEVELOPS-FROM DAG	No	Yes	Obo.sourceforge.net	Used in www.fruitfly.org for annotating gene expression
Mosquito (development + adult)	TGMA	+++	Fine		DAG		yes	Obo.sourceforge.net	From OBO.org, but does not seem to be used yet for annotation purposes
Xenopus	XAO	++			PART_OF DEVELOPS_FROM	Some	unclear		Under construction contact p.vize@
Zebrafish (development + adult)			Fine	Yes	PART_OF ... various forms available	Yes		Obo.sourceforge.net and ZFIN	Current version focuses on periods in which a tissue is present rather than providing an ID for each stage
Medaka (development + adult)	MFO	+++	Fine	Yes	PART_OF hierarchy	No	Yes	Obo.sourceforge.net	Used in ani.embl.de:8080/mepd/ for annotating gene expression

Mouse development	EMAP, EMAA (for abstract mouse)	+++	Fine	Yes	Now: PARTS_OF hierarchy Soon: DEVELOPS-FROM DAG	No	Yes	Obo.sourceforge.net	Designed to handle gene-expression in GXD (www.jackson.org)
Mouse adult	MA	+++	Fine	Yes	PART_OF IS_A (partial)	No	Yes	Obo.sourceforge.net	Designed to handle gene-expression in GXD (www.jackson.org)
Human development	EHDA EHDAA (for abstract human)	++	Good	Yes	PART_OF hierarchy	No	Yes	Obo.sourceforge.net	Only covers up to E50 (Carnegie stages) but is available in an abstract format from www.xspan.org
FMA [Foundation Model of [Human] Anatomy]	FMAID	+++	Fine	Yes	IS_A, PART_OF, CONTINUOUS-WITH etc. HAS_CELL_TYPE Complex DAG	No	Yes	only available online	Ontology is used in the FMA (sig.biostr.washington.edu/projects/fm/FME/index.html), but is not externally available yet, even as a simplified part-of hierachy
Evoc anatomy		+	Coarse	No	Controlled vocabulary	No	No	Evoc.org	One of a set of small ontologies (at www.evocontology.org/) to be used for human data annotation, although much is appropriate for all vertebrates
SAEL	SAEL	+	Coarse	No	Controlled vocabulary	No	No	sofg.org/sael	Intended for coarse cross-species annotation, particularly for microarray data
Cell types	CL	++	Good	No	IS_A, DEVELOPS-FROM DAG	No		Obo.sourceforge.net	

Table 1.2. The anatomy ontologies of the main fungal and animal model organisms. [1]Upwards propagation means that information associated with a lower-level tissue is associated with higher level tissue linked through a part_of relationship - this is the key property need to handle gene expression data.

References

1. J.B.L. Bard. Anatomics: the intersection of anatomy and bioinformatics. *J. Anatomy*, 206:1–16, 2005.
2. J.B.L. Bard and S.Y. Rhee. Ontologies in biology: design, applications, and future challenges. *Nature Rev. Gen.*, 5:213–222, 2004.
3. M. Bownes. A photographic study of development in the living embryo of drosophila melanogaster. *J Embryol Exp Morphol.*, 33:789–801, 1975.
4. A. Burger, D. Davidson, and R. Baldock. Formalization of mouse embryo anatomy. *Bioinformatics*, 20:259–267, 2004.
5. D. Davidson and R. Baldock. Bioinformatics beyond sequence: mapping gene function in the embryo. *Nat.Rev. Gen.*, 2:409–418, 2001.
6. G.V. Gkoutos, E.C. Green, A.M. Mallon, J.M. Hancock, and D. Davidson. Building mouse phenotype ontologies. In *Proceedings of Pac Symp Biocomput. 2004*, pages 178–189, 2004.
7. J. Hendler, T. Berners-Lee, and E. Miller. Integrating applications of the semantic web. *J. Inst. Elec. Eng. Jap*, pages 676–680, 2002.
8. A. Hunter, M.H. Kaufman, A. McKay, R. Baldock, M.W. Simmen, and J.B. Bard. An ontology of human developmental anatomy. *J Anat.*, 203:347–355, 2003.
9. H. Parkinson. et al. The SOFG Anatomy Entry List (SAEL): an annotation tool for functional genomics data. *Comp. Func. Genom.*, 5:521–527, 2004.
10. C. Rosse and J.L. Jr. Mejino. A reference ontology for biomedical informatics: the foundational model of anatomy. *J Biomed Inform.*, 36:478–500, 2003.
11. P. Simons. *Parts: a study in ontology*. Oxford University Press, Oxford, UK, 1987.
12. B. Smith. Mereotopology: A theory of parts and boundaries. *Data and Knowledge Engineering*, 20:287–303, 1996.
13. M.E. Winston, R. Chaffin, and D. Herrmann. A taxonomy of part-whole relations. *Cognitive Science*, 11:417–444, 1987.

Plant Structure Ontology (PSO) — A Morphological and Anatomical Ontology of Flowering Plants

Katica Ilic, Seung Y. Rhee, Elizabeth A. Kellogg and Peter F. Stevens

Summary. The Plant Structure Ontology (PSO) is a controlled vocabulary of anatomy and morphology of a generic flowering plant, developed by the Plant Ontology Consortium (POC) The main goal of the POC was to reduce the problem of heterogeneity of terminology used to describe comparable object types in plant genomic databases. PSO provides standardized set of terms describing anatomical and morphological structures pertinent to flowering plants during their normal course of development. Created as a tool for annotation of gene expression patterns and description of phenotypes across angiosperms, PSO is intended for plant genomics databases and broad plant genomic research community. Currently, this ontology encompasses diverse angiosperm taxa; further development will include new model organisms and important crop species. This chapter describes the rationales for creating PSO and discusses the guiding principles for its development and maintenance. The content of the PSO and the ontology browsing functionalities are outlined. The PSO can be browsed and downloaded at www.plantontology.org.

2.1 Introduction

Terminology-based application ontologies, or controlled vocabularies [21], have become increasingly important tools in biological and medical fields. This is largely due to two factors: they facilitate standardization of terminology of a given domain, and they allow for acquisition, integration and computation of large amount of biological information (i.e., data annotated to terms in the ontology). The best known and most widely used application ontology in biology, the Gene Ontology (GO), was initiated by a few model organism databases several years ago [9, 10]. Over the years it has become an established standard for describing functional aspects of genes and gene products and is used by a number of genomic databases, as well as by the research community at large. GO was the first generic controlled vocabulary that described three well-defined and distinct biological domains - cellular component, biological process and molecular function. As each of the three aspects is taxon-independent, that is, applicable to any given gene in any organism – GO has succeeded in facilitating consistent functional characterization of gene products in many species, spanning all kingdoms.

Since GO does not describe morphological and anatomical structures above the level of a cell, anatomical controlled vocabularies have been created for animal model organisms, e.g., fruit fly [8], mouse [4, 11], zebrafish [24], and humans [12]. Anatomical vocabularies were developed for a few plant species too, such as *Arabidopsis* [3], maize [25] and cereals [26]. Plant anatomical ontologies were either species-specific (*Arabidopsis* and *Zea* vocabularies) or applicable to a small number of closely related cereal crops (e.g., Cereal Plant Anatomy Ontology). No attempts were made to map the existing plant ontologies to each other, conceivably due to apparent variation in nomenclature and different organizational principles on which these vocabularies were built. Following the GO paradigm and embracing the idea of a generic, standardized terminology that would ultimately encompass many flowering plants, and that would allow for comparison across species, the Plant Ontology Consortium (POC) developed the first controlled vocabulary of anatomy and morphology of flowering plants, the Plant Structure Ontology (PSO) [13]. The primary goal of the POC was to create a shared descriptive set of terms that can be consistently applied across many angiosperms, and be used to associate and compare gene expression data and phenotypic descriptions across several plant genomic databases.

In this chapter we describe the PSO. This ontology represents the morphological-anatomical aspect of Plant Ontology (PO); the temporal aspect, Plant Growth and Development Stages Ontology, is described elsewhere [19]. This chapter primarily focuses on why and how we developed PSO, its content and comparison with other anatomical ontologies and basic ontology browsing functionalities. The applications of PSO as a tool for functional annotations are demonstrated briefly with examples from plant genomic databases. Possible future directions and further development of this ontology are briefly discussed at the end.

2.2 Objectives and Scope of Plant Structure Ontology

To our knowledge, PSO is currently the only morphological-anatomical ontology in the public domain that is pertinent to more than one organism. The initial public release of the PSO (July 2004) integrated existing species-specific ontologies for *Arabidopsis*, maize and rice; subsequent releases have encompassed terms for other cereal crops (Triticeae), Fabaceae, Solanaceae, and a number of terms for *Populus* (poplar), a recently sequenced model woody plant. The long term goal of the POC is to keep expanding the PSO by adding terms for other angiosperms, keeping pace with whole-genome sequencing efforts and large-scale functional genomics projects. Development and active maintenance of the PSO eliminate the need for creating species-specific anatomical ontologies for each plant whose genome sequence will be determined by large scale genome sequencing projects [e.g., tomato (*Lycopersicon esculentum L.*), potato (*Solanum tuberosum L.*), barrel medic (*Medicago trun-*

catula L.) and grapevine (*Vitis vinifera L.*)].

The main practical purpose of this ontology is to provide a standardized, biologically sound and computationally tractable set of terms describing a distinct domain (i.e., plant structure) as a tool for facilitating annotation of genes and germplasms in angiosperms. Therefore, PSO can be characterized as an annotation-centric controlled vocabulary. The level of detail in PSO is determined pragmatically – it is limited (i.e., very granular terms are generally excluded) but should be sufficient to make possible the description of tissue samples and experimental data, such as mRNA expression patterns, protein localization, description of mutant phenotypes and natural variants. PSO is not designed as a botanical glossary or as a vocabulary for taxonomy for use in taxonomic databases. This is because descriptors (attributes) of the component terms are, to a large extent, avoided in the PSO. Also, PSO does not address phylogeny of angiosperms and is neutral on the questions of organ homology in different angiosperm clades.

2.3 Organizing Principles of the PSO

At the inception of collaborative work on the ontology, a set of organizing principles for the PSO was established by the POC. While keeping in mind the main practical purpose of this ontology (gene and germplasm annotations), the POC members agreed on the crucial importance of preserving biological accuracy of the descriptions and relationships of the domain of knowledge this ontology encompasses, that is, anatomical and morphological structures of flowering plants during the normal course of their development, from zygote to an adult organism. A virtual, generic flowering plant would consist of anatomical and morphological parts that have been described in a range of angiosperm species, and so would comprise terms applicable to diverse angiosperms. The decision was made first to create an extensible backbone for the ontology, initially encompassing only a few species, namely Arabidopsis, rice and maize [18]. We would then proceed gradually to add terms for other angiosperms, keeping up with whole-genome sequencing projects and expanding the POC to include individual research communities involved in these large genomic projects. Retrospectively, the lessons we learned from each of the three predecessors of the PSO were invaluable, and have greatly influenced our decisions on the principles and design for the PSO. The following shortcomings in the anatomical ontologies created by the three plant genomic databases were discovered:

- Terms with no definitions or with definitions that were difficult to understand;
- Terms referring to a developmental stage rather than anatomical structure (such as *seedling* or *tetrad of microspores*);
- A number of non-botanical, mainly agronomical or colloquial terms (such as *whole plant*, *crown* or *shank*);
- Terms that did not describe plant structure but rather some qualitative features and/or descriptors (e.g., *leaf blade color*, *ligule appearance* or *ligule consistency*).

Based on our analysis of these three plant ontologies, several decisions were made in the initial stages of development of the PSO:

1. Every term in the PSO would be defined as concisely as possible.
2. Non-botanical and crop-specific agronomical terms would be avoided as much as possible.
3. Terms describing developmental stages would be excluded from the PSO.
4. Attributes of terms would be avoided.
5. Most importantly, synonymy would be used whenever possible, to group species-specific terms.

We also established criteria of what would constitute a valid term in PSO. Terms in PSO are morphological and anatomical structures of a flowering plant, from a cell to the whole plant level. Unlike in botanical or taxonomical glossaries, descriptors (also called attributes or qualifiers) are intentionally omitted. For example, the term *leaf margin* exists in the PSO (PO:0020128), but leaf margin shapes, such as dentate or serrate are not included. Very few exceptions have been made, such as cases where positional attributes of terms are included to accommodate accurate gene annotations (for instance, terms like *terminal bud*, PO:0004713, and *axillary bud*, PO:0004709). Each term in the PSO has a term name, a unique identifier, i.e., an accession number that always starts with the PO prefix followed by seven digits (PO:nnnnnnn), a textual definition, and a specified relationship to at least one other term. While term names can be modified to some extent (only if absolutely necessary), and may or may not be unique in the PO, the accession number associated to each term is always unique - it does not change or get reassigned to another term. Textual definitions are brief and sufficiently broad to reflect the position of a term in the ontology. They are often adopted from standard references, such as textbooks [7] and glossaries [Angiosperm Phylogeny Website (APWeb) (URL: http://www.mobot.org/MOBOT/research/APweb/)]. These definitions are conventional botanical definitions; at present, they do not follow formal ontological rules such as transitivity or reflexivity [22].

One of the most important organizational principles in the PSO is the use of synonymy. Extensive use of synonyms was acknowledged to be critical to keep the PSO relatively straightforward and easy to understand, and also to avoid problems with term multiplication (see below). Therefore, we chose a generic form of a plant organ as a term name, while various specific types of that organ were created as its synonyms. This was particularly effective for the *inflorescence* and *fruit* nodes. Since both structures occur in a range of morphological forms in angiosperms, instead of creating multiple terms for different specific types of inflorescence and fruit (such as the several terms for fruit types in the original Cereal Plant Anatomy Ontology [26], we decided to introduce a single generic term for each entity, and place all specific types of inflorescence and fruit as their synonyms (Fig. 2.1). Although this practice of creating synonyms in PO differs from the botanical usage of synonymy, to some extent, it is similar to the GO concept of narrower synonyms (see URL: http://www.geneontology.org/GO.usage.shtml#synonyms). As a result, the hierarchy of some higher-level nodes was considerably simplified, and excessive term multiplication was, to some degree, alleviated. Also, users could search the PSO using either generic terms, for instance, fruit, or its taxon-specific synonyms, silique, caryopsis or kernel.

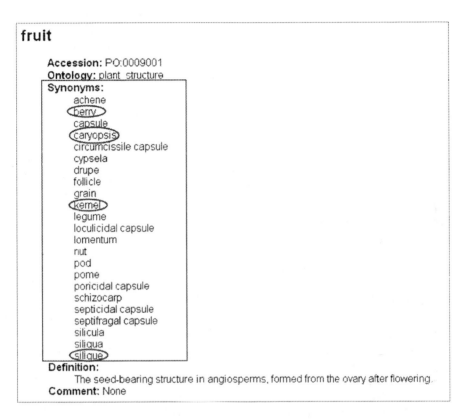

Fig. 2.1. Term detail page for *fruit* (PO:0009001), showing multiple synonyms of the generic term, all of which are particular fruit types in angiosperms. From the top, encircled are specific fruit types for Solanaceae (including tomato), rice, maize and Brassicaceae (including *Arabidopsis*).

Terms in PSO (as in GO) are linked in a hierarchical network structure called a Directed Acyclic Graph (DAG) (URL: http://www.nist.gov/dads/HTML/directAcyc Graph.html). The DAG structure of the ontology allows a term to be placed in multiple branches of the ontology - a term can have multiple parents and consequently have relationships to more than one parental (higher-level) node [1]. Parent-child relationships are connecting links between the nodes in the ontological hierarchy. However, a general term cannot appear as a child of a more specific (lower level) term. This is because the relationships in the DAG structure are directional; no path can start and end at the same node. Most importantly, the position of any given term in the ontology and the type of its relationship to a parental term conveys information about that term beyond its name and textual definition (see examples in text).

The PSO has three types of parent-child relationships, as illustrated in Fig. 2.2. The relationships is_a and **part_of** (Fig. 2.2a and Fig. 2.2b) are adopted from GO, and are the principal relationships used in PSO. The third relationship, **develops_from** (Fig. 2.2c), is an additional type, a modification of **derived_from** that is sometimes used in anatomical ontologies [22]. The **is_a** relationship represents a generalized relationship where a child is a subclass or a type of its parent. The **is_a** relationship is transitive. For example, a *cambial initial* **is_a** (a type of) *initial cell*, which **is_a** *meristematic cell*. Therefore, a *cambial initial* **is_a** *meristematic cell* (Fig. 2.2a). Mainly for computational purposes (e.g., automated reasoning and error checking), each term in the ontology is required to have at least one parent with an **is_a** relationship. This specifies that every term in PSO is at least a subclass of *plant structure*. Several terms with missing **is_a** relationship were detected in PSO by using the software tool Obol [17], and we are currently trying to assign a parent with an **is_a** relationship to every term that is missing such a relationship. The **part_of** relationship represents a component or subset relationship. In this ontology, **part_of** is used in a non-restrictive manner, where a parent does not have to be composed of all of its children. However, the child must be a **part_of** the parent to exist in the ontology. For instance, *stamen* is necessarily a **part_of** *androecium*, which is a **part_of** *flower*; therefore, whenever a stamen exists, it is part of a flower, but not all flowers have stamens (Fig. 2.2b). Like the **is_a** relationship, the **part_of** relationship is transitive. The **develops_from** relationship indicates that structure A develops from structure B, meaning that the structure A begins to exist at the same time as structure B ceases to exist. Unlike the other two relationship types, **develops_from** is not transitive. For example, a guard cell develops as a result of asymmetric division of a guard mother cell. Therefore, in the ontology, *guard cell* **develops_from** *guard mother cell*, which in turn **develops_from** the *epidermal initial* (a type of meristematic cell), as shown in Fig. 2.2c. However a guard cell does not develop from a meristematic cell. Consequently, annotations to the term *guard cell* or *guard mother cell* should not be propagated up in the ontology tree to any of their parental terms.

2.4 Content of the PSO

An important decision in the design of the PSO was the organization of the top-level nodes. The goal was to make a robust and extensible backbone of the ontology, which would allow regular updating without a need for significant changes of the top-level hierarchy of the ontology. The PSO describes flowering plant structures spanning cell types to organs and organ systems, from zygote to adult organism and including both sporophytic and gametophytic generations. Hence, the high level nodes in the ontology are *plant cell*, *tissue*, *organ*, *sporophyte* and *gametophyte* (Fig. 2.3a). Because of the need for more accurate gene annotations, an additional top level node,

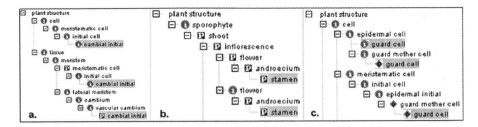

Fig. 2.2. Relationship types in PSO: **a.** The relationship **is_a** - a term is a subclass of its parent, i.e., *cambial initial* **is_a** *initial cell*, which **is_a** *meristematic cell*. **b.** relationship type, **part_of**, e.g., *stamen* is **part_of** *androecium*, which is **part_of** *flower*, therefore, *stamen* is **part_of** *flower*. **c.** Non-transitive, temporal relationship **develops_from** is used to make derivation assertion, e.g., *guard cell* **develops_from** *guard mother cell*, which in turn **develops_from** *epidermal initial*.

in vitro cultured cell, tissue and organ was added. The term *whole plant* was also introduced at the top level of the PSO (it was previously used in annotations by the three databases). This term is not a botanical term and is intentionally left without children terms. Thus, we recommend that the term *whole plant* is used as a last option, only when precise annotation to any other term in the PO is not possible.

Plant cell types are included in the PSO, but subcellular structures are not, since GO describes these in the cellular component ontology. The POC has made efforts to eliminate overlaps of PO with other bio-ontologies under the Open Biomedical Ontologies (OBO) umbrella (http://obo.sourceforge.net), and has been looking for the best solutions to eliminate an apparent overlap of the *plant cell* node in PSO with the Cell Ontology [2].

At the top-level of PSO, terms under *plant cell* (PO:0009002) and *tissue* (PO: 0009007) nodes constitute all cell and tissue types found in flowering plants. Many cell and tissue types are located in different plant organs and organ systems, during different stages in development (often with slight modifications). In such cases, introduction of a number of more granular terms in the ontology was necessary, with corresponding terms in every organ (and position within the organ) where the cells or tissues are located. To help avoid a massive proliferation of terms in PSO, which would make ontology navigation and browsing very difficult, (discussed in more detail by Ilic et al. [13]), a decision was made to instantiate cell and tissue terms on a selective basis – only as a response to annotation requirements. Exceptions were made in cases where a cell or tissue type was not present everywhere in a plant, but was localized to specific organs, such as *stem periderm* or *leaf mesophyll*.

At the top level of PO, parallel to the node *plant structure* (PO:0009011) is the *obsolete_plant_structure* node. A term that has been removed from the ontology is never permanently deleted. Instead, the term and its assigned identifier (accession)

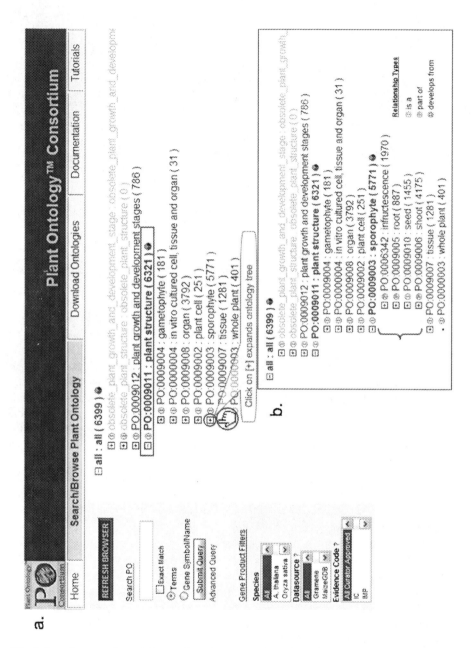

Fig. 2.3. a. A screenshot of the ontology browser - top nodes of the PSO. Clicking on the [+] or [–] sign in front of a term vertically expands or collapses the ontology tree, respectively. **b**. Expanded *sporophyte* node is indicated by curly bracket. A mouse click on a term itself opens a term detail page (see Fig. 2.1). Numbers in parentheses next to the term name indicate the number of the annotations for unique object types associated to a term (including annotations to all children terms). Relationship type icons are shown at the bottom right.

are kept in the ontology file for the record. The definition is always prefaced by the word with OBSOLETE. Because many obsoleted terms in PSO are valid botanical terms, the Comment section always provides an explanation as to why a term was removed (e.g. term *filiform apparatus* was made obsolete because it depicted a sub-cellular structure).

2.5 Comparison to other Anatomical Ontologies

Three plant anatomical ontologies that paved the way to the PSO have been super-seded by it and are no longer actively maintained. Shortly after the PSO was publicly released, TAIR and the Gramene database retired their respective anatomical ontolo-gies for *Arabidopsis* and cereals, respectively, and began using the PSO. The original *Zea mays* Plant Structure Ontology [25] has been partially integrated with the PSO terms, and both controlled vocabularies are currently in use by the MaizeGDB.

The *Arabidopsis* anatomical ontology [3] had just over 300 terms that were per-tinent to the model core eudicot plant, *Arabidopsis thaliana*, with an organization of the top-level nodes that was similar to the other plant species-specfic ontologies. The *Zea mays* Plant Structure ontology [25] had 136 terms that described anatomi-cal and morphological structures of maize, a member of Poaceae. The Cereal Plant Anatomy Ontology [26] had 360 terms, mainly describing anatomical structure of cereal crop species (also Poaceae). However, it also contained many terms that were specific to core eudicotyledonous families, such as Asteraceae, Fabaceae, Cucur-bitaceae and others. For example, unlike the two other vocabularies, the Cereal Plant Anatomy Ontology contained several terms describing different fruit types in an-giosperms. Fruit types were divided into main classes: dry and fleshy, and also into monocarpous and syncarpous fruits. Each specific fruit type was a separate term in this ontology. After assessing the scope of terms that would be required for each fruit type, and also taking into consideration that a number of fruit part terms needed to be added for each fruit type, we realized that the increase in the total number of terms under the fruit node in the PSO would grow exponentially, resulting in a massive term proliferation. Our solution to this problem was to create multiple synonyms of a single generic term *fruit* – this became one of the founding principles for the PSO.

Unlike vertebrate anatomical ontologies, the three original plant anatomical on-tologies were never mapped to each other. Because of historical differences in the ways these ontologies were constructed, term-to-term mapping of the three con-trolled vocabularies would be difficult at best, and a number of terms would be left unmapped. Furthermore, since the contributing databases were each commit-ted to adopting the PSO, and retiring their own anatomical ontologies, the mapping of species-specific ontologies to PSO was not necessary.

Compared to animal anatomical ontologies or to GO, PSO is a relatively small ontology, comprising just 727 terms (in the release PO_0906) that are descendants of the root node, *plant structure* (PO:0009011). There are 384 (or 53%) leaf terms, also called terminal nodes (the most specific terms with no children terms below), and 342 (47%) interior nodes (terms with children). Currently, PSO also has 304 synonyms assigned to 149 terms. The length of the longest path (i.e., from the root node to the leaf node in the path) is 15 nodes, while an average ontology depth is 5 nodes. More detailed analysis of the ontology structure is provided elsewhere [13].

2.6 Search, Browse and Download PSO

POC has shared software resources with GO and adopted the GO database schema and software infrastructure for storing, editing (OBO-editor and its preceding version, DAG-editor, http://www.oboedit.org) and displaying ontologies and annotations (ontology browsing tool, AmiGO). AmiGO is a web-based tool for searching and browsing ontologies and associated data and is freely available, open source software (http://www.geneontology.org/GO.tools.shtml#in_house). The browser has been slightly modified to suit specific requirements of the PO and its association files. Some browse and search functionalities of the PO AmiGO are displayed in Fig. 2.3a, which shows the top-level nodes in the screenshot of the browser. The *obsolete_plant_structure* node (grayed out) contains terms that are no longer in use in PSO, and are kept in the ontology file for historical and record keeping purposes. This node is placed at the same level as the PSO root node, *plant structure* (PO:0009011).

For browsing the PSO, a click on the [+] or [-] sign in front of a term vertically expands or collapses the ontology tree, respectively (Fig. 2.3a). Each horizontal line corresponds to a term (i.e., node), and consists of an icon depicting the specific relationship (edge), term accession number, term name, and the number of associated annotations in parentheses. The icon represents the relationship between the term and its immediate parental term, e.g., *shoot* is a **part_of** *sporophyte*, which **is_a** *plant structure* (Fig. 2.3b). The accession number starts with the PO prefix, identifying the Plant Ontology database. Next to the term name, the number in parentheses specifies the number of annotations of unique object types, i.e., genes and germplasms, which are associated directly to the term and also to its children terms. A mouse click on any term name opens a new page, the term detail page. This page consists of two parts. On the top, elements of a term are displayed, such as accession, aspect of the ontology, synonyms, definition, comment and term lineage (see example of the term *fruit* in Fig. 2.1), and the second part contains a list of annotations associated to the term, i.e., gene and germplasm annotations (see annotations to the term *embryo* in Fig. 2.4).

The output of a query can be filtered using several options. If a user clicks on the term *embryo* (PO:0009009), and is primarily interested in mutant phenotypes in *Arabidopsis*, maize and rice, he or she can apply species-specific filters, choosing particular species under Gene Product Filters, and selecting a specific Evidence Code (Fig.

2.3a, left-hand panel, and also top of Fig. 2.4, Filter Associations). The Evidence Code is the type of experimental or computational evidence used to support the annotation. For example, the evidence code IMP (*Inferred from Mutant Phenotype*) is used to support a phenotype annotation. The resulting term detail page will have a list of all IMP annotations for *Arabidopsis*, maize and rice, with several hyperlinks which provide quick access to additional information about each annotation entry (for details see Fig. 2.4).

To search the PSO, the left-hand panel on the browser offers a choice of searching for terms (and synonyms) or gene symbols. When searching for terms in the PSO, users are advised to apply the Ontology filter, choosing the Plant Structure aspect; otherwise the search result page contains hits to both aspects of the Plant Ontology, the Plant Structure (PSO) and Plant Growth and Developmental Stages. For example, a search with the term *mesophyll* (not shown) resulted in 9 terms in the PSO that contained word mesophyll in a term name or in a synonym, including the term *leaf mesophyll*. Clicking on the tree icon right below the check box in the first column, multiple term lineages for the *leaf mesophyll* in the PSO will be displayed. This term is placed under *tissue* (as a type of a parenchyma tissue), and also under *organ* and *sporophyte* nodes (as a part of a leaf). Children terms of the term *leaf mesophyll* can be viewed by expanding the ontology tree downward, after clicking on the [+] sign in front of the term accession (PO:0005645). The filtering options described previously for browsing options are also available for searches in the Advanced Queries in AmiGO, and also on the term detail page (see top of Fig. 2.4, Filter Associations is encircled).

The PSO ontology files (in OBO and GO flat file formats) and annotations from each contributing database are freely available and can be downloaded from the POC Concurrent Versions System (CVS) repository (http://www.plantontology.org/download/database/). Users can also download copies of the entire POC database as MySQL dumps from POCs CVS repository (http://www.plantontology.org/download/database/). The POC continues to maintain the CVS repository as the central place for the plant ontology files, associations, and mapping files. The Plant Ontology Browser accesses the MySQL database located at Cold Spring Harbor Laboratory, Cold Spring Harbor, NY. The structure of the POC database, CVS repository and the main features of the web site are described in more detail elsewhere [14].

2.7 Application of Plant Structure Ontology

PSO has been used for annotations of gene expression data and phenotype descriptions of mutants and natural variants in plant genomic databases such as TAIR [20], Gramene [14], NASC (URL: http://arabidopsis.info) and MaizeGDB [16], and also in databases specializing in large scale gene expression data such as GENEVESTIGATOR [27] NASCArrays [6] and ArrayExpress (H. Parkinson, personal communication). As of August 31, 2006, the POC database has over 4,400 unique object

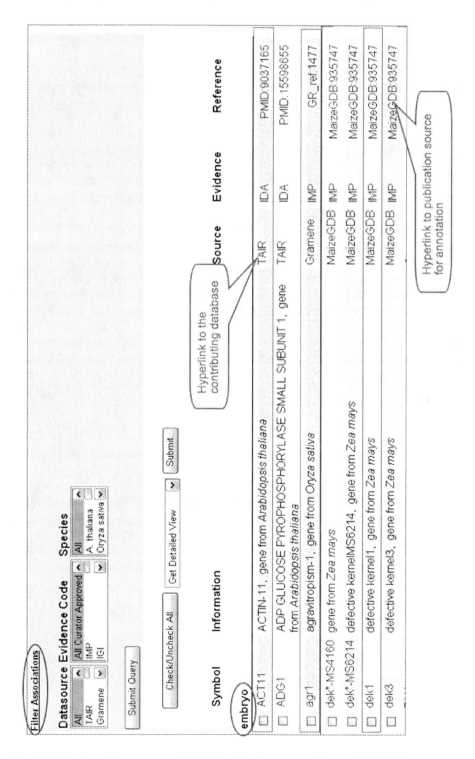

Fig. 2.4. Ontology browser term detail view for the term *embryo* (PO:0009009), showing selected annotations for this term.

types (genes and germplasms) annotated with PSO terms amounting to over 10,000 annotations. These annotations were contributed by TAIR, Gramene, MaizeGDB and NASC.

Functional annotation of genes and gene products can be described as a process of extracting information about gene function at the molecular level, its biological role(s), protein localization, and its spatial/temporal expression patterns during plant development [3]. Manual annotations are made by biologists - scientific curators who either extract the information from published literature, or record phenotype descriptions directly by observing plants (natural variants and mutants) in the greenhouse or field. Based on the type of collected information, curators make short statements by creating gene-to-term associations [3, 5]. Any object type can have multiple controlled vocabulary terms associated to it; however, each gene-to-term association is a separate annotation entry (Fig. 2.4). Indispensable components of each annotation entry are the unique identifier of the object type (gene or germplasm/stock), an appropriate, most granular controlled vocabulary term that describes the object type, a database reference number of the original paper (or another type of source) from which the annotation is extracted, and finally, a specific evidence code - a defined type of experimental or computational evidence that was used to support the annotation. Details on evidence codes and evidence descriptions can also be found online (URL: http://www.plantontology.org/docs/otherdocs/evidence_codes.html). More details on the process of literature curation using controlled vocabularies can be found elsewhere [3, 5].

The POC database is set up as a gateway through which the data curated using PO terms can be easily acquired and downloaded; researchers can quickly retrieve annotation data for multiple species. For example, a user who is interested in all genes that have mutant phenotypes affecting embryo development and all genes that are expressed in the embryo, can search for the term *embryo* in the AmiGO browser and will quickly retrieve all gene annotations and phenotypic descriptions associated to this term (and its children terms) for *Arabidopsis*, rice and maize, as shown in Fig. 2.4. Hyperlinks to the individual databases that contributed the annotation provide a quick access to additional information about annotated genes and germplasms. More detailed description of applications of PSO is provided by Ilic et al. [13].

2.8 Conclusion and Future Directions

PSO has been in the public domain since 2004, and has been actively maintained by the POC. Ontologies are work in progress and this ontology is far from being finished; rather, it is still in the early phase of development. As a long term goal, we envision PSO as a continuously expanding ontology that will gradually encompass many angiosperms. The ultimate measure of the success of this ontology will depend directly on how widely it is used by plant genomic databases as well as bench scientists.

The team that created PSO consisted of database curators and plant scientists, all experimental biologists by training. As expected, our main focus was the biological content of the ontology and the very pragmatic goal of providing a practical framework for annotations. The formal philosophical ontological rules and software implementations have been, to a large extent, neglected. The range of problems this approach created is similar to those described for some other informal bio-ontologies such as GO [23]. In the next phase, the POC will need to make an effort to become more compliant with formal ontology rules by applying formal ontological characteristics for term definitions such as transitivity and reflexivity [22], creating **is_a** relationship type for every term in the PSO, and improving other aspects of this ontology to make it computable by automated reasoners [17].

A persistent problem with term multiplication [13] has limited our ability to further expand the PSO by adding new terms for other angiosperm species. New visualization and ontology-editing software, and perhaps a different approach in designing broad anatomical ontologies, will be needed to reflect the modularity of biological reality more accurately and comprehensively, and to allow PSO to be more accessible and easier to comprehend for end-users with limited ontological expertise.

Acknowledgements

We greatly acknowledge Pankaj Jaiswal, Felipe Zapata, Leszek Vincent and Shuly Avraham for their invaluable participation in development of Plant Structure Ontology, and also Anuradha Pujar, Mary Schaeffer, Leonore Reiser, Doreen Ware, Susan McCouch and Lincoln Stein for their contribution. We also thank POC collaborators, Quentin Cronk, Rex Nelson, Naama Menda, Victoria Carollo and William Friedman for their participation in the ontology development. We are grateful to numerous researchers and curators who have reviewed the Plant Ontologies, who are listed individually online (URL: http://www.plantontology.org/docs/otherdocs/ acknowledgment_list.html). We acknowledge the Gene Ontology Consortium for software infrastructure and technical support. This project is supported by National Science Foundation Grant (No. 0321666) to the Plant Ontology Consortium.

References

1. J. Bard. Ontologies: formalizing biological knowledge for bioinformatics. *BioEssays*, 25:501–506, 2003.
2. J.B. Bard, S.Y. Rhee, and M. Ashburner. An ontology for cell types. *Genome Biology*, 6(R21), 2005.
3. T.Z. Berardini, S. Mundodi, L. Reiser, E. Huala, M. Garcia-Hernandez, P. Zhang, L.A. Mueller, J. Yoon, A. Doyle, G. Lander, N. Moseyko, D. Yoo, I. Xu, B. Zoeckler, M. Montoya, N. Miller, D. Weems, and SY. Rhee. Functional annotation of the *Arabidopsis* genome using controlled vocabularies. *Plant Physiol*, 135:1–11, 2004.

4. A. Burger, D. Davidson, and R. Baldock. Formalization of mouse embryo anatomy. *Bioinformatics*, 20:259–67, 2004.

5. J.I. Clark, C. Brooksbank, and J. Lomax. It's all GO for plant scientists. *Plant Physiol*, 138:1268–1278, 2005.

6. D. J. Craigon, N. James, J. Okyere, J. Higgins, J. Jotham, and S. May. NASCArrays: a repository for microarray data generated by NASC's transcriptomics service. *Nucleic Acids Res*, 32:D575–7, 2004.

7. K. Esau. *Anatomy of Seed Plants*. Wiley & Sons, Inc., 2nd edition, 1977.

8. FlyBase Consortium. The FlyBase database of the Drosophila genome projects and community literature. *Nucleic Acids Res*, 30:106–8, 2002.

9. Gene Ontology Consortium. Gene Ontology: tool for the unification of biology. *Nature Genetics*, 25:25–29, 2000.

10. Gene Ontology Consortium. Creating the gene ontology resource: design and implementation. *Genome Res*, 11:1425–1433, 2001.

11. T.F. Hayamizu, M. Mangan, J.P. Corradi, J.A. Kadin, and M. Ringwald. The Adult Mouse Anatomical Dictionary: a tool for annotating and integrating data. *Genome Biol*, 6(R29), 2005.

12. A. Hunter, M.H. Kaufman, A. McKay, R. Baldock, M.W. Simmen, and J.B. Bard. An ontology of human developmental anatomy. *J Anat*, 203:347–55, 2003.

13. K. Ilic, E.A. Kellogg, P. Jaiswal, F. Zapata, P.F. Stevens, L. Vincent, S. Avraham, L. Reiser, A. Pujar, M.M. Sachs, N.T. Whitman, S. McCouch, M. Schaeffer, D.H. Ware, L. Stein, and S.Y Rhee. The plant structure ontology - a unified vocabulary of anatomy and morphology of a flowering plant. *Plant Physiology*, 143:587–599, 2007.

14. P. Jaiswal, S. Avraham, K. Ilic, E.A. Kellogg, S. McCouch, A. Pujar, L. Reiser, S.Y. Rhee, M.M. Sachs, M. Schaeffer, L. Stein, P. Stevens, L. Vincent, D. Ware, and F. Zapata. Plant Ontology (PO): A controlled vocabulary of plant structures and growth stages. *Comp Funct Genom*, 6:388–397, 2005.

15. P. Jaiswal, J. Ni, I. Yap, D. Ware, W. Spooner, K. Youens-Clark, L. Ren, C. Liang, W. Zhao, K. Ratnapu, B. Faga, P. Canaran, M. Fogleman, C. Hebbard, S. Avraham, S. Schmidt, T. M. Casstevens, E.S. Buckler, L. Stein, and S. McCouch. Gramene: a bird's eye view of cereal genomes. *Nucleic Acids Res*, 34:D717–23, 2006.

16. C.J. Lawrence, Q. Dong, M.L. Polacco, T.E. Seigfried, and V. Brendel. MaizeGDB, the community database for maize genetics and genomics. *Nucleic Acids Res*, 32:D393–397, 2004.

17. C. Mungall. Obol: Integrating language and meaning in bio-ontologies. *Comp Funct Genom*, 5:509–520, 2004.

18. Plant Ontology Consortium. The Plant Ontology Consortium and Plant Ontologies. *Comp Funct Genom*, 3:137–142, 2002.

19. A. Pujar, P. Jaiswal, E.A. Kellogg, K. Ilic, L. Vincent, S. Avraham, P. Stevens, F. Zapata, L. Reiser, S.Y. Rhee, M.M. Sachs, M. Schaeffer, L. Stein, D. Ware, and S. McCouch. Whole plant growth stage ontology for angiosperms and its application in plant biology. *Plant Physiol*, 142:414–428, 2006.

20. S.Y. Rhee, W. Beavis, T.Z. Berardini, G. Chen, D. Dixon, A. Doyle, M. Garcia-Hernandez, E. Huala, G. Lander, M. Montoya, N. Miller, L.A. Mueller, S. Mundodi, L. Reiser, J. Tacklind, D.C. Weems, Y. Wu, I. Xu, D. Yoo, J. Yoon, and P. Zhang. The Arabidopsis Information Resource (TAIR): a model organism database providing a centralized, curated gateway to *Arabidopsis* biology, research materials and community. *Nucleic Acids Res*, 31:224–228, 2003.

21. C. Rosse, A. Kumar, J.L. Mejino(Jr), D.L. Cook, L.T. Detwiler, and B. Smith. A strategy for improving and integrating biomedical ontologies. In *AMIA Annu Symp Proc.*, pages 639–643, 2005.

22. B. Smith, W. Ceusters, B. Klagges, J. Kohler, A. Kumar, J. Lomax, C. Mungall, F. Neuhaus, A.L. Rector, and C. Rosse. Relations in biomedical ontologies. *Genome Biol*, 6(R46), 2005.

23. B. Smith, J. Williams, and S. Schulze-Kremer. The ontology of the gene ontology. In *AMIA Annu Symp Proc*, pages 609–613, 2003.

24. J. Sprague, L. Bayraktaroglu, D. Clements, T. Conlin, D. Fashena, K. Frazer, M. Haendel, D.G. Howe, P. Mani, S. Ramachandran, K. Schaper, E. Segerdell, P. Song, B. Sprunger, S. Taylor, C.E. Van Slyke, and M. Westerfield. The Zebrafish Information Network: the zebrafish model organism database. *Nucleic Acids Res.*, 34:D581–5, 2006.

25. P.L. Vincent, E.H. Coe, and M.L. Polacco. *Zea mays* ontology - a database of international terms. *Trends in Plant Sci*, 8:517–520, 2003.

26. Y. Yamazaki and P. Jaiswal. Biological ontologies in rice databases. an introduction to the activities in Gramene and Oryzabase. *Plant Cell Physiol*, 46:63–68, 2005.

27. P. Zimmermann, M. Hirsch-Hoffmann, L. Hennig, and W. Gruissem. GENEVESTIGATOR. *Arabidopsis* microarray database and analysis toolbox. *Plant Physiol*, 136:2621–32, 2004.

3

Anatomy for Clinical Terminology

Alan L Rector

Summary. Anatomical notions provide the foundations for much clinical terminology. However, clinicians' concerns are notoriously practical. The goal in clinical terminology is not to represent anatomy for its own sake, but rather to facilitate faithful communication by clinicians about what they have heard, seen thought and done in their care of patients. The focus of clinical anatomy is therefore the interaction between the locus of disease and partonomy - that diseases of parts are, with specific exceptions, diseases of the whole. Clinical anatomy must also accommodate the clinical/functional view as well as the structural view basic to classic anatomy. One means of doing so in current ontology formalisms is through a hierarchy of relations. There are a number of problems where there is no consensus on solutions. Of particular importance are the relations of tissues and substances to structures and representation of characteristics collective effects of entities such as cells, One approach to this issue is discussed here. Finally, there is currently no expressively adequate and computationally tractable means for expressing abnormal and variant anatomy. All resources are explicitly about "normative anatomy". In dealing with all of these problems it is essential to separate issues of "terminology" - i.e. the labeling of entities from their formal definitions and relations. Clinicians frequently use terms that anatomist now consider obsolete or deprecated. However, a distinction should be drawn between the mere renaming of an entity - however well motivated - and more fundamental revision as to its structure or function.

3.1 Introduction

Anatomical notions provide the foundations for much of clinical terminology. Anatomy is a key part of the fundamental vocabulary, and it is not possible to discuss disorders and treatment of the body without using anatomical notions.

However, clinicians' concerns are notoriously practical. They might recognise the same underlying foundations in anatomical structure embodied in the Foundational Model of Anatomy [6, 15]. They might assent to the formal descriptions of part-whole relations put forward by Smith et al. on behalf of the Open Biomedical Ontologies (OBO) consortium [19]. However, what they use day-to-day is a mixture of structure, function, and convenience. The *Open*GALEN anatomy schemata were created to address issues this clinical level of abstraction. They are well documented in several publications [10, 11, 14]. They have also been compared with the Foundational Model of Anatomy (FMA) [7, 8, 22]. We will not describe them in detail here. Rather this chapter will use them to illustrate key issues that must be addressed by any formal representation of anatomy that seeks to be used in clinical applications.

The goal in clinical terminology is not to represent anatomy for its own sake, but rather to facilitate faithful communication by clinicians about what they have heard, seen, thought and done concerning patients and their conditions. Because clinicians are concerned primarily with disorders and dysfunctions, any representation of anatomy for clinical purposes must deal with function as well as structure and development. Because a significant part of clinical terminology deals with pathology, any terminology for clinical anatomy must deal with the arrangements of tissues as well as with organs and other structures.

Our goal is a logical foundation for clinical terminology which is sufficient as a reference point for clinical terminologies. With respect to anatomy, there are five key goals, each of which will be discussed in turn:

- *"Locus" and partonomy.* The most important relationship between clinical medicine and anatomy is in specifying the locus for disorders and procedures. The fundamental internal relation in anatomy is partonomy. The interaction between partonomy and locus is therefore critical to using any formal ontology of anatomy in clinical medicine. This includes issues of understanding when partonomy does not appear to be treated as transitive.
- *Reconciliation of clinical/functional and structural views.* How to reconcile the classical structural view of anatomy and the more functional and pragmatic view used for reasoning in clinical medicine. The challenge in a more generic anatomy is to accommodate the clinical view without distorting the "pure" view of the structural anatomist and developmental biologist.
- *Tissues, substances and structures.* Clinical descriptions of organs of the tissues that constitute them are performed by different specialists in attending to different features using different vocabulary. Developmental anatomists tend to focus on tissues whereas clinicians and anatomists concerned with adult organisms tend to focus on structure.
- *Abnormal anatomy and congenital anomalies.* Classical anatomy is "normative". It deals with what is typically true. Clinical practice must deal with the myriad variations and distortions that occur in practice. Representation of normal and abnormal anatomy and the relationships between various abnormalities is a major unmet challenge.
- *Synonyms and variant terminology.* Issues of terminology and synonyms is as prevalent in anatomy as in any other biological field. Differences occur over time and between different academic communities and between different geographical locales. In particular, clinical usage is often different from formal anatomical

terminology. Hence any anatomical terminology, needs to be able to cater for alternative terminology.

A set of ontologies formulated in OWL demonstrating these principles can be found at http://www.co-ode.org/ontologies under "Sample Top Bio".

3.2 Locus and Partonomy – The Basis of Clinical Reasoning about Anatomy

The most important function for anatomy in clinical terminology is as the locus for disorders. Many diseases clearly manifest at an anatomical location or in an anatomical structure – e.g. "fracture of femur", "tumour of lung" – or in anatomically related notions – e.g. "bronchogenic carcinoma". Other notions notions are so closely tied to anatomy that they are usually described as localised to some particular organ even though the manifestations may occur remotely or affect the body globally – e.g. "hypothyroidism" or "cerebellar ataxia". Signs and symptoms are almost by definition related at least loosely to anatomical notions, even when they are vague terms such as in "indigestion". The pattern is so ubiquitous that many clinical nomenclatures, including some versions of the original SNOMED pathology codes, require an anatomical locus[1]; where there is no specific locus, some construct such as "body as a whole" is often used.

However, since the relation between disorders and anatomy is not strictly speaking one of physical location, it is preferable to use a label such as "locus" for the relation rather than any term which implies physical location. Nor is it useful to define the notion of "locus" too tightly. For example the exact relation between disorder and anatomy is different in "pneumonia", "pleuridinia", "intestinal obstruction", and "cholelithiasis"; however, the differences are not consistently reflected in language, and each condition would normally be classified under disorders of the anatomical structures involved. In effect the "locus" relationship is defined pragmatically as the disjunction of all of these relations and others, which are not distinguished in common clinical parlance.[2]

The fundamental relation of anatomy is the part-whole relation – the foot is part of the lower extremity; the ventricle is part of the heart, etc. The key pattern for clinical anatomy relates "locus" and "partonomy".

The fundamental pattern is that, with few exceptions, disorders of parts are disorders of the whole. For example, a fracture of the neck of the femur is considered a kind of fracture of the femur; stenosis of the aortic valve is considered as a heart disease; cataracts, which occur in the lens, are considered a disorder of the eye, etc.

---

[1] now the "site" qualifier in SNOMED-CT

[2] *OpenGALEN* used the label "location" and came to regret the confusion caused.

The same is typically true of procedures. Fixation of the neck of the femur is an operation on the femur; repair of the aortic valve is a heart operation; removal of a cataract is an operation on the eye, etc.[3]

Put another way, when we use the phrase "Disorder of X", what we usually intend is "disorder that has locus X and/or any of its parts". Likewise, when we speak of "Procedure on X", we usually mean a "procedure that has locus X and/or or any of its parts".

There are two principled exceptions:

- *Effects on entites "as a whole" – e.g.* amputations, where we would not consider the amputation of a finger to be an amputation of the hand although we would consider an injury to the finger to be an injury to the hand . There is a limited but important set of constructs such as "Amputation of X" that should not be interpreted as applying to "X and/or its parts".
- *Collective effects – e.g.* the collective failure of the pancreatic islet cells in type one diabetes. Collectives are a more complex case that is discussed under 3.4.2 below.

The standard pattern of a "disorder a structure and/or any of its parts" gives rise to standard transformation whereby separate hierarchies of kinds and parts give rise to a combined hierarchy of disorders as shown in Figure 3.1. Because this transformation is so common, it is also common to confuse the hierarchies of parts and wholes with the kind-of or subsumption hierarchy. The two are distinct, but related through the common patterns shown in Figure 3.1.

These notions are easily represented in any logic based formalism supporting disjunction (the logical operator "or"), including the new web ontology language, OWL.[4] Alternatively, even without disjunction, any system supporting inheritance hierarchies can achieve a similar effect by rewriting each structure as a "SEP triple" [4, 16, 17] consisting of: i) a "Structure" node for what we have called the reflexive parts of the entity, ii) an "Entire" node, for the entity itself, and iii) a "Part" node for the proper parts as shown in Figure 3.2. This transformation is possible even in relatively weak formalisms and, correspondingly, leads to highly computationally efficient representations. (Normally, the transformation is hidden so that users need not be aware of the extra entities). The SEP triple transformation makes it easy to see the interaction between the *locus* relation and the kind of relation as shown in Figure 3.3. If we interpret "Disorder of Heart" as "Disorder of Heart of its parts"; interpret "Disorder of Ventricle" as "Disorder of Ventricle or its parts", etc., then it follows

---

[3] This is not true just of medicine, but of our conceptualisation of the world in general. For example, a fault in a fuse is considered to be a fault of the electrical system, which is in turn considered to be a fault of the car.

[4] For a discussion of how to formulate such expressions in OWL see http://www.w3.org/2001/sw/BestPractices/OEP/SimplePartWhole/index.html.

HIERARCHY OF KINDS
Organ
  Heart
  Aorta
Organ_part
  Heart_valve
    Aortic_Valve
  Cusp_of_Valve
    Cusp_of_Aortic_Valve

HIERARCHY OF PARTS
Cardiovascular_system
  Heart
    Aortic_Valve
      Cusp_of_Aortic_Valve

RESULTANT HIERARCHY OF KINDS OF DISORDERS
Cardiovascular_disease
  Heart_disease
  Valvular_heart_disease
  Aortic_valve_disease
    Disorder_of_cusp_of_Aortic_valve

Fig. 3.1. Interaction of hierarchies of parts and kinds

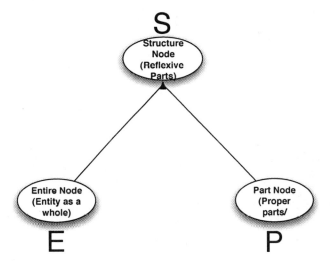

Fig. 3.2. SEP Triple

immediately that a disorder of a part of the ventricle, say the "Aortic Valve" can be inferred to be a disorder of the Ventricle and therefore of the Heart.

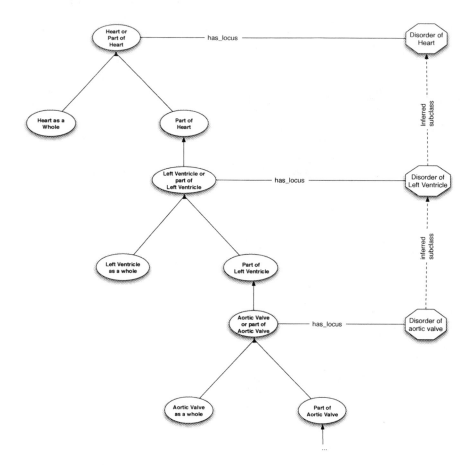

Fig. 3.3. Example hierarchy of SEP triples and interaction with locus and inferred hierarchy of disorders

3.3 Accommodating Clinical/Functional and Structural Views – using a Hierarchy of Relations.

Classical structural anatomy is defined in terms of physical structure and developmental morphology. However, reasoning with these structures in clinical medicine normally involves their functions. We want to preserve the notion that the disease hierarchy results from an interaction of locus and partonomy even when the part-whole relation does not conform, strictly, to the structural notions of classical anatomy.

The paradigmatic case is the pericardium. No clinician would dispute that, in terms of embryological development, the pericardium is a separate organ from the heart. However, few clinicians would expect to find pericarditis classified elsewhere than under cardiac disorders, and most clinicians would expect a comprehensive dis-

Clinical part-of (parthood as commonly used by clinicians)
Structural part-of (the relation used in classical anatomy and the FMA)
Functional part-of (functional partonomy - not yet well elaborated)

Fig. 3.4. A fragment of the hierarchy of partonomy relations showing clinical partonomy subsuming functional partonomy

cussion of malfunctions of the heart to include pericarditis[5] and cardiac tamponade[6].

For these clinical purposes, the pericardium behaves *functionally* as part of the heart. The challenge in a more generic anatomy is to accommodate the clinical view including function without distorting the "pure" view of the structural anatomist.

We can achieve this if we regard the notion of "clinical parthood" as a more general relation of classical structural parthood. Most logic based formalisms support a hierarchy of relations and subrelations.[7] In any such formalism we can define a hierarchy of relations as shown in Figure 3.4 [13]. We wish to say that anything which is structurally part of a whole is also clinically part of the whole, but not vice versa – *e.g.* that the pericardium is clinically part of the heart but not structurally part of the heart. On the other hand, we want to say that the ventricle is clinically part of the heart because it is structurally part of the heart. This is precisely the meaning of stating that structural parthood is a sub-relation of clinical parthoood. In general, to say that one relation is a sub-relation of another, is to say that any two entities related by the sub-relation are also related by the super-relation – in the example given, that any two entities related by structural parthood are also related by clinical parthood. More succinctly, to say that structural parthood is a sub-relation of clinical parthood is to say that structural parts are clinical parts.

This allows us two "views" of parthood, the first using the structual parthood relation is confined to just those things that would be represented by anatomist; the second allows a broader definition to take in those things that would be included in clinical expressions typically expressed, at least informally, as parthood.

In the above, and in our practical developments, we have left "functional parthood" as a placeholder because we do not wish to prejudge a complex discussion which should probably be left for a reference model of biological functions and processes. In particular, whether or not structural parthood always implies functional parthood is not a debate we wish to enter into here.

---

[5] inflammation of the pericardium
[6] restriction of cardiac function because of fluid in the pericardial sac
[7] "properties" and "subproperties" in OWL or "slots" and "subslots" in Protege

3.4 Tissues, Substances and Structures

3.4.1 Constituents: Tissues and Substances

While some clinicians are concerned primarily with gross anatomical structures, pathologists and clinical histologists are concerned primarily with tissues and micro structures. Much of clinical medicine, particularly with respect to cancer and surgery, revolves around the description of tissues. However, it is notable that there is a recognised line between gross and micro anatomy, both in delineation of clinical specialities and in medical education. In addition there are a variety of substances – blood, urine, sweat, etc. – that are also described by related specialists in related terms.

What should be the status of tissues and substances in a terminology of clinical anatomy? Clearly the information conveyed about them is different from that conveyed about gross structures. We say things like "Bone tissue is made up of matrix with osteocytes scattered sparsely throughout", which would make no sense of individual structures.

This issue is paralleled by a continuing debate amongst ontologists concerning how constituents should be represented. The argument is typically formulated concerning a "statue made of clay". The cognitivist or multiplicative approach taken by Guarino and Welty in DOLCE and OntoClean [1, 3, 20] maintains that the clay and the statue should be represented as two different entities, and that the clay "constitutes" the statue. The argument is that if the statue is damaged and looses some clay and then repaired with different clay, it is still the same statue but a different mass of clay. Smith in BFO maintains that, since there can only be one physical object occupying the same space and time, the clay and the statue must be the same entity [2, 18]. Smith resolves the problem of changing composition by indexing parthood by time and does not give identity the same central role in BFO that it has in DOLCE and OntoClean.

Translated to anatomy, the question is whether there is a distinction to be made between a lobe of the liver and the liver tissues that make up that lobe? Or more practically, whether there is a distinction to be made between a "piece of liver parenchyma" and "liver parenchymal tissue".

The philosophical dispute cannot be resolved here. What is clear from a clinical point of view, is that different sorts of clinicians have different things to say to about tissues and structures. There is different information to be represented concerning structures and concerning tissues and substances. Furthermore, a pathologist examining a slide will describe it as "liver parenchymal tissue" or perhaps tissue even if it is for some reason "ectopic" – *i.e.* not found in the liver. In fact the whole notion of "ectopic tissue" makes no sense unless we can speak separately of the tissue and the structure that it would normally constitute. Hence, for purposes of a clinical representation for information systems, it is simpler to treat tissues and the structures they constitute separately, as in DOLCE.

We then label the relation between the structure and what makes it up as the "constitutes" relation, whose inverse we shall label "constituent" and which corresponds, roughly, to the use of "consituent part" in the FMA. Tissues and other constituents are analogous to mass nouns whereas structures are analogous to count nouns – we can say that we have one liver and two kidneys but only that the liver consists of roughly two kilograms of some mixture of tissues. Whereas structures have shape and spatial relationships, mass constituents have composition and arrangements. We can say that the liver is below the diaphragm but that the liver cells are arranged in roughly hexagonal patterns around bile canniculi. Or we can say that the proportion of collagen in liver tissue is of a given value or specify the percentage by weight of sodium ions in serum.

For completeness it is useful also to distinguish a "portion-of" relation between substances and mixtures of substances. We can then say that "substances" form *portions of* "mixtures" which *constitute* "structures".

3.4.2 Collectives

The issue of patterns brings us to the issue of collectives. A pathologist might say, for example, that this pattern was either intact or disrupted. In doing so he or she is referring not to the individual cells, or even to one example of the pattern, but to the overall collective appearance.

The individual cells in tissue, like the individual molecules in water and other substances, are of little concern. Loosing or gaining a discrete structure such as a finger matters clinically; losing or gaining a liver cell or a molecule of water in the interstitial fluid does not.

We are concerned with the collective effects and patterns of cells and molecules, rather than their individual effects. Hence we need to represent the collectives explicitly in the ontology. Typically we use the relation "is grain of – by analogy to grains of sand on a beach – to link the individual grains of a collective to the collective as a whole. There is a recurring pattern. Typically, discrete grains at one level form collectives that act as mass entities at the next, e.g. cells form (portions of) tissue, collectives of molecules form substances, etc. [9, 21].

Putting these two patterns together we have the larger pattern that collectives of grains form mass entities which constitute discrete structures. There are different things to be said at each level, for example of the individual molecules of collagen, of their arrangement connective tissue, and of the ligament constituted by the connective tissue. Frequently different information sources and experts are involved at each level. Maintaining each level separately allows the information system to be factored

Generic part-of (parthood in its most general form which is always transitive)
  Clinical part-of (parthood as commonly used by clinicians - transitive)
    Structural part-of (the relation used in classical anatomy and the FMA-transitive)
    Functional part-of (functional partonomy - not yet well elaborated - transitive
      Constitutes (relation between substances that constitute structures - not transitive)
      Portion-of (relation between a substance and a mixture of substances - transitive)
  Grain-of (relation between a grain and a collective - not transitive)

Fig. 3.5. A fragment of the hierarchy of partonomy relations showing clinical partonomy subsuming functional partonomy

so that each group can deal with it separately and relatively independently.

Note that grains are not sets. Sets are defined by their members. However, the members of collectives change constantly. For example, blood cells are constantly created and destroyed even when all measures of the collective "full blood count" remain constant. Cells are constantly lost from the skin and muscosa, and indeed most tissues without are considering it an injury. Indeed the failure of the normal turn-over represents an abnormality. The same applies to grains of sand on a beach or flocks of birds. The loss or gain of a few members does not affect the identity of the whole until that loss or gain affects the collective behaviour or function or appearance.

Note also that the notion of collective also provides a means of setting limits on the transitivity of locus across partonomy. A disorder of "a liver cell" is not normally treated as a disorder of the liver whereas a disorder of "liver cells" – *i.e.* a collective of liver cells – is considered as a disorder of the liver. The issue of when partonomy is, or is not, transitive is one which has long been debated by mereologists [5]. Many of the cases of interest can be explained if we simply state that, for most purposes, the transitivity of parthood stops at collectives.

Hence in figure 3.5, the "grain of" relation is not a subrelation of the "clinical partonomy" relation because being a grain of a collective that constitutes an entity does not imply being a clinical part of that entity. However, we provide a still higher level notion of partonomy – corresponding to the union of classical mereological partonomy, functional partonomy, and clinical partonomy – which does include the "grain of" relation. (A more detailed example can be found in [12]).

3.4.3 Summary of Views on Structure, Function, Constituents and Grains: Extended Relations Hierarchy

The issues of the differing views of clinical and structural anatomy, the different modes of constituents and structural parts, and the need to query the anatomy sometimes in a sense in which partonomy is considered always transitive and sometimes in a sense at which transitivity stops at the boundaries of grains and collectives can all be captured using a hierarchy of relations as shown in Figure 3.5. Since anything true of any of the subrelations is also true of the super-relations, a query using the most generic part-of will include all structural, functional parts, constituents, portions, and grains transitively. Clinical partonomy includes all relations but grains; structural partonomy just classical structural parts. If there is a need to include a form of parthood that includes both structural parthood, constituents and proportions, this can be accommodated by further elaborating the hierarchy of relations. Although not all disputes can be resolved in this way, developing a hierarchy of relations is a powerful methodology for accommodating multiple views.

3.5 Abnormal Anatomy and Congenital Abnormalities

Most anatomy references deal primarily with normal anatomy. The FMA is explicit in being a reference of "normative anatomy." However, much of clinical medicine concerns abnormal anatomy. Dealing with abnormal anatomy raises a series of issues:

- The notion of "normal" and "abnormal" are difficult to define. In particular do they refer to whether or not a structure is "normative" or "typical", whether or not it performs its usual function, or whether or not it is clinically significant?
- The notion of "normative" is probably distinct from "normal" in the clinical sense. Most references of anatomy, notably the FMA describe themselves explicitly as references on "normative" anatomy.
- The notion of "normality" is applied both to structures as a whole and to features of those structures. If applied to structures as a whole, it may be questionable as to whether any real structure even corresponds completely to the "normative" ideal.
- Notions of "normative" are essentially meta-level notions which relate an individual to the type or class of such individuals. For example, whereas we can say that a hand has five fingers without reference to other hands, we can only say that it is "normal" in relation to other hands.
- Interrelations amongst abnormalities are common. Multiple congenital anomalies often spring from a single cause.

The simplest notion of normality, and perhaps the most useful from a clinical point of view, is pragmatic. A structure is "abnormal" if it is somehow clinically noteworthy and not otherwise. To a first approximation, it is abnormal if a clinician would consider it worth noting in a medical record. GALEN's distinguished two notions pragmatically, motivated by the need to make sense of the common clinical construct "abnormal but not pathological":

- *Normal vs NonNormal* – whether or not some feature or structure was worth noting in a clinical record.
- *Pathological vs NonPathological* – whether or not some feature or structure required medical management in some sense (including continued monitoring).

This pragmatic formulation has two immediate advantages. Firstly, it supports the expected inferences for clinical practice and statistics. Secondly, it is strictly first order.[8] It refers to the individual structure itself without relying directly on any notion of the class. It is a statement that a particular structure is noteworthy or in need of attention, without commitment as to why. In general it also follows the pattern that if a part is non-normal or pathological, then the whole is likewise non-normal or pathological.

Other notions of normality are more problematic and intrinsically higher order. A closely related notion is "typical" which illustrates the practical consequences of such notions being higher order rather than first order, i.e. referring to the class itself rather than to all of the individuals in the class. Such meta-notions are not inherited in the usual way. Consider, for example, the case of mammalian red blood cells. "Typical" mammalian cells have nuclei; typical mammalian red blood cells do not; therefore they are not kinds of typical mammalian cells. However, abnormally red cells may have nuclei, so that certain abnormal mammalian red cells may be classified as kinds of typical mammalian cells in this respect.

As the example of mammalian red blood cells illustrates, the notion of "normative" is intimately tied up with issues of defaults and exceptions. It is slightly easier to manage if we attach the notion of normality not to the structure but to the feature – in this example to "having a nucleus" rather than to the cell type. If we attach the notion of normality to the feature rather than the entity, then the standard frame style default and exception mechanism appears to work reasonably well, but to the best of our knowledge has not been carried through on any large scale for any anatomical reference resource.

In summary, at a pragmatic level, a notion of "abnormality" that amounts to flagging a specific entity as noteworthy is relatively simple and easy to implement. Deeper notions of normative anatomy and abnormal development present difficult conundrums and are beyond the bounds of current formalisms and the scope of this brief chapter.

3.6 Synonyms and Variant Terminology

As in any other terminology, a clear distinction is required between the linguistic "term" and the underlying "concept" or "entity" that it labels. The various clinical terminologies such as SNOMED and ICD each have idiosyncrasies of anatomical naming. There are marked differences between classic references such as *Nomina Anatomica*, the Foundational Model of Anatomy, and various model organism anatomical references. Furthermore, common clinical usage is often at variance with

---

[8] More accurately it is "epistemic", *i.e.* it refers to the clinician's understanding of the specific structure. However, most clinical usage is "epistemic" in this sense.

any official anatomic nomenclature and usually lags behind official changes.

Because the use of anatomical terms in clinical terminologies brings together different disciplines, problems of linguistic labelling often give rise to disputes. In resolving such disputes it is essential to distinguish whether the issue is the label or the underlying anatomical concept. For example, it is clear that the reclassification of the Thymus as an organ of the immune system rather than an endocrine organ is substantive. It is less clear that the renaming of the "left anterior descending artery" as the "anterior intraventricular artery" has any such substantive import or merely represents an improvement in linguistic consistency. Indexing and labelling of clinical resources need to provide both common clinical usage and various official preferred usages.

3.7 Summary

Anatomy is central to clinical terminology, but the notions of anatomy are pragmatic and functional. The clinician's pragmatic view is often at variance with the strict "biological" view of the anatomist or developmental biologist. The key issue for clinical terminology is that, with certain well defined exceptions, disorders and procedures of parts are considered as kinds of the corresponding disorders or procedures of the whole.

It seems likely that many, if not most, of these differences between clinicians on the one side and anatomists and developmental biologists on the other can be overcome by treating clinicians' pragmatic notion of partonomy as an abstract super-relation and the more biologically motivated notions of the anatomist and developmental biologist as subrelations. However, this has not yet been demonstrated on a large scale.

It is essential in clinical anatomy to distinguish between discrete parts and pieces common in gross anatomy and the patterns and arrangements in micro-anatomy. This is is not a matter of size but of whether individual or collective effects are of concern. Cells usually function collectively; hands, and hearts usually function individually. This remains an area of controversy in which no complete solution is forthcoming.

As long as these issues remain, it is likely that the anatomy for clinical resources will remain distinct from major reference anatomies. How much this is a matter of principle, and how much a matter of inertia only the future will tell.

References

1. Aldo Gangemi, Nicola Guarino, Claudio Masolo, Alesandro Oltramari, and Luc Schneider. Sweetening ontologies with dolce. In A Gmez-Perez and V Richard Benjamins, edi-

tors, *European Knowledge Aquisition Workshop (EKAW-2002)*, pages 166–181, Siguenza, Spain, 2002. Spring Verlag.

2. Pierre Grenon, B Smith, and L Golberg. Biodynamic ontology: Applying bfo in the biomedical domain. In D M Pisanelli, editor, *Ontologies in Medicine*, pages 20–38. IOS Press, Amsterdam, 2004.

3. Nicola Guarino and Chis Welty. An overview of ontoclean. In S Staab and R Studer, editors, *Handbook of Ontologies*, pages 151–159. Springer Verlag, 2004.

4. Udo Hahn, Stefan Schulz, and Martin Romacker. Partonomic reasoning as taxonomic reasoning in medicine. In *Proc. of the 16th National Conf. on Artificial Intelligence & 11th Innovative Applications of Artificial Intelligence (AAAI-99/IAAI-99)*, pages 271–276, Orlando FL, 1999. AAAI Press/MIT Press.

5. Ingvar Johansson. On the transitivity of parthood relations. In H. Hochberg and K. Mulligan, editors, *Relations and Predicates*, pages 161–181. Ontos Verlag, Frankfurt, 2004.

6. J L V Mejino and C Rosse. Conceptualization of anatomical spatial entities in the digital anatomist foundational model. *Journal of the American Medical Informatics Association*, (1999 Annual Symposium Special Issue):112–116, 1999.

7. Peter Mork and Philip Bernstein. Adapting a generic match algorithm to align ontologies of human anatomy. In *20th International Conference on Data Engineering*, Boston, 2004. IEEE.

8. Peter Mork, Rachel Pottinger, and Philip Bernstein. Challenges in precisely aligning models of human anatomy. In Marius Fieschi, Enrioco Coiera, and Yu-Chuan Jack Li, editors, *Proceedings of Medinfo 2004*, pages 401–405, San Francisco, CA, 2004. IMIA.

9. James J Odell. Six different kinds of composition. *Journal of Object Oriented Programming*, 5(8):10–15, 1994. A clear presentation, apparently taken from Winston et al 87. Distinguishes: component-integral object, material-object (made-of), portion-object (slice of bread, loaf), place-area (city-state), member-bunch, member-partnership (as a kind of member-bunch). These he distinguishes from non-partitive relations: topological inclusion (seems confussed to me with place-area), classification inclusion (i.e. kind-of), attribution, attachment, ownership. Then gives smashing examples of transitivity and shows how the transitivity of each type is distinct.

10. AL Rector, A Gangemi, E Galeazzi, AJ Glowinski, and A Rossi-Mori. The galen core model schemata for anatomy: Towards a re-usable application-independent model of medical concepts. In P Barahona, M Veloso, and J Bryant, editors, *Twelfth International Congress of the European Federation for Medical Informatics, MIE-94*, pages 229–233, Lisbon, Portugal, 1994.

11. A.L. Rector and J.E Rogers. Ontological and practical issues in using a description logic to represent medical concept systems: Experience from galen. In P Barahona, F Bry, E Franconi, N Henze, and U Sattler, editors, *Reasoning Web*, LNCS 4162, pages 197–231. Springer-Verlag, Heidelberg, 2006.

12. Alan Rector, Jeremy Rogers, and Thomas Bittner. Granularity, scale & collectivity: When size does and does not matter. *Journal of Biomedical Informatics*, 39(3):333–349, 2006. httxp://dx.doi.org/10.1016/j.jbi.2005.08.010.

13. Alan L Rector. Defaults, context and knowledge: Alternatives for owl-indexed knowledge bases. In Russ B Altman, A Keith Dunker, Lawrence Hunter, Tiffany A Jung, and Teri E Klein, editors, *Pacific Symposium on Biocomputing (PSB-2004)*, pages 226–238, Kona, Hawaii, 2004. World Scientific.

14. JE Rogers and AL Rector. Galen's model of parts and wholes: Experience and comparisons. *Journal of the American Medical Informatics Association*, ((Fall symposium special issue)):819–823, 2000.

15. C Rosse, I G Shapiro, and J F Brinkley. The digital anatomist foundational model: Principles for defining and structuring its concept domain. *Journal of the American Medical Informatics Association*, (1998 Fall Symposium Special issue):820–824, 1998.

16. Stefan Schulz and Udo Hahn. Mereotopological reasoning about parts and (w)holes in bio-ontologies. In *Formal Ontology in Information Systems (FOIS-2001)*, pages 210–221, Ogunquit, ME, 2001. ACM.

17. Stefan Schulz, Udo Hahn, and Martin Romacker. Modeling anatomical spatial relations with description logics. In JM Overhage, editor, *AMIA Fall Symposium (AMIA-2000)*, pages 799–783, Los Angeles, CA, 2000. Hanly & Belfus.

18. Barry Smith. The basic tools of formal ontology. In Nicola Guarino, editor, *Formal Ontology in Information Systems (FOIS)*, pages 19–28, Amsterdam, 1998. IOS Press (Frontiers in Artificial Intelligence and Applications).

19. Barry Smith, W Ceusters, B Klagges, J Kohler, A Kumar, J Lomax, CJ Mungall, F Neuhause, A Rector, and C Rosse. Relations in biomedical ontologies. *Genome Biology*, 6(5):46, 2006.

20. C Welty and N Guarino. Supporting ontological analysis of taxonomic relationships. *Data and Knowledge Engineering*, 39(1):51–74, 2001.

21. ME Winston, R Chaffin, and D Hermann. A taxonomy of part-whole relations. *Cognitive Science*, 11:417–444, 1987. find.

22. Songmao Zhang and Olivier Bodenreider. Comparing associative concepts among equivalent concepts across ontologies. In Marius Fieschi, Enrioco Coiera, and Yu-Chuan Jack Li, editors, *Medinfo-2004*, pages 459–464, San Francisco, CA, 2004. IMIA, IOS Press.

4

The Foundational Model of Anatomy Ontology

Cornelius Rosse and José L. V. Mejino Jr.

Summary. Anatomy is the structure of biological organisms. The term also denotes the scientific discipline devoted to the study of anatomical entities and the structural and developmental relations that obtain among these entities during the lifespan of an organism. Anatomical entities are the independent continuants of biomedical reality on which physiological and disease processes depend, and which, in response to etiological agents, can transform themselves into pathological entities. For these reasons, hard copy and in silico information resources in virtually all fields of biology and medicine, as a rule, make extensive reference to anatomical entities. Because of the lack of a generalizable, computable representation of anatomy, developers of computable terminologies and ontologies in clinical medicine and biomedical research represented anatomy from their own more or less divergent viewpoints. The resulting heterogeneity presents a formidable impediment to correlating human anatomy not only across computational resources but also with the anatomy of model organisms used in biomedical experimentation. The Foundational Model of Anatomy (FMA) ontology is being developed to fill the need for a generalizable anatomy ontology, which can be used and adapted by any computer-based application that requires anatomical information. Moreover it is evolving into a standard reference for divergent views of anatomy and a template for representing the anatomy of animals. A distinction is made between the FMA ontology as a theory of anatomy and the implementation of this theory as the FMA artifact. In either sense of the term, the FMA is a spatial-structural ontology of the entities and relations which together form the phenotypic structure of the human organism at all biologically salient levels of granularity. Making use of explicit ontological principles and sound methods, it is designed to be understandable by human beings and navigable by computers. The FMA's ontological structure provides for machine-based inference, enabling powerful computational tools of the future to reason with biomedical data.

4.1 Introduction

The Foundational Model of Anatomy (FMA) ontology is both a theory of anatomy and an ontology artifact. The theory defines anatomy and its content domain and thus provides a unifying framework for grasping the nature of the diverse entities that make up the bodily structure of biological organisms together with the relations that exist among these entities. In other words, FMA theory is a theory of structural phenotype. The FMA ontology artifact, on the other hand, is the computable implementation of the FMA theory. In this chapter we give an account of the FMA theory; the FMA ontology artifact, however, although readily comprehensible when accessed by computer, cannot be reproduced in its entirety on the printed page. We use portions of the artifact to illustrate both the theory and its implementation.

The ontology is designated as foundational for two reasons. First, the high-level nodes of the FMA's taxonomy generalize to vertebrates and, in several respects, to metazoa; second, the entities encompassed by FMA theory are the salient participants of all biological processes which ultimately become manifest as health or disease. Thus, ontologies designed to project to non-anatomical domains of biomedical reality must make explicit or implicit reference to anatomical entities.

The FMA conforms to the definition of an ontology advanced by Grenon et al.:

"An ontology grasps the entities which exist within a given portion of the world at a given level of generality. It includes a taxonomy of the types of entities and relations that exist in that portion of the world seen from a given perspective." [36]

A terminology or vocabulary, on the other hand, is a system of terms relying largely on linguistics and is established for coding or annotating particular kinds of data [78].

Unlike biomedical terminologies and vocabularies, and most extant ontologies, the FMA is not intended to meet the needs of any particular user group or support any particular task, such as the learning of anatomy or the annotation of biomedical data of different sorts. Rather, the FMA ontology is being developed as a *reference ontology*, intended to be reused in *application ontologies* designed to support any computational tool - with or without advanced inference capabilities - which calls for anatomical information. In this sense, the FMA is, in fact, the first of biomedical *reference* ontologies. Consistent with its foundational nature, it is providing the basis not only for several evolving application ontologies, but also for reference ontologies in other basic biomedical sciences, such as physiology, pathology, developmental biology and neuroscience.

The developers of the FMA have greatly benefited from extensive and substantial collaboration with leading investigators in knowledge representation and ontological methodology. The need for depicting the complexity of anatomy in the FMA has served as a motivation for refining such methods and enhancing knowledge representation systems and reasoners. Largely as a consequence of these interactions, the FMA has come to be regarded by the biomedical informatics community as an example of a principled ontology constructed with sound ontological methods [80, 99].

We first introduce a case study to illustrate the kinds of distinctions an anatomy ontology has to make. We shall see that these distinctions are diverse and complex, making the sorting of anatomical entities into types quite challenging. We derive from this case study the need for a theory of anatomy, and then we illustrate the implementation of this theory in the FMA ontology artifact. Before concluding, we illustrate the realization of the FMA's potential as a reference ontology in the basic and applied biomedical sciences.

4.2 Case Study: The Esophagus

We present the esophagus of the human species as a case study to illustrate the challenges for developing an ontology and to provide a consistent cohort of examples in subsequent sections of this chapter. The following account would fit well in a textbook of anatomy intended for biomedical education.

The esophagus connects the pharynx, located in the neck, to the stomach in the abdomen. Its cervical part begins at the level of the 6th cervical vertebra and its abdominal part ends at the level of 10th thoracic vertebra. Its cervical and abdominal parts are connected by a thoracic part, which is located in the posterior mediastinum. Much of the esophagus runs more or less vertically in front and to the left side of the vertebral column. The esophagus is part of the upper gastrointestinal tract and is derived embryologically from the foregut.

The esophagus has the shape of a tube, the lumen of which is surrounded by a multi-laminar wall: innermost is the mucosa, succeeded concentrically by a layer of submucosa, a muscle layer (tunica muscularis) and, on the outside, the adventitia. Each of the inner three layers has its own layers: the mucosa, for example, has epithelium, muscularis mucosae and lamina propria; the tunica muscularis has circular and longitudinal layers. All of these layers consist of portions of different types of tissue. The character of these tissues varies along the length of the esophagus because of differences in cellular composition: the muscle tissue, for example, is striated muscle in the upper part and smooth muscle in the lower part. To support assertions in the last sentence, this account would need to be extended to the types of muscle cells and their respective parts, including some macromolecular complexes, by virtue of which the cell types are distinguished from one another.

The lumen of the esophagus contains portions of swallowed air, saliva and mucus secreted by the mucosa, which cover the luminal surface of the esophageal epithelium; from time to time, it also contains a bolus of food. On its external surface, the esophagus is loosely attached to several of its neighboring structures by extensions of its adventitia, and to the diaphragm by the phreno-esophageal ligament. The structures adjacent to – in other words, touching – the circumference of the esophagus vary from vertebral level to vertebral level and include the trachea and aorta.

A comprehensive account of the anatomy of the esophagus would also include the nerves, arteries, veins and lymphatic vessels which supply or drain (are distributed to) the esophagus. Much anatomical information about the esophagus, however, remains unspecified in available sources, because it is taken for granted; for example, no mention is made in textbooks of the plexuses of blood and lymphatic capillaries which pervade all layers of the esophageal wall, except its epithelium. On the other hand these texts routinely make reference to function.

4.3 Challenges for an Anatomy Ontology

The account of the esophagus is replete with anatomical terms but omits to specify whose esophagus the talk is about. Ontology developers seem to assume that anatomical terms point to plural entities, usually understood to be classes instantiated by individual objects or entities in the real world, such as my or your esophagus and their lumina. Although anatomists – and our case study – may implicitly share this assumption, no explicit reference is made to classes or types in anatomy texts or *Terminologia Anatomica*, the international standard of anatomical nomenclature [29]. Only case reports of anatomical variants or abnormalities make it clear that their accounts pertain to one or a few particular individuals. If the intent of our case study is to describe the "normal" type of esophagus, then the bounds of normality and the meaning of the term type call for specification.

If one is to respect the definition of ontology cited earlier [36], then we must sort the entities which exist in the anatomy domain of biomedical reality into a taxonomy of types, choosing a perspective in which we selectively see these entities. This particular perspective or context will constrain the kinds of entities and relations that come under the ontology's purview. How can the boundaries of anatomical reality be decided and which of the contexts prevalent in anatomical sources and discourse should we choose? Clearly, we must resort to different methods, approaches and even ways of thinking than those employed in text-based artifacts of communication.

In devising an ontological account of the esophagus we must consider the great variety of material entities such as the neck, the esophagus and its various parts; also cells, mucus and a bolus of food; as well as immaterial entities such as a lumen, surfaces, levels or coordinates, the shape of the esophagus and of its cells. Moreover, the particular arrangement of these entities entails diverse relations, such as location, containment, continuity, adjacency, attachment and implied boundaries. Such plurality of properties is not unique to anatomical entities of the size of the esophagus. For example, the anatomy of a pyramidal cell in the cerebral cortex or one of its mitochondria is manifest through similar kinds of properties and relations as those of the esophagus. It seems therefore that anatomical entities of different sorts, size, appearance and complexity share a number of fundamental properties or qualities, whereas other properties distinguish them from one another. We must first focus on those properties which, according to Aristotle, determine their *essence* [10, 58].

We can then begin the sorting with the intent to assure inheritance of general properties by entities of more and more specialized type distinguishable from one another by some properties they do not share. The result should be an inheritance class subsumption hierarchy or taxonomy. The nodes of the taxonomy should be marked by appropriately descriptive anatomical terms, many of which exist in the anatomy literature. We shall see, however, that terms for denoting many types have to be newly introduced. Thus established, such a taxonomy forms the essential part of an ontology which distinguishes it from a term list. The ontology itself should, however, take account also of nonessential or non-definitional properties of the entities under its purview. The association of the latter with nodes of a taxonomy will assure that the ontology can provide anatomical information which will inevitably be called for by different sorts of reasoners and applications.

The essence of the entities on the basis of which a taxonomy is to be established depends largely on the context or view in which its developers regard the domain of their interest. In the scientific literature anatomical entities are, more often than not, viewed in several parallel contexts. Textbooks of human anatomy fall into two main categories: those that subdivide the body into so-called functional systems such as the respiratory or reproductive system, and those that treat it according to so-called regions or body parts, such as the upper limb and abdomen; in each category, however, reference is usually made to structure and function, and even dysfunction, as well as embryonic development. The purpose of these artifacts, however, is not the sorting of the entities under their purview, but rather to serve the needs of particular groups of students and professionals.

Which of these contexts to choose as the predominant one may not be self-evident. The functional orientation is proclaimed by many time-honored sources of human anatomy and some anatomy terminologies. A taxonomy of functional anatomy has in fact been proposed [43]; however, it has not been exploited by developers of anatomy terminologies/ontologies. As the case study may suggest, it is problematic to sort many anatomical entities by virtue of their function. In fact, function tends to be used only for classifying and naming the major systems of the body; whereas the components of these systems are often viewed in structural contexts (see for example *Terminologia Anatomica* [29] and a number of other anatomy terminologies).

Single inheritance, a desirable feature of a sound taxonomy, is more often than not disregarded in anatomy terminologies/ontologies; it presents formidable problems in a functional context. For example, functions of the kidney include the disposition for excreting urine and secreting erythropoietin and renin. Should the kidney be classified both as an excretory and an endocrine organ? Likewise, should a bone such as a vertebra or humerus be classified in three ways: a support organ, a hematopoietic organ and an electrolyte-regulating organ, since in addition to supporting a part of the body, a vertebra and humerus also accommodate bone marrow, and store and release calcium into the circulation to be used in a variety of cellular processes? How could such a classification account for the anatomical differences between a humerus and a vertebra? This is not to deny the fact that functional anatomy is well-nigh indispensable in applications of anatomy such as those for education, biomedical research and clinical practice.

Sorting of anatomical entities into types in a predominantly structural context, however, is not without its challenges either. For example, if the taxonomy is to have one node as its root, the following kinds of questions call for an answer: At one end of the size scale, what are the structural properties shared by the thorax and the lumen of the esophagus? At the other end of the scale, what are the structural properties that distinguish a myofilament in a striated muscle fiber from that in a smooth muscle cell found in different regional parts of the esophageal wall? What properties are shared by the portion of mucus that coats the internal surface of the esophageal wall and the wall itself, and what properties distinguish them? The same questions should be asked about a bolus of food and the esophagus, or any other pair of entities mentioned in the case study. Similar questions arise pertaining to distinctions between relations. Are containment, parthood, adjacency and coordinates, such as 'level of,' different kinds of locations? If not, how are they distinct from one another? Likewise, is attachment a kind of continuity, and if not, how are the two different?

Choosing a predominant context for sorting anatomical entities should be influenced by the essential properties of anatomy itself. These properties should distinguish anatomy from its sister domains, such as physiology and pathology. It follows therefore that before addressing the foregoing problems and questions, we have to answer the question 'what is anatomy'. The answer should assist us in distinguishing essential properties of anatomical entities from incidental or nonessential properties. Coherence of a taxonomy for a domain as large and complex as anatomy can only be assured if decisions about essential (definitional) and nonessential (non-definitional) properties are guided by a unifying theory of the domain.

4.4 Theory of Anatomy

By the term theory we mean a tentative explanation of a portion of reality, derived from principles independent of the phenomena to be explained. The principles for guiding the establishment of an anatomy ontology artifact must be rooted in such a theory. The quality of this artifact will depend to a great extent on the distinctions its underlying theory makes about the portion of reality to which the artifact projects. These distinctions are of two sorts: those made by top-level ontologies, which generalize to any domain of reality, and those specific to a particular domain, such as anatomy. The theory of anatomy propounded by the FMA is rooted in high-level ontology; in particular, the FMA adopts and extends into the anatomy domain *Basic Formal Ontology (BFO)* [36], a domain-independent, spatio-temporal theory which provides a rigorous ontological framework. Some of the challenges of establishing an anatomy taxonomy will be met from the distinctions made by BFO; others by the FMA's unifying theory.

4.4.1 Basic Formal Ontology

BFO deals with the philosophy of reality; the distinctions it makes between the following pairs of entities are fundamental to the FMA:

1. **Reality and knowledge.** Instead of transcribing the content of textbooks, the FMA regards anatomy as a domain of biological reality and comprehends this reality at a more general or higher level than textbooks.

2. **Instances and universals.** Instances exist as discrete, specific individuals, also called tokens or particulars (e.g., my and your esophagi and their lumina); they instantiate universals, such as 'esophagus' referenced in the case study, which implies any esophagus that is presumably "normal". BFO distinguishes between instances and universals by virtue of their location in space and time. Instances exist in particular places at particular times; universals, on the other hand, are multipley located in space and time (all entities conforming to the notion of the term esophagus, which exist in any place at any time: past, present and future).

   Synonyms of the term universal are kind, category, class and type. The FMA selects type as the preferred name for universal; the more widely used term 'class' in some contexts implies the extension of the class (the sum of individuals which instantiate the corresponding universal at a particular time). In the FMA instances of a type are marked out by the fact that, in Aristotelian terms, they share a common essence [10, 58].

3. **Continuant and processual entities**. A continuant is an entity which endures *in toto* while it undergoes changes during the period of its existence; it is bound with respect to space and has spatial parts. A process – designated in BFO as an occurrent – is an entity which does not endure in time; rather it unfolds from its beginning in successive temporal phases to its ending; it is bound with respect to time and it has temporal parts. The instances of the type 'esophagus' as well as the universal they instantiate qualify as continuants, as do the respective surfaces, lumina and their contents.

4. **Dependent and independent entities.** In addition to the orthogonal continuant and process categories, BFO draws distinctions between dependent and independent entities. Processes depend for their existence on their participants. The act of swallowing cannot exist without some esophagus; nor can the process of peristaltic contraction proceed without the muscle layers of the esophageal wall. Such processes are all dependent on some continuant entity, which in an organism is an anatomical structure or a portion of some body substance. BFO also

draws distinctions between dependent and independent continuants. The lumen of the esophagus or its surfaces cannot exist without some esophagus also existing. Function is likewise a dependent continuant, rather than a process. The function of a sperm endures while the sperm exists even if it is never realized through engaging in the process of fertilizing an ovum.

The adoption of such fundamental distinctions by any domain ontology is a requirement for the soundness of the ontology. As far as we are aware, the FMA is the only extensively populated ontology which takes advantage of the theoretical foundations of a top-level ontology and extends the latter into the biomedical domain.

4.4.2 The FMA Theory

Theories are conspicuous by their absence in the biomedical domain. The cell theory advanced in 1838 and 1837 by Schleiden [73] and Schwann [75], respectively, seems to be the only one that has been proposed in the field of anatomy. None of the time-honored textbooks or sources declares a theory for sorting into types cellular and acellular entities, which together make up the anatomy of the human body or that of a metazoan organism. The FMA ontology was the first attempt to fill this gap [71, 72]: it is a theory of anatomy which provides the rationale for implementing the FMA as an ontology artifact. Since its first inception more than ten years ago, the theory has been refined substantially by virtue of the insight its authors have gained during the implementation of the corresponding artifact, and – not the least – as a result of guidance by, and interactions with, leaders in the fields of knowledge representation and ontological theory.

Adoption of the foregoing distinctions made by BFO means that FMA theory should apprehend the anatomy domain of reality by sorting independent and dependent continuants into types. The theory should then account for relations prevailing between these entities such that the relations capture the anatomical characteristics of these entities. The tasks are to 1) draw the boundaries of this domain and demarcate it from other domains; 2) specify distinctions between independent continuants of anatomy and other domains; 3) select a predominant context for viewing anatomical reality; 4) comprehend essential properties of anatomical entities on the basis of which they may be grouped together and distinguished from one another as types; 5) establish a taxonomy of anatomical types supported by Aristotelian definitions that assert the essential properties of instances subsumed by increasingly specific types; 6) define anatomical relations and link a given node of the anatomy taxonomy with other nodes.

4.4.3 What is Anatomy?

The term anatomy commands a plurality of meanings. A recent addition is the one which refers to anatomy ontologies simply as anatomies. Regarding an ontology artifact as an anatomy is a new permutation of an established use of the term for a textbook of anatomy such as Gray's anatomy [93]. Before the 20th century, the term anatomy was also used for the public demonstration of dissections of executed criminals, and as a *mise en scène* for group portraits of surgical societies, exemplified by the so-called 'Anatomy Lesson' of Rembrandt, the correct title of which is 'The Anatomy of Dr. Nicolaas Tulp'.

Dictionaries, on the whole, view anatomy as a branch of biological science. The FMA distinguishes this meaning from the one implying the structure of a biological entity. Anatomy of the hand and anatomy of the mouse are expressions that imply the structure of these biological entities, whereas the human activity primarily concerned with investigating, recording and comprehending the structure of biological organisms and their parts is the science of anatomy, distinct for example from the science of physiology. In other words, anatomy as structure exists as a portion of biological reality and is independent of the way human beings analyze it or create artifacts depicting it.

Despite the fact that *morphe* is shape in Greek and anatomy is a contraction of *ana* and *temnein* – meaning in Greek apart and to cut, respectively – the FMA regards the term morphology as synonymous with anatomy, pointing to both anatomy-science and anatomy-structure. The justification for the synonymy is that 1) *form* is dependent on the *structure* of biological organisms and their parts; 2) the study of form is not the principal component of contemporary anatomical knowledge, whereas structure is; 3) investigators who profess to be morphologists primarily study structure not just form; and so on.

The noun 'structure' is a homonym for a material object composed of parts and for the spatial arrangement and interrelation of the parts of a material or immaterial entity within a whole. In the FMA these two meanings of structure are conjoined, which means that the FMA takes account of anatomical entities and their mereotopological relations. In other words, the essence of anatomy is structure; whereas the essence of physiology is processes in which at least one salient participant is an anatomical structure. Although the terms process and function are often used interchangeably in biomedical discourse, their ontological distinction is fundamental. The demarcation of anatomy and pathology is contingent on the definition of anatomical structure. The association of the dependent continuants functions and pathological entities, as well as processes, with appropriate anatomical entities must be accomplished by inter-ontology relations once an ontology of each of these non-anatomical entities has been established.

4.4.4 Independent Anatomical Continuants

Extrapolating from the adopted meanings of the terms 'anatomy' and 'structure', FMA theory regards each instance of the type anatomical structure as an independent anatomical continuant. We introduce anatomical structure here independent of its taxonomic position, because it stands as the cornerstone of FMA theory. The definition of an instance of anatomical structure is key to comprehending all other types of anatomical entities by virtue of the relations they bear to anatomical structures.

Anatomical structure:

> *material anatomical entity which is generated by coordinated expression of the organism's own genes that guide its morphogenesis; has inherent 3D shape; its parts are connected and spatially related to one another in patterns determined by coordinated gene expression.*

The first and foremost essential property that distinguishes anatomical structures from other material objects is the involvement of genes in their generation or morphogenesis. By morphogenesis we mean the development of an organism's structure or that of any of its parts. A bullet or a swallowed coin is excluded. A prosthesis of a cardiac valve, or one transplanted from another individual, be that a member of the same or different species, does not qualify as a particular individual's own anatomical structure; nor do parasites and bacteria that invade the organism. Similarly, tumors, granulomas and other so-called space occupying lesions are excluded from the type *anatomical structure*.

On the basis of this first essential property, a boundary may be drawn for excluding biological continuants from anatomy and – more importantly – including continuants in this domain. A pathological formation such as a carcinoma of the esophagus is excluded, because gene expression patterns implicated in its generation are distinct from those involved in the morphogenesis of the esophagus or any of its parts. The largest and smallest structures to be included in anatomy may also be specified by virtue of possessing this property. At one end of the spectrum is the body or carcass of the organism itself, and at the other end are the macromolecules synthesized as a consequence of DNA-RNA transcription. Subatomic particles, oxygen and carbon atoms, and carbon dioxide and water molecules are also parts of an organism and participate in biological processes; they are ignored by the theory, however, since they are not gene products. Embryos, fetuses, their parts, and other gestational structures such as the placenta and yolk sac are embraced by FMA theory, because they satisfy the definition of anatomical structure.

The second essential property, inherent 3D shape, distinguishes anatomical structures from other anatomical entities illustrated by the example of the esophagus: the esophagus, its wall, layers of the wall and muscle fibers qualify as anatomical structures because of the shape they possess; whereas the esophageal lumen, portions of mucus and swallowed air, which assume the shape of their container, do not.

The third essential property, the gene-dependence of the arrangement of an anatomical structure's parts, distinguishes *bona fide* anatomical structures from *ad hoc* collections of cells or molecules that may come about within an organism. For example a rouleau, consisting of erythrocytes adherent to one another like a stack of coins, has its inherent 3D shape, but its members are not connected and their ordered arrangement is not influenced by genes, whereas both requirements are fulfilled by the esophagus and the layers and cells of its wall. Likewise the arrangement of cells within a carcinoma does not conform to those established during morphogenesis.

The second and third properties together stipulate that each anatomical structure is a maximally connected entity and that – except for the organism itself – some complement entity must exist for any one of its parts in order to account for the whole.

The FMA theory assigns a dominant role to anatomical structure for three reasons: first, the definition of anatomical structure distinguishes material objects that are alive at some phase of their existence from inanimate objects and thus sets the boundaries of the theory; second, the types of anatomical structures determine the salient levels of organization – also known as levels of granularity – within biological organisms on which distinct levels of biological processes depend; and third, dependent continuants that come under the purview of the theory can be best systematized by virtue of their relation to anatomical structures at various levels of granularity. The elements of the theory discussed thus far should be of assistance in establishing an anatomy taxonomy.

4.4.5 The Anatomy Taxonomy (AT) of FMA

A taxonomy is a tree in the mathematical sense and has the following properties: 1) it has a unique root which serves as maximal universal or type, and 2) the *is_a* relation connects all other types and instances to this root in conformity with the principle of single inheritance. We use the *is_a* relation in accord with its formal definition which includes both the *subtype_of* and *instance_of* relations [83].

The challenges for establishing a taxonomy of anatomy are recounted in Section 4.3. The elements of FMA theory discussed thus far solve several of these challenges: 1) the AT traces over instances, and its nodes point to types of anatomical structures and other entities that depend for their existence on anatomical structures; 2) structure is the predominant context, since structure is the essence of anatomy; 3) types are established on the basis of shared structural properties of instances; processes and functions are excluded.

Several decisions, however, remain to be made: 1) the sense in which the terms entity and type are used; 2) the criteria of normality and deviations from it; 3) the principles for formulating definitions; and 4) selecting the root of the taxonomy.

Entity and Type

Dictionary definitions of the term entity assign it the most general meaning, including things that have real existence and those that do not, such as beliefs and thoughts; some distinguish entities from relations, as seems to be the case also in BFO. The definition of ontology cited earlier [36], however, includes relations along with entities. FMA theory adopts the meaning consistent with the latter definition and includes anatomical relations in the AT.

The term type is generally meant to imply a plural entity which encompasses the majority of the members of a species (see, for example [40]). As we have seen, BFO regards type – synonym of universal – as the entity instantiated by instances or individuals. FMA theory restricts the meaning of the term anatomical type by introducing the factor of *idealization*. The morphogenetic process is subject to a variety of (micro-) environmental influences and consequently its fidelity varies to a greater or lesser extent from individual to individual. Qualitative observations of members of the human and other species, which have been refined and sanctioned by generations of scientists and recorded in textbooks and atlases, however, have resulted in an implicit consensus about the ideal or prototypical anatomy to which each individual and its parts should conform. The nodes of FMA's AT point to such idealized types.

The introduction of idealization has several consequences, in that the theory 1) can sidestep the need for defining the normal; 2) establishes a benchmark with reference to which anatomical variants can be specified and represented as types of anatomical variants (distinct from pathological structures and formations); and 3) makes a distinction between canonical and instantiated anatomy [72].

Canonical anatomy ranges over those types which are idealizations of an organism's body and its component parts. The case study deals with canonical anatomy and the esophagus it describes is an idealized type. *Instantiated anatomy* comprises anatomical data obtained by invasive and non-invasive methods of clinical practice or experimentation about individual organisms which can be documented in clinical and other records and stored in databases. Instantiated anatomy does not come under the purview of the FMA; however, the FMA may provide the framework or schema for storing anatomical data in computable form.

Definitions

The FMA formulates its definitions consistent with Aristotle [58], exemplified by the definition of anatomical structure in section 4.4.4.

The first assertion in the definition specifies the *genus* as *is_a* 'material anatomical entity', which is the immediate taxonomic ancestor or super-type of 'anatomical structure'; the subsequent assertions are the *differentiae*, which, as discussed earlier (section 4.4.4), assert the essential properties shared by instances of the type and by which they may be distinguished from those of other types. It will be observed that only the last differentia conforms to the predominant context of the FMA; the first refers to a process and the second to a physical property indirectly related to structure (e.g., the shape of a cell is determined primarily by the arrangement of its cytoskeleton). These exceptions to the ontology's predominant context, however, are justified for defining high-level types by reference to properties that are independent of the domain of the theory.

The definition does not account for all properties inherent in anatomical structures. Through a lineage of broader ancestor types, anatomical structure inherits additional distinguishing properties asserted by the differentiae of its successive ancestors (Taxonomy 1 [Figure 4.1] and Appendix Table 4.1). This means that the definition of a type is incomplete without those of its taxonomic ancestors. The line of this inheritance becomes evident when anatomical structure is inserted as a node of a taxonomic tree along with its ancestors.

```
= anatomical entity
   = physical anatomical entity
      = material anatomical entity
         ⊕ anatomical structure
         ⊕ portion of body substance
      ⊕ immaterial physical anatomical entity
   ⊕ non-physical anatomical entity
```

Fig. 4.1. Taxonomy 1. High-level nodes of the anatomy taxonomy; here, as in subsequent taxonomies, displayed through the Foundational Model Explorer – FME [32]. Each indentation signifies the *is_a*, or more specifically, the *subtype_of* relation.

The properties inherited from successive taxonomic ancestors may be illustrated by the esophagus, which – we will agree – is an anatomical structure. Material anatomical entities have mass (e.g., muscle fiber, portions of mucus and swallowed air), whereas immaterial anatomical entities (e.g., lumen and surfaces of the esophagus) do not. Both material and immaterial anatomical entities, however, qualify as physical anatomical entities because they have spatial dimensions including the imaginary plane at the level of the 6th cervical vertebra; whereas non-physical anatomical entities such as the longitudinal and circular patterns in the layers of the tunica muscularis and the relations *has_part* and *surrounds* lack this property. All these entities, however, are anatomical entities by virtue of the definition of the root of the taxonomy: anatomical entity (Appendix Table 4.1).

The genus of the type 'anatomical entity' links the AT to the higher-level domain ontology OBR – Ontology of Biomedical Reality [70]; bona fide boundary as a differentia is defined in section 4.4.8 and Appendix Table 4.4.

Thus, all anatomical structures are anatomical entities, possess three-dimensions, their own inherent shape, and are the products of those genes of an individual organism which encode its structure. Although the lumen and surfaces of the esophagus and a portion of mucus do not qualify as anatomical structures, they are anatomical entities. Anatomical entity, therefore, fulfills the requirement for the root of the AT.

Although definitions must be consistent with the understanding of a domain by its experts, in an ontology their primary purpose is to marshal arguments in support of the ontology's taxonomy. The order of succession of the nodes of the taxonomy is established by the genus of successive definitions; instead of relying on mere opinions, the differentiae explicate the justification for including an instance in a given type or excluding it from the type.

4.4.6 Types of Anatomical Structure

The Challenge

As far as we are aware, the entities represented in Taxonomy 1 (Figure 4.1) have not been recognized in the published legacy of anatomy science. With the exception of anatomical structure, the terms pointing to these entities are not to be found in these publications or in the versions of terminologies which predate the FMA. Anatomy science has been primarily concerned with anatomical structures; however, the treatment of these structures by established sources is problematic and presents a number of challenges for the developers of ontologies.

As noted in section 4.4.3, the international standard of anatomical nomenclature [29] and many textbooks organize their content according to so-called functional systems, but do not make the meaning of this term clear. What kinds of anatomical structures constitute such systems? For example, why does the conducting system of the heart not qualify as one of them? Some sources include joints in the skeletal system; in others they are regarded as a separate system, and in yet others, bones and joints are grouped together with muscles as the musculoskeletal system. What type of anatomical structures are bones, joints and muscles? The same question may be asked about the components of body parts or regions such as the upper limb or the back.

Likewise, although the term organ is omnipresent in biology and medicine, *Terminologia Anatomica* provides no indication as to which of its terms point to organs. In fact, the general term organ is omitted from TA altogether. While there is implicit consensus that the liver and uterus are organs, opinions would vary widely about whether or not a nerve such as the vagus, a bone such as the femur, or the knee joint, should be regarded as organs. Although most anatomical terms are defined in dictionaries, the term organ serves as an example to illustrate that such definitions often compound rather than resolve ambiguities.

Organ:
> *a fully differentiated structural and functional unit in an animal that is specialized for some particular function [94].*

a somewhat independent part of the body that is arranged according to a characteristic structural plan and performs a special function or functions; it is composed of various tissues, one of which is primary in function [26].

These definitions mirror the perceptions of anatomists and probably also zoologists; the definitions, however, are satisfied by a number of anatomical structures cited in our case study. The esophagus conforms perhaps best to the most widely held intended meaning of the term. But the definition also holds just as well for the esophageal mucosa, the muscularis and any one of the other laminae of the esophageal wall, as well as the wall itself. The mucosa and the circular and longitudinal muscle layers are arranged in distinctive patterns characteristic for each, have specific functions and are composed of one primary tissue (epithelium and muscle tissue respectively), along with subsidiary connective tissue. The WordNet definition is satisfied by the esophageal wall just as well as by the thorax, the knee joint and the big toe.

A similar difficulty is presented by the current use of the term tissue and its implied meanings. The terms muscle and bone illustrate the difficulty. Depending on their context, these terms may project to macroscopic entities such as the biceps or the humerus, respectively, or to specialized cohorts of cells which are the respective parts of the biceps and humerus. Unfortunately, developers of some ontologies have enhanced rather than improved on these ambiguities. For example, the Adult Mouse Anatomy Dictionary, an ontology of the OBO library [66], classifies connective tissue as an organ system, along with the cardiovascular and nervous systems [1, 38]. The same kind of confusion between tissues and organs pervades another ontology of anatomy [41]. In a treatise on the computational representation of anatomy, a joint, a participating phalanx, its hyaline cartilage, and also the atrial septum and the right and left ventricles of a developing mouse, are all regarded as candidates for the class tissue, because "a distinct name such as right ventricle ... is cumbersome ... and not really required in a computational context" [7].

Such examples highlight the need for sound ontological methods as an approach to eliminating ambiguity prevalent in scientific discourse; an ambiguity which presented no serious problems while human experts were its primary participants. We should not only introduce specificity about the context in which a given ontology views anatomy, but also guard against the injudicious use of anatomical terms by assigning them meanings which make sense only in the context of a particular application domain, such as the annotation of gene expression maps or computational models of physiological function. Such practices will hamper interoperability between computational resources which target anatomical entities at different levels of granularity, discussed in the next section.

Units of Structural Organization

FMA theory approaches the task of sorting anatomical structures into types by considering salient structural units of an organism's corporeal framework. We look for precedence to the cell theory, which established the cell as the fundamental organizational unit of plants and animals [73, 75].

All multicellular organisms begin their existence as a single cell. This cell is destined not only to multiply, but also – governed by the regulation of groups of genes – its descendants become more and more specialized and aggregate into more or less distinct anatomical structures of increasing levels of complexity. Levels of the resulting structural organization have long been recognized by biologists. Since its earliest iterations, the FMA adopted these levels and specified the organizational unit of each [72]. A formal theory has been propounded about granular partitions which correspond to levels of structural organization in biological organisms [13].

Modern biology has focused attention on the products of DNA-RNA transcription. Such *biological macromolecules* are viewed by the FMA as the elementary units of structural organization for three reasons: they satisfy the definition of anatomical structure; they are essential components of all cells; and, suspended in body substances, they exist as discrete anatomical structures. The unit of the level of complexity beyond cell, if properly defined, is *tissue*. We saw in examples cited earlier that supra-cellular units and levels are much more open to opinion and interpretation than molecule and cell, which calls for applying sound ontological methods for the definition and classification of such complex structures. The FMA was the first to propose *organ* as the unit of organization at the macroscopic level in human anatomy, because the units at higher levels can be best defined in terms of the organs which constitute them. These units are *cardinal body part* and *organ system*, which, unlike units at lower levels, overlap each other. The meronymic sum of either cardinal organ parts or organ systems, respectively, is the maximal structural unit, namely the *body* or carcass of an entire vertebrate organism. Some organisms, of course, exist at the cellular or tissue levels and lack organs and cardinal body parts.

Two cautions must be raised. The first is proclaimed by the third differentia in the definition of anatomical structure: as implied by the name 'units of structural organization,' the assembly of units of a lower level into units at higher levels must be governed by genes implicated in morphogenesis. Secondly, in more highly evolved organisms, it seems necessary to define subdivisions of their tissues, organs, body parts and organ systems and treat them as types of anatomical structure for two reasons: the great variety and specialization of anatomical structures at each of these levels; and the prevailing elaborate detail and specificity in which anatomical structures are analyzed and treated in biomedical research and clinical practice. Such subdivisions and so-called cardinal parts of the salient units of organization are *bona fide* types in their own right and serve a transition to the next higher level.

The salient units of granularity levels are highlighted in Taxonomy 2 (Figure 4.2). In the first part of the taxonomy, nodes are aligned in decreasing order of structural complexity, starting with the whole vertebrate body and ending with biological macromolecule, followed by types which span more than one granularity level. The definitions of most types – in reverse order from simple to complex – are shown in Appendix Table 4.2; others may be retrieved from the FME [32].

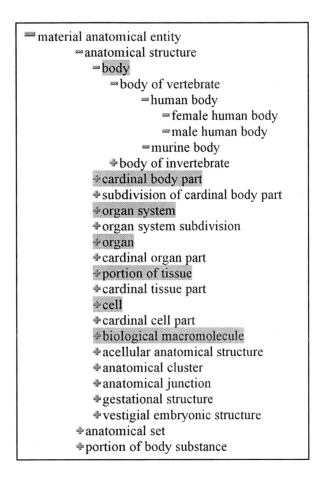

Fig. 4.2. Taxonomy 2. Types of anatomical structure. Salient units of structural organization are highlighted.

It is in the definitions of types of anatomical structure that the structural context is most strictly applied. Some qualifying comments about some of these types, however, are in order.

Cell

All biologists profess to know what a cell is. Quite surprisingly, however, it is not a simple matter to find a satisfactory definition; most sources leave it undefined. A key reference textbook for cell biology asserts: "All living creatures are made of cells – small membrane bound compartments with a concentrated aqueous solution of chemicals." [3]. The assertion is problematic, not only because as the authors demonstrate such compartments can be created by agitating a vessel containing some lipid admixed with an aqueous solution, but also because the assertion is true for several cell parts, such as a cistern of the Golgi apparatus or a mitochondrion, as well as macroscopic entities such as a cyst. WordNet's definition captures the meaning from several hard copy dictionaries: "the basic structural and functional unit of all organisms" [94]. The Cell Type ontology (CL) adopts the definition from MeSH: "Minute protoplasmic masses that make up organized tissue, usually consisting of a nucleus which is surrounded by protoplasm which contains the various organelles and is enclosed in the cell or plasma membrane. Cells are the fundamental, structural, and functional units of living organisms" [18].

The latter definition excludes so-called solocytes which exist independent of any organized tissue; as a consequence of its neglect to distinguish protoplasm from cytoplasm, it also excludes non-nucleated eukaryotic cells, such as erythrocytes, reticulocytes and lens fibers, which lack any nuclear material. In order to include bacterial and plant cells, the Gene Ontology (GO) extends the definition: "The basic structural and functional unit of all organisms. Includes the plasma membrane and any external encapsulating structures such as the cell wall and cell envelope" [34]. If the cell wall and cell envelope are an integral part of the cell, what is the complement of wall or envelope; namely the one bound by the outer surface of the plasma membrane within the wall or envelope; or a cell which lacks such external casings?

All types of cells – including prokaryotic and eukaryotic cells – share two essential properties: a maximally connected cytoplasm surrounded by a maximally connected plasma membrane. The FMA's definition of cell is dependent on the definitions of cytoplasm and plasma membrane, shown in Appendix Table 4.3, along with those of some other cell parts and the whole cell. The latter definition regards the outer surface of a maximally continuous plasma membrane as the boundary of the cell; hence it distinguishes from a cell as a whole cell appendages such as a dendrite or pseudopodium (which consist of less than maximal parts of the plasma membrane and cytoplasm). The distinction of protoplasm and cytoplasm assures that cells lacking nuclear material are classified as *bona fide* cells. This definition holds for cells in interphase and any phase of mitosis and meiosis, as well as their enucleated progeny; it may also be elaborated to include prokaryotic cells.

While some organisms consist of only one cell, in the human body several hundreds of cell types have been distinguished. CL – employing multiple inheritance – classifies cells along the parallel axes of function, histology and lineage [6]; whereas the cell section of the FMA, which antedates CL, sorts cells into a rich hierarchy of 665 types adhering to single inheritance in a predominantly structural context. The high level nodes of the resulting ontology are shown in Taxonomy 3 (Figure 4.3):

```
≡anatomical structure
    ≡cell
        ≡nucleated cell
            ≡haploid nucleated cell
                ✤oocyte
                ✤sperm cell
            ≡diploid nucleated cell
                ≡somatic cell
                    ✤epithelial cell
                    ✤muscle cell
                    ✤connective tissue cell
                    ✤neural cell
                    ✤hemal cell
                    ✤stem cell
                ≡diploid germ cell
                    ✤primordial germ cell
                    ✱ primary oocyte
                ✱ zygote
        ≡non-nucleated cell
            ≡non-nucleated solocyte
                ✤erythrocyte
                ✱ platelet
            ≡non-nucleated colligocyte
                ✱ corneocyte
                ✱ non-nucleated lens fiber
    ✤cardinal cell part
```

Fig. 4.3. Taxonomy 3. The major categories of FMA's cell types. Except for five leaf nodes, a hierarchy of subtypes in most categories remains unopened.

Tissue

Tissues are usually referred to in the biomedical literature in the context of types (e.g., columnar epithelium, mesenchyme), whereas in reality, tissues exist as concrete portions within an organism, rather than as mass objects – a requirement for them to qualify as anatomical structures; hence the corresponding organizational unit is *portion of tissue*; the types in FMA's AT point to such portions. In Taxonomy 4 (Figure 4.4) and the FMA artifact, however, the phrase 'portion of' is omitted in the name of subtypes of the four major tissues types, taking it for granted that all these subtypes exist also in portions. The FMA's definition of portion of tissue honors the definition of tissue in established textbooks of histology (Appendix Table 4.2); however, as noted earlier, it is at variance with the use of this term in some fields of biomedical discourse and some terminologies or ontologies.

```
⇒anatomical structure
    ⇒portion of tissue
        ⇒portion of epithelium
            ⇒ unilaminar epithelium
                ⇘simple squamous epithelium
                ⇘simple cuboidal epithelium
                ⇘simple columnar epithelium
            ⇒ multilaminar epithelium
                ⇘stratified squamous epithelium
                ⇘stratified cuboidal epithelium
                ⇘stratified columnar epithelium
                ⇘mixed stratified cuboidal and columnar epithelium
                ⇘transitional epithelium
            ⇘ atypical epithelium
        ⇒ portion of connective tissue
            ⇘ regular connective tissue
            ⇘ irregular connective tissue
        ⇒ portion of muscle tissue
            ⇘smooth muscle tissue
            ⇒striated muscle tissue
                ⇘skeletal muscle tissue
                ⇘cardiac muscle tissue
        ⇘ portion of neural tissue
        ⇘ portion of heterogeneous tissue
            ⇘lymphoid tissue
            ⇘myeloid tissue
    ⇘portion of cardinal tissue part
```

Fig. 4.4. Taxonomy 4. Types of portion of tissue.

The classification extends to several tiers of subtypes beyond most of the nodes shown, mirroring the specialization of tissues by virtue of the cells of which they are predominantly composed. Depending on the tissue type, there is a varying degree of anatomical – or morphological – similarity among its predominant cells. Unilaminar epithelia are more homogeneous; the stratification of several multilaminar varieties reflects the structural differences in the cohort of cells as they move in unison from a basal to a superficial stratum through a maturational gradient. The heterogeneity is most startling in the epidermis, a keratinized subtype of stratified squamous epithelium. Although there is a direct developmental lineage between cells of the basal and keratinized strata, arguments could be advanced for regarding each stratum as a distinct tissue type. To respect traditions, however, the FMA classifies such subdivisions of a tissue as *cardinal tissue part* (see below), a sibling node of portion of tissue.

The definition accommodates incidental cells in a portion of tissue of a particular type (Appendix Table 4.2). Although, in most cases, they are too few to be visible, there is experimental evidence that stem cells are present in the majority of, if not all, tissues. Also, various types of macrophages and other immunologically competent cells normally pervade loose connective and other tissues. Portions of connective tissue provide the bulk of such anatomical structures as ligaments and bones; however, with the exception of epithelia, some type of connective tissue is also present in portions of various types of muscle, neural and heterogeneous tissue. During development, this incidental connective tissue is largely responsible for establishing the patterns in which the principal cells are arranged. It will be recalled from the definition of anatomical structure (Appendix Table 4.1), that such a characteristic pattern is a requirement for a cohort of cells to qualify as portion of tissue. Moreover, the incidental connective tissue component of portions of tissue of all other types contains nerve fibers, blood and lymphatic capillaries, and larger pre- and post-capillary vessels, essential for the survival and functioning of the principal cells of a tissue other than epithelia. Thus it is erroneous to think of tissues as mere aggregates of similar cells.

Organ

In fully developed vertebrates, portions of types of tissue are not found outside the confines of organs, except for a subtype of connective tissue – areolar connective tissue – which loosely connects organs around their circumference, allowing them to move and modify their shape independent of one another. Such connections are illustrated by the extensions of the esophageal adventitia to neighboring organs. The breaking of such tenuous connections permits ready separation and demarcation of one organ from another (e.g., esophagus from trachea or vertebrae) without the use of sharp tools in either the living or dead body of most vertebrates, which is one factor that justifies regarding organ as the unit of macroscopic anatomy.

Although continuity may prevail between portions of two types of tissue (e.g., non-keratinized stratified squamous epithelium of esophagus and microvillous columnar epithelium of the stomach), as a rule, it is also connective tissue that secures portions of two or more types of tissue to one another as they contribute to structures of a higher order, such as the wall of the esophagus.

As the definition of organ suggests (Appendix Table 4.2), multi-tissue complexes may qualify as *simple organ* by virtue of their relative independence from other similar structures - particularly the case in organisms of lower orders. In vertebrates, however, as a rule, multi-tissue complexes are united to form anatomical structures which by themselves do not qualify as organs; they need to be 'welded' to other similar multi-tissue complexes to form *compound organs*. The FMA classifies structures intermediate in complexity between portion of tissue and organ as *cardinal organ part*. At one end of the size and complexity scale is, for example, the circular muscle lamina of the cervical esophageal wall – consisting of striated muscle and connective tissue; at the other end of the scale is the wall of the entire esophagus or the cervical part of esophagus; all these cardinal parts connected together qualify as the compound organ esophagus. It follows, therefore, that unless the requirements asserted by the differentiae in the definition of organ are fulfilled, a structure should not be classified as an organ, even though this term may be part of its conventional name (e.g., organ of Corti, which in the FMA is an organ component).

```
▬anatomical structure
    ▬organ
        ▬simple organ
            ✤parathyroid gland
            ✳ paraaortic body
            ✳ paraganglion
            ✳ coccygeal body
            ✤carotid body
        ▬compound organ
            ▬solid organ
                ▬parenchymatous organ
                    ✤lobulated organ
                    ✤corticomedullary organ
                ▬nonparenchymatous organ
                    ✤muscle organ
                    ✤ligament organ
                    ✤cartilage organ
                    ✤membrane organ
                    ✤neural tree organ
                    etc.
            ▬cavitated organ
                ✤organ with organ cavity
                ▬organ with cavitated organ parts
                    ✳ heart
                    ✤cavernous organ
                    etc.
    ✤cardinal organ part
```

Fig. 4.5. Taxonomy 5. The high-level types of organ.

Definitions of organ types may be retrieved via the FME. These definitions employ as differentiae salient structural properties of instances of each type: whether or not organs have a cavity; if they are solid, whether their cardinal parts are arranged in lobules and lobes or as an inner core, the medulla, capped by a cortex; or whether they conform to neither of these typical arrangements; and if a cavity is present, whether it occupies the entire organ (e.g., lumen of the esophagus, cavity of urinary bladder), or major parts of the organ (e.g., a cardiac chamber, an air cell of the ethmoid bone). At a lower nodal level, the differentiae for cavitated organs, for instance, are specified by the organs with which each connects. For example instead of the conventional definition of the human heart by its functioning as a pump, in the FMA the heart *is_a*

> organ with cavitated organ parts which has as its parts chambers continuous with the systemic and pulmonary arterial and venous trees;

and the liver *is_a*

> lobular organ which has as its parts lobules connected to the biliary tree.

No anatomical structure other than the heart and liver satisfy these definitions, respectively, whereas major arteries also pump blood, though perhaps less forcefully than the heart.

Even parenchymatous organs, such as liver, lung or kidney, are not truly solid; they resemble a sponge in that some of their parts contain spaces of a lesser dimensional magnitude than the organs themselves. These spaces include the bile canaliculi, air filled cavities of alveolar sacs and the lumina of renal tubules, as well as the lumina of blood and lymphatic capillaries. Hence, for an organ to qualify as a cavitated organ, it must have as its parts one or more anatomical spaces of the same dimensional order of magnitude as the organ itself.

Cardinal Parts and Subdivisions

Rather than implying any part transitively removed from the whole (e.g., as an epithelial cell as part of the esophagus), the intent with the use of the terms cardinal part and subdivision in Taxonomy 2 is to point to anatomical structures of distinctive types. The corresponding definitions specify the criteria for assigning instances to these types (Appendix Table 4.2).

Cardinal Organ Part

Although *cardinal organ parts* are also composed of more than one portion and type of tissue, they are distinguished from simple organs by virtue of their continuity with their complement in constituting a compound organ. As a rule, they cannot be demarcated from one another by blunt dissection as most organs can.

Each of the circular and longitudinal layers of the tunica muscularis, for example, qualifies as a cardinal organ part, because its direct parts include not only portions of muscle tissue, but also portions of connective tissue. The latter is essential for packaging portions of muscle tissue into bundles and sheets, and arranging these bundles within a sheet in circular or longitudinal patterns. During development the establishment of these distinctive patterns of muscle fiber arrangement seems to be mediated by gene products in embryonic connective tissue (mesenchyme). Such arrangements of portions of striated or smooth muscle account for the emergent structural and functional properties of these cardinal organ parts and distinguish them from portions of tissue. For example, contraction of any portion of muscle tissue results in its shortening; whereas the summation of the contraction of muscle laminae in esophageal wall is manifest as a peristaltic action; a similar summation of contraction of portions of muscle tissues in the muscle belly and long and short heads of the biceps results in flexion of the elbow and supination of the hand. The explanation for these rather startling differences in muscle action in these two organs is to be found in the distinct patterns in which quite similar muscle fibers are arranged in their cardinal organ parts.

In addition to the wall of cavitated organs and their laminae, characteristic cardinal organ parts are the cortex and medulla of the human kidney, the head and shaft of the humerus, a cardiac chamber and a lobule or lobe of the liver.

Lobes of lung and liver may seem to be exceptions to the requirement for continuity among cardinal organ parts. In some species lobes of these organs have free surfaces whereas in others they do not. Lobes of the human lung and murine liver can be freely separated from one another; those of the human liver and mouse lung cannot. Continuity between these distinct lobes is always present at their root or the hilar region of the organ they constitute. The fundamental architecture of lobulated organs is established by the branching pattern of the hollow tree responsible for their morphogenesis; fissures which carve the lobular or acinar parenchyma into the larger chunks of lobes are inconsequential and variable among and within species.

Cardinal Body Part and Organ System Subdivision

Terminologia Anatomica lists a number of body parts and body regions which overlap to a great extent in both name and meaning [68]. The FMA's definition distinguishes *cardinal body part* and admits only four types of anatomical structures in this category: head, neck, trunk and limb. Likewise, only those body systems qualify as instances of *organ system* which have as their direct parts organs connected to one another. Thus, the gastrointestinal and respiratory systems qualify, whereas the conducting system of the heart, cited earlier, does not, since its parts are portions of specialized tissue; neither do functional systems made up of unconnected organs (e.g., endocrine and immune systems), the sum of which does not constitute one anatomical structure.

Both cardinal body parts and organ systems are subdivided into structures larger than organs which are designated in the FMA as their respective *subdivisions* (Taxonomy 2 [Figure 4.2]). For example, thorax and abdomen are classified as subdivisions of trunk; forearm and hand subdivisions of the upper (pectoral) limb; the upper and lower gastrointestinal and respiratory tracts as subdivisions of the alimentary and respiratory systems, respectively. Both types of subdivisions have entire organs as their direct parts (e.g., lungs and stomach or discrete bones and muscles); whereas other organs cross the boundary between subdivisions as does the tracheobronchial tree between upper and lower respiratory tracts and the tendons of several muscles between forearm and hand.

Miscellaneous Anatomical Structures

A theory of anatomy must also account for anatomical structures that do not fit naturally into the units of structural organization. The last five nodes of Taxonomy 2 (Figure 4.2) project to such structures.

Acellular anatomical structure is the type that subsumes, for example, basement membrane (on which an epithelium is supported), collagen fiber, zona pellucida of ovum, an otolith in organs of balance or an intracellular crystal.

The root of the lung and the renal pedicle consist of cardinal parts of several organs (principal bronchus, a pulmonary artery, pulmonary veins, lymphatic vessels and nerves in the lung root) grouped together in a predetermined manner; the collection, however, dos not qualify as either the cardinal part of an organ or the body, nor as a subdivision of an organ system. To account for such structures, we introduce the type *anatomical cluster*. A joint, such as the interphalangeal joint, is an anatomical cluster made up of the joint capsule, synovial sac (each an organ) and the proximal and distal ends of phalanges (cardinal organ parts) covered by articular cartilage (a portion of tissue, part of the articulating bones). Anatomical clusters exist at various levels of granularity exemplified by the juxtaglomerular complex (made up of the macula densa and juxtaglomerular and mesangial cells) or a nerve fasciculus, the parts of which are zones of a number of axons surrounded by a perineurial sheath.

Anatomical junctions, such as the pharyngo-esophageal and csophagogastric junctions mentioned in the case study, and others at the cellular level such as synapses, neuromuscular junctions and desmosomes, establish continuity between two or more anatomical structures. Each type of junction has its own characteristic components and structure.

The embryo, fctus and their parts, qualify along with the placenta, amnion and umbilical cord as anatomical structures. They are grouped together under the type *gestational structure*. Some embryonic structures persist postnatally in a vestigial state and assume a different character. In human beings, examples of such *vestigial anatomical structures* include a lateral umbilical ligament (distinct from ligaments of the musculoskeletal system) which, before birth, was an umbilical artery, and the appendix of the testis, which persists as the fibrous transformation of some mesonephric ductules.

All the above miscellaneous structures qualify as anatomical structure because they possess an inherent 3D shape and come about, directly or indirectly, as the result of morphogenetic processes regulated by particular groups of genes.

Summary of Anatomical Structure

In conclusion, anatomical structures at each level of granularity share some structural properties inherited from their taxonomic ancestors, and also exhibit additional properties specific to their own level. These inherited and level-specific attributes account for the emergent properties of anatomical structures at levels of increasing structural complexity. One of these emergent properties is the potential they manifest for participating in higher level biological processes than those at a lower level, illustrated earlier by different actions exerted by portions of muscle tissue and those of the esophagus and biceps.

Thus, as a result of designating not only cell but also biological macromolecule, portion of tissue, organ, cardinal body part and organ system as units of granular partitions, the human body, or the body of any vertebrate, can be stratified into seven salient levels of structural organization. Five transitional levels provide the connection between the salient levels (Taxonomy 2 [Figure 4.2]).

Such a structural stratification of the vertebrate organism advanced by theories of the FMA and granular partitions is by no means original. Notions similar to the levels here propounded are implicit in many time-honored accounts of anatomy. However, the notable distinction is that the types of anatomical structures, body substances and boundary entities encompassed by each organizational level are explicitly defined in the context of FMA's taxonomy, whereas in other sources such entities, notably tissue and organ, remain more or less ambiguous.

4.4.7 Other Material Anatomical Entities

Anatomical Set

Singular material objects forming part or the whole of an individual organism are classified as the type anatomical structure. Such singular structures need to be distinguished from plural material objects which exist as collections, distinct from types. Such collections are the referents of terms such as ribs and spinal nerves. These terms as used in anatomical discourse do not point to any number of ribs or spinal nerves, but rather to their maximal number in a canonical member of a given species; for example 12 pairs of ribs and 32 pairs of spinal nerves in a human being. We designate such maximal collections as the type *anatomical set*, which is a sibling, rather than a subtype of anatomical structure in the ontology (Taxonomy 1 [Figure 4.1]; Appendix Table 4.1). Set of ribs and set of cranial nerves are two subtypes of anatomical set.

These particular sets consist of organs; anatomical sets, however, exist at all levels of granular partitions. For example, all skeletal muscle fibers innervated by a single alpha motor neuron are members of the anatomical set myone; these members intermingle with members of other myones, an arrangement which offers significant functional advantages.

The foregoing examples illustrate that, unlike anatomical clusters, anatomical sets have members rather than proper parts in that sets lack one maximal boundary; no direct continuity or spatial adjacency prevails between the members; and members are of the same type which is not the case for parts of anatomical clusters.

Members of an anatomical set are distinct from elements of a mathematical set in at least two respects: 1) indirect connections exist between them since, with a few notable exceptions, all anatomical structures of an organism are interconnected directly or indirectly; and 2) as a rule, the members are ordered in accord with genetically determined patterns. For example, the oculomotor nerve is the third pair in the row as cranial nerves emerge from the brainstem; the second rib on the right is not interchangeable with the left second rib or with the right third rib. To our knowledge, the pattern of intermingling between members of particular myones within a muscle has not been analyzed; it is, however, unlikely to be random.

Portion of Body Substance

Anatomical structures, including clusters and members of anatomical sets, have their own inherent 3D shape. Material anatomical entities which lack this property adopt the shape of cavities and spaces within or among anatomical structures (e.g., swallowed air, saliva and mucus in the esophagus) or, like enamel, are molded to the surfaces of anatomical structures. To designate this type of entity at the highest level, we borrow the term body substance from current clinical usage, which is distinct from substance in Aristotle's categories [4]. Portion of body substance is defined in Appendix Table 4.1 and its subtypes are shown in Taxonomy 6 (Figure 4.6).

Like tissues, body substances exist in biological organisms as distinct portions rather than as mass entities. The differentiae for distinguishing subtypes include composition and containers. For example,

portion of blood:
> *portion of body substance which has as its direct parts blood cells suspended in a portion of plasma.*

Blood cell and portion of plasma are defined independently of blood.

Leaf nodes in this taxonomy point to specific anatomical spaces which contain the corresponding instances of portion of body substance. As for tissues, the phrase 'portion of' is taken for granted in most cases. For example, portions of blood can be

```
═material anatomical entity
    ═portion of body substance
        ✦secretion
        ✦excretion
        ✦transudate
        ✦blood
        ✦blood plasma
        ✳ semen
        ✳ vitreous humor
        ✦aqueous humor of eyeball
        ✳ colloid of thyroid follicle
        ✦intercellular matrix
        ✦substance of tooth
        ✦cell substance
        ✦ingested food
        ✦gas in anatomical space
```

Fig. 4.6. Taxonomy 6. Types of portion of body substance.

distinguished by the vascular trees or the blood vessels that contain them. For example, the portion of blood in a pulmonary arterial tree is distinct not only from that in a pulmonary venous tree, but also from the portion of blood in a coronary arterial tree. Such distinctions and their refinements have practical importance in physiology and clinical medicine. For example, during coronary angiography, oxygen saturation is assessed separately for zones and branches of a coronary artery distal and proximal to a partial blockage in order to inform therapeutic decisions. The annotation of such detailed clinical data calls for corresponding resolution in the parts of the coronary arterial tree and their contents, levels of specificity not to be found in any current anatomy textbook. Similar levels of specificity are called for also by computational mathematical models of physiological processes (e.g., [45]).

The taxonomy of portion of body substance in the FMA is as yet tentative: many subtypes have not been defined and some of the definitions rely for differentiae on the structures that synthesize or filter the particular substances (e.g., secretions, exudates, transudates), which, though sensible and useful, is not strictly consonant with a structural context.

One of the salient distinctions made by FMA theory is that between portion of body substance and portion of tissue. Time-honored textbooks of anatomy and histology have for long been regarding such body substances as blood and lymph as subtypes of connective tissue. Because of the fundamental properties they share, portions of blood and lymph are classified in the FMA together with the substances that fill the cavities of anatomical structures at all levels of granularity. They all lack a defining property of portion of tissue: a predetermined pattern of their architecture.

4.4.8 Immaterial Anatomical Entities

Unlike portions of swallowed air, saliva and mucus, the lumen of the esophagus which contains portions of these substances has no mass, although it has spatial dimension. By virtue of these properties lumen is classified as immaterial anatomical entity, a subtype of physical anatomical entity. The external and internal surfaces of the esophagus fall also into the same category, as do the virtual planes which demarcate the esophagus from the pharynx (plane of pharyngo-esophageal junction) and the stomach (plane of esophagogastric junction). Immaterial entities are categorized on the basis of whether they have three or fewer spatial dimensions (Taxonomy 7 [Figure 4.7]); the former are anatomical spaces and the latter operate as boundaries. These and further distinctions are captured by the definitions (Appendix Table 4.4).

Anatomical Space

The first criterion for sorting anatomical spaces into cavities and compartment spaces is whether or not their boundary is provided by the surface of one or more anatomical structures (Appendix Table 4.4). The second criterion is the content of the spaces: anatomical cavities contain portions of body substances, whereas compartment spaces contain anatomical structures.

The lumen of the esophagus and the lumina of blood vessels qualify as cavities, as do the spaces enclosed by the pleura, peritoneum, stomach and right ventricle. Compartment spaces contain cells or organs such as members of thoracic viscera (e.g., space of mediastinum). Despite its name, the abdominal cavity is classified as a compartment space because it is bound by the surfaces of a number of muscle organs and it is filled by organs such as the kidneys and the peritoneal sac, rather than by body substances; whereas the space within the peritoneal sac is a cavity, because it is surrounded by the wall of the sac and contains a portion of peritoneal fluid. Likewise, the space bound by the internal surface of the plasma membrane contains a maximal portion of cytosol; therefore, the FMA classifies it as an anatomical cavity, although this space is spoken of by biologists as a compartment [3].

The FMA also distinguishes anatomical conduits, which connect two or more spaces with one another and may contain either portions of body substances (e.g., ostium of coronary artery, median aperture of fourth ventricle known also as the foramen of Magendie, atrioventricular orifice), or anatomical structures (e.g., foramen magnum, space of inguinal canal, pulmonary hilum).

Spaces in a developing organism can be categorized according to these three types. Such spaces are, however, left in gestational space, a category of their own for the time being, mainly for the purpose of drawing attention to them by ontology

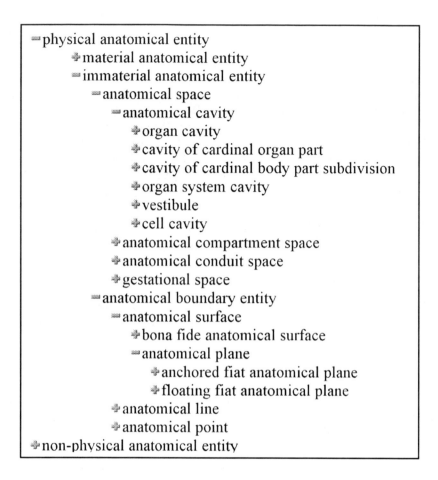

Fig. 4.7. Taxonomy 7. A selection of high-level types of immaterial anatomical entity.

developers concerned primarily with embryonic development (e.g., [27, 37]).

The distinction between anatomical cavities and compartment spaces is not a matter of gilding the lily. It is called for by the need to forestall erroneous conclusions by reasoners. For example, the presence of portions of body substance in compartment spaces, such as blood in the space of posterior mediastinum, would signal a medical emergency.

Anatomical Boundary Entity

Boundaries exist in reality in that they mark a natural discontinuity between objects. They are also employed extensively to subdivide an organism and its components into parts where natural discontinuities may not exist. In human anatomy, only spaces are said to have a boundary, which is equated with the wall or walls of the space. The FMA makes a distinction between boundary entities and walls. A physical anatomical entity must have fewer than three spatial dimensions to qualify as a boundary. For example, because it has two dimensions, the internal surface of the esophagus is a boundary entity, and as we shall see, it may be associated with either the wall of the esophagus or its lumen as the second relatum linked by the *has_boundary* relation (section 4.4.9). Although biomedical discourse makes routine reference to anatomical surfaces (e.g., cell surface, body surface, diaphragmatic surface of lung, abluminal surface of epithelium), they are not usually regarded as boundaries. Since they are taken for granted by traditional sources, boundaries are in general ignored by ontologies in the biomedical domain, although they have been thoroughly treated in ontology theory [77]. Boundaries are implicit in systems of categorization or sorting; they operate in the decomposition of an entity into its parts, notwithstanding the fact that some theories of mereology do not account for boundaries explicitly. FMA theory adopts Smith's treatment of boundary [77] and with his guidance extends it.

There is a distinction between the surfaces of the esophagus and the planes that demarcate the esophagus from the pharynx and stomach. The surfaces mark a discontinuity between the wall and the lumen, and also the neighborhood, of the esophagus; whereas the pharyngo-esophageal and esophagogastric planes, which demarcate the esophagus superiorly and inferiorly, respectively, are imposed by consensus across continuities which exist between the walls and spaces of the pharynx, esophagus and stomach. The surfaces are natural or *bona fide* boundaries, such as the one which demarcates an organism from its external environment, or a red blood cell within the portion of blood in which it is suspended. The planes are *fiat* or *virtual* boundaries, across which natural continuity prevails. The FMA extends these distinctions by designating some fiat boundaries as anchored and others as floating fiat boundaries. The position of the plane of the thoracic inlet which demarcates the cervical from the thoracic part of the esophagus is anchored by the level of the first pair of ribs. No comparable fixed reference exists, however, for the plane that demarcates the upper part of the esophagus from the lower part (in which the muscularis has distinct properties), the apical and basal parts of the lung or the apical and basal parts of a columnar epithelial cell. All the latter planes fall into the category of floating fiat boundary. Anatomical planes, both anchored and floating, are widely used for subdividing the body and other anatomical structures in anatomical and clinical descriptions of the human body.

Both bona fide and fiat boundaries operate also in demarcating 2D surfaces and planes by 1D anatomical lines. The sharp anterior border of the somewhat semi-cone-shaped human right lung is a bona fide boundary because it is an anatomical line formed by the intersection of the lung's costal and mediastinal surfaces; whereas the so-called posterior border is rounded and the demarcation between the two surfaces posteriorly is a floating fiat boundary. The intersection of the line of the horizontal fissure with the anterior border of the right lung marks an anatomical point, which is a bona fide boundary between the anterior borders of the upper and middle lobes.

A number of anatomical points and anchored fiat lines, such as McBurney's point and Nelaton's line, serve as useful landmarks and guides for clinical diagnosis and surgical procedures.

4.4.9 Anatomical Relations

The term relation has many meanings. In ontology theory, relation is a primitive which asserts some kind of association between two or more entities, such as A *is_a* B or A *contains* B. Relations in anatomy assert associations between anatomical entities. Relations between anatomical entities and those of other domains (physiology, pathology) do not come under the purview of a theory of anatomy or of anatomy science as defined here. Since they have no spatial dimension and cannot be quantified, the FMA classifies anatomical entities as one of the three subtypes of non-physical anatomical entity (Taxonomy 8 [Figure 4.8]).

The case study (section 4.2) illustrates the indispensable role relations play in taking account of the structure – i.e., anatomy – of anatomical entities. Such relations figure extensively in anatomical and clinical descriptions, but except for the part relation they have for long been largely ignored or inadequately treated by anatomy terminologies. For example, the Adult Mouse Anatomical Dictionary limits structural relations to parthood and seems to use the *is_a* relation for specifying location and containment, as in heart *is_a* thoracic cavity organ [1, 38]. The hierarchies of the international standard of anatomical nomenclature fail to specify the nature of links between their terms and only those familiar with human anatomy can imply that in a given hierarchy a link may be intended to mean *is_a*, *part_of* or *branch_of* [68]. A notable exception to these unsatisfactory practices is GALEN [33], the anatomy module of which predates the FMA, and employs several anatomical relations.

The challenge for a theory of anatomy is illustrated by the following kinds of questions related to the case study: Are the surfaces, wall, lumen and portions of mucus and swallowed air all part of the esophagus? Is the nature of the connection between the stomach and the esophagus the same sort as the one that anchors the esophagus to the diaphragm? Are the arborizations and networks of nerves and blood vessels embedded in the esophagus part of its wall, or part of the respective neural and vascular trees, or both? How can the location and position of the esophagus be specified with respect to the posterior mediastinum and the anatomical structures that surround it? And so on.

Adopting some of the precedents in GALEN and UMLS (Unified Medical Language System; [90]) – which also includes and defines several anatomical relations – evolving versions of the FMA have incorporated an increasing number and kinds of relations. Not only the number but also the expressivity and specificity of relations pertaining to anatomy has been extended and refined. As a result, the FMA has motivated much of the recent interest in relations by biomedical ontologists

[25, 64, 65, 79, 82, 83]. Taxonomy 8 (Figure 4.8) shows the salient relations employed by the FMA, which are defined in Appendix Table 4.5.

FMA theory distinguishes between two major categories of relations: taxonomic and structural. The former generalize to any domain; while none of the latter is unique to anatomy, they are particularly apt for specifying the arrangement of the physical parts of an organism.

```
⚌ non-physical anatomical entity
     ⚌ anatomical relation
          ⚌ taxonomic anatomical relation
               ⚌ is_a
                    * has_instance < > instance_of
                    * has_type < > type_of
          ⚌ structural anatomical relation
               * has_spatial_dimension < > spatial dimension of
               * has_shape < > shape_of
               * has_boundary < > boundary_of
               ⚌ has_part < > part_of
                    * has_generic_part < > generic_part_of
                    * has_constitutional_part < > constitutional_part_of
                    ⚌ has_regional_part < > regional_part_of
                         * has_branch < > branch_of
                         * has_tributary < > tributary_of
                    * has_member < > member_of
               * has_orientation
               ⚌ connected_to < > connected_to
                    * continuous_with < > continuous_with
                    * attached_to <> receives_attachment_of
               ⚌ has_location < > location_of
                    * contained_in < > contains
                    ⚌ adjacent_to < > adjacent_to
                         * surrounds < > surrounded_by
                    ⚌ has_anatomical_coordinate < > anatomical_coordinate_of
                         * has_qualitative_anatomical coordinate
                         * has_geometric_coordinate
               * has_organizational_pattern < > organizational_pattern_of
               * has_segmental_innervation < > segmental_innervation_of
     ❀ organizational pattern
     ❀ segmental innervation
```

Fig. 4.8. Taxonomy 8. Anatomical relations. The symbol <> designates inverse relations.

Taxonomic Relations

As noted earlier, the FMA employs the *is_a* relation strictly in accord with its formal definition [83] and implements its specifications along with their inverses (Taxonomy 8 [Figure 4.8]). Although instances are excluded from the anatomy taxonomy implemented in the FMA artifact, the theory conforms to high-level ontology in that it adopts the distinction between instances and types (Section 4.4.1). Consistent with the distinction between canonical and instantiated anatomy, the FMA takes account of the *instance_of* relation between individuals and types, as well as the *subtype_of* relation between types.

Structural Anatomical Relations

Structural relations can be defined primarily with reference to instances. The type esophagus has no parts - only your and my esophagi do. Instance to instance relations, however, are extrapolated to obtain between types under the constraints propounded elsewhere [12, 25, 82]. Taxonomy 8 (Figure 4.8) presents such type to type relations, which are defined in Appendix Table 4.5.

Several published accounts about the FMA deal with these structural relations and justify the need for introducing them [53, 56, 57, 62, 82]. They also explicate some of the rules and principles for distinguishing between relations of different sorts. Here it should suffice, as an illustration, to provide answers to some of the foregoing questions about the esophagus.

By virtue of the definitions of the relations, the wall and lumen qualify as parts of the esophagus because, although each entity is of a different type, they all have three dimensions; moreover wall and lumen are complements of one another in that together they account for the whole of the esophagus. The case study, however, also refers to the cervical, thoracic and abdominal parts of the esophagus; together they also account for the whole organ. None of the latter can substitute either for the wall or the lumen and each has its own wall and lumen. Such overlapping partitions of an anatomical structure highlight the need for specifying different kinds of part relations: an entity in one partition cannot qualify as part or complement in another partition of the whole. The distinction between constitutional and regional part relations – which obtain for anatomical structures at all levels of granularity – resolves such conflicts and ambiguities (Figure 4.9).

Yet another distinction is called for when considering the surfaces of the esophagus. Because they have two, rather than three dimensions, the surfaces must be associated with the wall, lumen and the whole of the esophagus through the *boundary_of*, rather than the *part_of*, relation [62]. The internal surface of the esophagus is the boundary of both the wall and the lumen. Such a specific view invalidates the one prevalent in anatomical discourse, in which the wall of the esophagus is generally regarded as the boundary of its lumen.

To clarify the relation of portions of air and mucus to the esophagus and its parts, location – and in particular – containment relations need to be considered, the need

A

has_constitutional part

Esophagus	wall of esophagus	lumen of esophagus
cervical part of esophagus	wall of cervical part of esophagus	lumen of cervical part of esophagus
thoracic part of esophagus	wall of thoracic part of esophagus	lumen of thoracic part of esophagus
abdominal part of esophagus	wall of abdominal part of esophagus	lumen of abdominal part of esophagus

B

has_constitutional part

Neuron	plasma membrane of neuron	cytoplasm of neuron
soma of neuron	plasma membrane of soma of neuron	cytoplasm of soma of neuron
axon	plasma membrane of axon	cytoplasm of axon
dendrite	plasma membrane of dendrite	cytoplasm of dendrite

Fig. 4.9. Regional and constitutional part relations shown for the esophagus (A) and a neuron (B).

for distinctions between which will soon become apparent. Location is the most general relation which associates objects, substances and spaces with spatial regions into which the universe is divided by mereotopology [12, 25, 82]; some of these regions are enclosed by an organism's maximal boundary. Thus, not only portions of mucus but also the esophagus, its lumen, bacteria or a swallowed coin, are located within the human body. The FMA distinguishes parthood from location and further specifies the latter as containment, adjacency and having anatomical coordinates [53, 57]. Parthood in biological organisms must meet a number of other criteria [74, 82], the pertinent one in the current context enforced through the rule of dimensional consistency [53, 57].

Whereas part relations can be asserted between instances of two types of physical anatomical entity of the same dimension, the *contains* relation associates anatomical cavities with portions of body substances, and compartment spaces with anatomical structures. By virtue of these constraints, the valid assertions are: lumen of esophagus *contains* portion of mucus; lumen of esophagus *part_of* esophagus; space of posterior mediastinum *contains* thoracic part of esophagus. Imposing such restricted meaning on the *contains* and *contained_in* relations may seem pedantic. The purpose of such specificity, however, is to assure that the role of container is constrained to anatomical structures which have anatomical space as their part.

The formal properties of these relations in the FMA have been analysed [11, 25, 64, 79]. It deserves emphasis, however, that whereas part relations are transitive within their regional and constitutional categories, containment relations are not. To assert that a portion of esophageal mucus is contained in the lumen of the esophagus must not imply that such mucus is also contained in the space of the posterior mediastinum, in which the esophagus itself is contained.

In addition to containment relations, the location of the esophagus can also be specified by its adjacencies and anatomical coordinates. For example, the adjacencies of the thoracic vertebral column and trachea give an approximate location for the esophagus, which can be specified by attributing the adjacency with anatomical coordinates illustrated in Figure 4.10. For a particular regional part of the esophagus we may assert

'trachea' *adjacent_to* 'esophagus' *left_anterior, right_anterior*
'apex of left lung' *adjacent_to* 'esophagus' *left posterolateral.*

These qualitative coordinates refer to the standard 'anatomical position' of bipedal erect posture and therefore hold regardless of the position an individual assumes. They translate into a quadrupedic orientation of non-human species if – for example – anterior and posterior are equated with ventral and dorsal, respectively; and superior and inferior correspond with rostral and caudal. When anterior is used to mean rostral, however, as often is the case, it becomes problematic to identify interspecies homologies such as those between lobes of the human prostate and members of the murine prostate anatomical set [88].

In clinical medicine, not only qualitative but geometric coordinates are also employed (e.g., anteroposterior diameter of thoracic inlet, conjugate diameter of pelvis). In an individual human being, such as the Visible Human [92], location of an anatomical structure can be stated by a set of numerical coordinates, which, however, need to be translated into qualitative anatomical coordinates to be meaningful to human beings. Orientation provides additional information relevant to location, and like adjacency, is also attributed. For example, esophagus *has_orientation* pharyngo-esophageal junction *superior*, esophagogastric junction *inferior*.

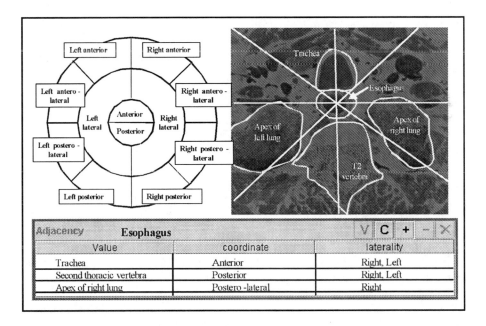

Adjacency	Esophagus		
Value	coordinate	laterality
Trachea	Anterior	Right, Left
Second thoracic vertebra	Posterior	Right, Left
Apex of right lung	Postero-lateral	Right

Fig. 4.10. A system of qualitative anatomical coordinates superimposed on the esophagus in a transverse section of the male Visible Human cadaver specimen [92]. The lower part of the figure shows the attributed adjacency relations of the esophagus implemented in the FMA artifact.

4.5 The FMA Ontology Artifact

Selected parts of FMA theory have been implemented as ontology artifacts in a variety of terminology and ontology authoring and editing environments. The master copy, populated and maintained by the FMA's curators, is in Protégé and is stored in a relational database [65, 71]. It is the largest anatomy ontology or terminology and one of the largest ontologies in the biomedical domain: its more than 135,000 terms point to 75,500 types, which are interrelated by over 2.5 million iterations of 198 kinds of specific relations. (We shall see that a number of these relations, such as *has_FMA_ID*, *has_synonym*, are not anatomical or even ontological relations.) A main reason for such extensive data entry was to test and validate the theory, which, as a consequence, has been revised and enhanced through several cycles, an activity which continues to this day. Whereas the objective with the FMA theory is to treat the anatomy domain comprehensively, for several of its subdomains the artifact is populated merely with examples to illustrate a particular aspect of the theory. For example, although we have proposed a high-level ontological scheme (theory) for developmental continuants and relations [71], they have not been introduced in the FMA artifact in any detail. The main focus has been the macroscopic and microscopic anatomy of the entire body, including neuroanatomy [50]. Cell and its parts are extensively covered (a feature not widely appreciated); with surprisingly little overlap with the GO [5] and CL, and substantial differences in their ontological perspectives.

Such extensive population of an ontological framework required the selection of a particular species as model organism. For a variety of reasons, the FMA artifact is concerned with the canonical anatomy of *Homo sapiens*. Its nodes and relations become the more specific to this species the further removed they are in the taxonomic tree from the root node. This circumstance accounts for the prevailing view that the FMA is an ontology of human anatomy. Except for 'human body', the ontology's terms do not specify that they point to parts of the human body; it is taken for granted that the types esophagus and stomach, for example, are instantiated by organs of human canonical anatomy.

We use the frame of esophagus, the subject of our case study, to illustrate the expressive machinery of the Protégé system for representing aspects of the theory.

A frame is a data structure which displays all the information in the ontology about a named anatomical type, including the properties which its instances share and the relations they have with instances of other types. The left panel (Figure 4.11) shows the node esophagus along with its taxonomic ancestors and siblings. Related information is displayed in the right panel in slots that bear the name of a particular property or relation. The contents of the slots are its values, and are admitted into a slot only if they point to a node of the anatomy taxonomy or one of the ancillary taxonomies of the FMA. Exceptions are the slots for numerical identifiers, preferred name, synonyms and foreign language equivalents associated with the taxonomic node of the frame, the definition of the corresponding type and comments about it (the latter not shown in Figure 4.11). Other slots cannot be filled unless the terms are imported from one of the taxonomies of the FMA. For example, the Dimensional Ontology provides the values for the slot *has_shape* (e.g., cylinder, polyhedron, which

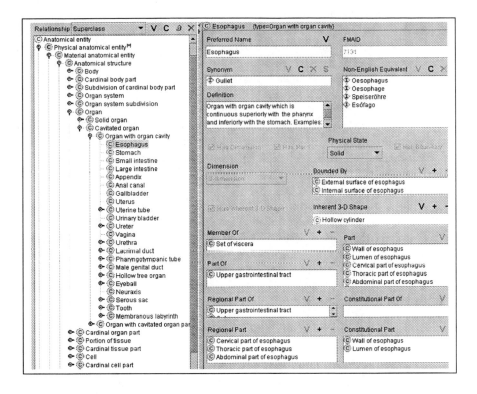

Fig. 4.11. The Protégé frame of esophagus in the FMA artifact.

are subclasses of 3-D volume), whereas the values for the part and adjacency slots in a frame are derived from the AT.

4.5.1 Identifiers and Terms

In addition to identifiers built into Protégé, each node has its unique numerical FMA identifier. When the corresponding type has an accepted name, it is adopted as the preferred name from *Terminologia Anatomica* [29] or established textbooks of subdomains of anatomy [3, 17, 28, 39, 49, 69, 93]. The FMA is the only anatomy ontology or terminology which comprehensively incorporates the approximately 10,000 terms comprising the international standard of anatomical nomenclature, accommodating also plural terms through the type anatomical set (section 4.4.7) [47].

In addition to those for most of the high-level types in the AT, new descriptive terms are also associated with a large number of leaf nodes; these point either to the unnamed complement of previously named parts, or are more specific than the terms in extant sources. The construction of new compound terms follows the rule of progression from the most specific to the most general component of the phrase (e.g., 'apex of heart' – not 'heart apex'; 'upper lobe of right lung' – not 'right upper lobe of lung'; 'left third rib' – not 'third left rib'; 'meningeal branch of left eighth thoracic spinal nerve' – not 'eighth thoracic spinal nerve meningeal branch'). Where not all parts of the entity have been named, a descriptive name is assigned to the complement (e.g., 'upper segment of uterus' where only the lower segment had been named previously).

We use the term 'proper' to designate the major unspecified part of an anatomical structure to distinguish it from lesser parts; for example, we distinguish 'epithelium proper of esophagus' from 'epithelium of esophageal gland' – both *part_of* 'epithelium of esophagus'; 'cytoplasm proper of neuron' from 'axon hillock' – both *part_of* 'cytoplasm of neuron'.

An audit is maintained of the terms adopted from other sources. For example the English language synonym of the preferred name 'esophagus' is 'gullet'; its non-English language equivalent in German is 'Speiseröhre', and in Latin 'oesophagus.' The audit for the latter records the term's derivation from *Terminologia Anatomica* (Figure 4.12). The audit can also indicate when a term is erroneous or outdated, as is the case for example for 'Botallo's ligament', the preferred name of which is 'ligamentum arteriosum'.

These examples are intended to illustrate that although the FMA is primarily ontologically rather than terminologically oriented, it is more inclusive, specific and comprehensive for terms of human anatomy than are other sources that we are aware of. The inclusion of such a spectrum of terms pointing to a node of the taxonomy enables searches of the FMA by a variety of users.

4.5.2 Properties and Relations

The machinery Protégé provides for distinguishing between inherited slot values and "own" slot values is explained elsewhere [65]. In the frame 'esophagus', the values for the slots of dimension, mass, physical state and 3-D shape are inherited from the frames of a hierarchy of taxonomic ancestors (Figure 4.11); so are the kinds of slots the esophagus frame can have (e.g., preferred name, definition, part of, adjacency, nerve supply, but, among others, not *has_branch*). A particular feature of the FMA, for which Protégé makes special accommodation, is attributed relations, illustrated for the kinds of adjacencies the esophagus has (Figure 4.10).

Protégé imposes constraints on the values of a slot. For example, the *part_of* slot in the frame of organ specifies that there can be multiple values for the slot and that the values can be derived only from AT types organ system, organ system subdivision, cardinal body part and cardinal body part subdivision. Since esophagus is a subtype of organ, the allowed values for its *part_of* slot include upper gastrointestinal tract, which is a subdivision of the GI tract, in turn a subdivision of the alimentary system. Another example is the restriction for the *nerve_supply* slot; values for this

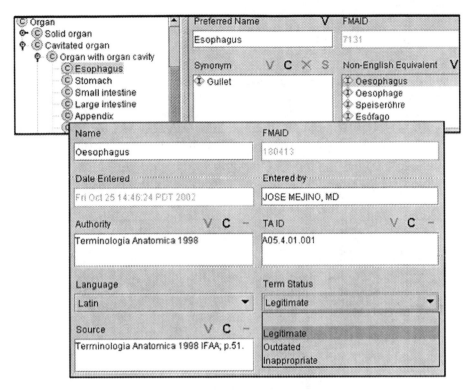

Fig. 4.12. A record of a term entry for a non-English equivalent of the preferred name esophagus in the FMA artifact.

slot may only be derived from AT types cranial nerve, spinal nerve and peripheral nerve. Also, the value 'lumen of esophagus' is allowed for the part slot, because lumen is a kind of organ cavity, and having a cavity as a part is inherited from the frame of 'cavitated organ'. Selecting a particular value in a slot, automatically opens the frame of the corresponding node of the taxonomy, both in Protégé and the FME [32].

We cite these examples to illustrate the discipline the Protégé ontology authoring environment has imposed on the FMA artifact and thereby significantly enhanced its ontological soundness.

4.5.3 Automatic Derivation of Hierarchies

The implementation of the FMA in Protégé enables the automatic generation of hierarchies linked by a uniform transitive relation, such as *has_part*. A number of the current anatomy terminologies employ this relation as a primary link within their hierarchies, as noted earlier, and may enter both *has_part* and *is_a* relations in the same directed acyclic graph, an approach promoted by some tools for terminology authoring.

The partonomy of the esophagus, based on the *has_generic_part* relation and illustrated in Figure 4.13, was automatically derived from the frame-based representation. Selected nodes of the hierarchy have been opened up at all levels of granularity, starting with the whole human body, moving onto organ system, its subdivision, an organ, cardinal organ parts, portions of tissue, cell, organelle, organelle part and biological macromolecule, as well as an acellular anatomical structure, the basement membrane and its molecular components; all seamlessly included in the same tree. Similar trees can be automatically generated on the fly for other transitive relations [60].

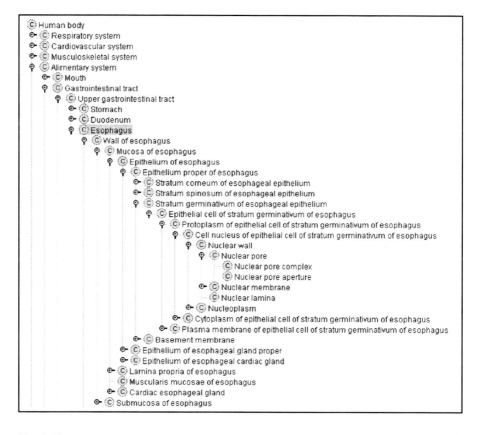

Fig. 4.13. A part hierarchy automatically generated from FMA's Protégé's frame-based representation. The partonomy spans all levels of granularity from the whole body to biological macromolecules.

4.5.4 Artifact Alternatives

The initial iteration of the Digital Anatomist vocabulary, the first incarnation of the FMA, was implemented in a simple terminology editing tool designed in-house at the University of Washington. This tool delivered the data for incorporation in the Unified Medical Language System to the National Library of Medicine; the vocabulary was then merged into the concept-based system of UMLS. At a later stage, the vocabulary was migrated to earlier versions of Protégé and its evolution into an increasingly complex ontology was a significant motivating factor for the realization of the current Protégé system.

As noted in Section 4.1, more recently the FMA also motivated a number of avenues of research in ontology and biomedical informatics, a notable one being to serve as a substrate and case-study for solving the problem of migrating complex frame-based systems to Web Ontology Language (OWL). Most comprehensive is the approach led by Golbreich at the University of Rennes, France and the National Library of Medicine [35], but other investigators at Stanford University [20], and at the University of Mannheim, Germany [30] have also undertaken similar tasks. Work is also in progress on migrating the FMA to the OBO edit modeling environment [61] and a new format of UMLS [63].

A simplified web browser, the Foundational Model Explorer or FME, has also been developed for providing ready access to a streamlined version of the Protégé-based artifact [22, 32].

Thus the FMA grew and was transformed from a simple terminology into one of the most complex and disciplined ontologies without having to discard any of the data entered over more than a 10 year period.

4.6 The FMA as Reference Ontology and Bioinformatics Resource

Since anatomy pervades essentially all subdomains of biology and medicine, the FMA was designed and developed as a general-purpose resource to fill a need for a unifying ontological framework of biological structure. It was the lack of such a reference, or standard, which had led to the creation of overlapping and often inconsistent representations of human anatomy in clinical terminologies. For example, in addition to SNOMED [85], GALEN [33], the Medical Entities Dictionary [52] and others, there are at least six terminologies in UMLS with a substantial anatomical content. Each of these terminologies was designed to support some task, application or activity in clinical medicine.

Unlike these terminological resources, the FMA was not tailored to the needs of any particular user group; rather, it was designed to serve as a resource for the developers of application ontologies in any specialized biomedical field. Indeed, the FMA is the first example of a *reference ontology* in the so-called basic biomedical sciences, and as such it has contributed to an awareness of a necessary distinction between application and reference ontologies in given domains of biomedicine. Knowledge pursued and represented by the basic biomedical sciences – anatomy, physiology, pathology, biochemistry, pharmacology, and the more recent additions to the list such as molecular and developmental biology and genome science – is not only indispensable and reused in such application domains as clinical medicine and experimental biology, but also often remains unarticulated, sinks to subconscious levels or is taken for granted. These basic sciences are the very fields of biomedicine that call for the establishment of their own reference ontologies as a solid backing to application ontologies.

Currently some tensions prevail between the promoters of sound ontological methods and many practitioners of clinical medicine and biomedical research. The demand for ontologies grows as knowledge-based applications gain increasing deployment. Practitioners in biomedical domains, however, look for knowledge organization schemes in these applications which mirror the ones they absorbed during their training. Our case study and many cited examples illustrate that such schemes are often not compatible with ontological principles and may not be suited for supporting nontrivial inference. These tensions can be resolved if developers of application ontologies 'mine' relevant reference ontologies, reuse segments appropriate for targeted tasks and design interfaces which accommodate the expectations of particular users. Such an agenda is reflected in a number of uses for which the FMA has already been exploited.

With the collaboration of investigators in computer science, members of the FMA's team have experimented with the development of knowledge-based applications and interfaces to facilitate access to the FMA. Although an application ontology for anatomy education has not yet been derived from it, the FMA has been used for the annotation of radiographs [54] and the 3D anatomical atlases of Digital Anatomist [23], which experience many thousands of hits per day from 95 countries. The FMA is the ontology back-end to a client-server anatomy information system which supports the semi-automatic generation of such atlases and also enables the interactive disassembly and assembly (virtual dissection and its reverse) of 3D computer graphics models of the human body [16]; it was also a key component in an open source toolkit for building biomedical web applications [42]. The FMA served as a test bed for developing the query agent OQAFMA for large semantic networks, which classifies and processes different types of queries [60]. In addition to its own interface for database queries, OQAFMA is also the agent behind a prototype interface to the FMA, which served for experimenting with the formulation of natural language queries [24]. The problem of constraining queries to entities and relations present in the FMA was solved by 'Emily', another "intelligent" interface [21]. The evaluation of 'Emily' revealed that correct answers, matching the key, could be generated to multiple-choice questions used in anatomy exams, which were culled from published compendia [76]. Since many of the answers were not hard-coded and had to be generated on the fly by traversing several paths, the results provide assurance that the FMA's ontological structure and content can support nontrivial inference comparable to the reasoning expected of medical students.

A number of ontological questions of general nature have been addressed in the course of the development of the FMA by its authors and independent investigators. Traditional representations of anatomical entities in terminologies have been influenced [2] and proposals have been advanced for assuring consistency in such representations [55]. Using the FMA as a reference, similar objectives were pursued for enriching the UMLS semantic network [95] and for designing metaschemas for it [96].

A particular topical problem is the development of methods for correlating or mapping ontologies with overlapping content to one another. The FMA has been used as one of the test ontologies in several such projects. Different investigators and approaches have compared the FMA to the anatomy module of GALEN [59, 97]. The rather surprising result with each method was that only around 5% match could be demonstrated between approximately 60,000 and 23,500 nodes in the FMA and GALEN ontologies, respectively. This match could not be improved substantially by combining the two independent methods [100]. The explanation of the divergence has not been analyzed systematically; however, it is likely related to the fact that the anatomy module of GALEN is primarily intended as an application ontology for diagnostic and therapeutic procedures; a substantial number of anatomical entities are classified in terms of their accessibility to such procedures, exemplified by some non-canonical structures designated by conjunctions pertaining to the esophagus: 'GITractFromEsophagusToDuodenum', 'EpibronchialPartOfEsophagus' and 'UnamedTractOfEsophagus'. The level of correspondence between the FMA and SNOMED's current version was found to be only somewhat better [14], despite the fact that – in contrast with SNOMED's earlier versions – the schemes of representation had much similarity in the two ontologies.

A comparison of a narrower scope was made, using yet a different approach, between cell parts in the FMA and the cell component section of GO [5]. After taking synonymy into account, 972 of 1,172 cell part terms remained unique to the FMA and 1,479 of 1,807 GO's cell component terms could not be aligned with the FMA. The two ontologies were comparable in their scope of breadth and depth and were found to be largely complementary rather than overlapping.

A finding suggestive of the advantage reference ontologies offer for improving alignment between application ontologies comes from the mapping of the human anatomy subset of the NCI Thesaurus and the Adult Mouse Anatomical Dictionary of Jackson Labs [98]. The correlation was improved when each terminology was first aligned with the FMA compared to when they were directly aligned with each other.

In addition to proving to be a substrate in biomedical informatics research, the FMA has also been exploited for some clinical informatics applications. A number of investigators at the National Library of Medicine have made use of segments of the FMA in systems designed for analyzing arterial branching patterns in cardiac catheterization reports [67], evaluating anatomical terminology in medical texts [84], facilitating integration of endoscopy terminology into the UMLS [90] and automating the interpretation of anatomical spatial relations in medical reports [8]. The FMA has provided the anatomical information for a system of radiation treatment planning in cancer therapy [44] and the development of a related application ontology for fields of lymphatic drainage and regions of predicted cancer spread [9, 86]. The anatomy component of another evolving clinical application ontology, *RadiO*, designed for radiology task reporting, is derived from the FMA [51].

In addition to its substantial section which takes account of neuroanatomical entities and relations [50] – a domain often treated as distinct from other anatomical entities – the FMA has also influenced bioinformatics ontology research in other fields of the basic sciences. An information system has been developed for the comparative anatomy of vertebrate species with the FMA serving as its reference ontology [87]. A high-level Ontology of Biomedical Realty – OBR – has been proposed as a framework for linking to one another the basic biomedical sciences [70]. The guiding principle of OBR is the designation of anatomical structures as independent continuants on which other continuants such as pathological lesions, functions, malfunctions and also processes depend. Actually, OBR is an explicit iteration of our long-held opinion that a sound conceptual framework of anatomical entities is at the root of sorting and ontologically organizing entities in other biomedical domains [15]. Reference has been made already to CARO, a common anatomy reference ontology which extends the FMA's orientation to vertebrate anatomy to all animals and developmental entities in particular [37]. The current version of CARO adopts from the FMA nearly half of its nodes and definitions, with or without modifications appropriate for its expanded scope.

OBR will realize its potential once basic science reference ontologies beyond anatomy become available. Examples of such ontologies include an evolving physiology reference ontology which integrates the FMA as the participants in physiological processes [19], and a reference ontology for pathology which adopts anatomical structures from the FMA as the continuants on which pathological entities are dependent [48, 81]. Although not reported in the literature, or noted in the artifacts, we hear from developers and curators that without adopting the FMA as such, they develop new terminologies/ontologies or update existing ones with reference to the FMA as a template. It is indeed rewarding to see the FMA reflected in these evolving resources. Access to the FMA as open source [31] should facilitate and enhance the role of the FMA as a reference ontology.

4.7 Concluding Remarks

The dual purpose of this chapter is to assist ontology developers only superficially familiar with biology in gaining some appreciation of the complexities of anatomy; and introduce anatomists unfamiliar with, but interested in, ontology – or "anatomical informatics" as it is currently designated [89] – to a new paradigm for viewing their discipline. Biologists, anatomists, health care professionals and students should not be more than peripherally concerned with high level types in the FMA, such as physical, non-physical, material and immaterial anatomical entities: they are necessary for an all-encompassing domain theory and for linking ontologies in different domains through anatomy to one another. There is a great need for application ontologies in anatomy tailored to diverse curricula in the basic science and clinical disciplines in order to raise web-based education and training to new levels [46]. The FMA should prove to be a useful resource for filling this need.

Soon after its initiation as a terminology, the FMA became a research project in biomedical informatics concerned with the development of methods for ontologically representing a fundamental and complex domain of biomedicine. As a consequence, its objectives are quite distinct from those of GO, GALEN or the Adult Mouse Anatomical Dictionary, for example, which were developed to support targeted tasks. We regard the FMA as an ongoing experiment in the evolving science of ontology and anticipate that it will continue to change and improve as it has during its ten year history. In addition to the examples cited, several chapters in this book attest to the influence the FMA has exerted on the thinking of ontologists about anatomy; some illustrate as well the challenges the FMA continues to pose for its own curators and others in ontological research. Although the FMA as yet has no substantial penetration in anatomy science and education, several professional societies and international organizations are in the process of considering its adoption as the standard for human anatomy.

In summary, the FMA has broken new ground in the science of anatomy, as well as in biomedical ontology and informatics, in that it has 1) defined anatomical structure and proposed it as the independent continuant of biomedical reality; 2) made the notion of canonical anatomy explicit and distinguished it from instantiated anatomy; 3) distinguished anatomy as structure from anatomy science; 4) drawn the boundaries for the scope of anatomy and demarcated it from the other biomedical basic sciences; 5) introduced Aristotelian definitions for the types of anatomical entities based predominantly on their structural properties; 6) proposed a unifying theory of anatomy; 7) distinguished this theory from its representation in a computable artifact; and 8) populated this artifact with types of anatomical entities such that its content is both more generalizable and detailed or specific than contemporary hard-copy or computable resources of human anatomy.

References

1. Adult mouse anatomical dictionary browser.
   http://www.informatics.jax.org/searches/AMA_form.shtml.
2. A. Agoncillo, J.L.V. Mejino, and C. Rosse. Influence of the digital anatomist foundational model on traditional representations of anatomical concepts. In *AMIA Symposium Proceedings*, pages 2–6, 1999.
3. B. Alberts, A. Johnson, J. Lewis, M. Raff, K. Roberts, and P. Walter. *Molecular Biology of the Cell*. Garland Science, New York, 4th edition, 2002.
4. Aristotle. *The categories*. Harvard University Press, Cambridge, Mass., 1973.
5. A. Au, X. Li, and J.H. Gennari. Differences among cell structure ontologies: FMA, Go and CCO. In *AMIA Symposium Proceedings*, pages 16–20, 2006.
6. J. Bard, S.Y. Rhee, and M. Ashburner. An ontology for cell types. *Genome Biology*, 6(R21), 2005.
7. J.B.L. Bard. Anatomics: the intersection of anatomy and bioinformatics. *J Anat*, pages 1–16, 2005.
8. C.A. Bean, T.C. Rindflesh, and C.A. Sneiderman. Automatic semantic interpretation of anatomic spatial relationships in clinical text. In *AMIA Symposium Proceedings*, pages 897–901, 1998.
9. N. Benson, M. Whipple, and I.Kalet. A markov model approach to predicting regional tumor spread in the lymphatic system of the head and neck. In *AMIA Symposium Proceedings*, pages 31–35, 2006.
10. J. Berg. Aristotle's theory of definition. In *AATI del Convegno Internationale di Storia della Logica San Gimignano*, pages 19–30, Bologna, 4-8 December 1982 1983. CLUEB.
11. T. Bittner. Axioms for parthood and containment relations in bio-ontologies. In *First International Workshop on Formal Biomedical Knowledge Representation*, pages 4–11, Bethesda MD, 2004. American Medical Informatics Association.
12. T. Bittner, M. Donnelly, and L.J. Goldberg. Modeling principles and methodologies spatial representation and reasoning. In Burger A., Davidson D., and Baldock R., editors, *Anatomy Ontologies for Bioinformatics: Principles and Practice*, New York, In press. Springer.
13. T. Bittner and B. Smith. A theory of granular partitions. In Duckham D., Goodchild MF, and Worboys MF., editors, *Foundations of Geographic Information Science*, pages 117–151, London, 2003. Taylor & Francis.
14. O. Bodenreider and S. Zhang. Comparing the representation of anatomy in the FMA and SNOMED CT. In *AMIA Symposium Proceedings*, pages 46–50, 2006.
15. J.F. Brinkley, J.S. Prothero, J.W. Prothero, and C. Rosse. A framework for the design of knowledge-based systems in structural biology. In *Proc. 13th Annual Symposium on Computer Application in Medical Care*, pages 61–65, 1989.
16. J.F. Brinkley, B.A. Wong, K.P. Hinshaw, and C. Rosse. Design of an anatomy information system. *IEEE Comp Graphics Applic*, 3:38–48, 1999.
17. M.B. Carpenter and J. Sutin. *Human Neuroanatomy*. Wilkins & Wilkins, Baltimore, 8th edition, 1983.
18. Cell type ontology.
    http://www.sanbi.ac.za/evoc/ontologies_html/latest/celltype.html.
19. D.L. Cook, J.L.V. Mejino, and C. Rosse. Evolution of a foundational model of physiology: symbolic representation for functional bioinformatics. In *Proceedings of MedInfo*, pages 336–340, 2004.

20. O. Dameron, D.L. Rubin, and M. Musen. Challenges in converting frame-based ontology into OWL: the foundational model of anatomy case-study. In *AMIA Symposium Proceedings*, pages 181–185, 2005.
21. L.T. Detwiler, E. Chung, A. Li, J.L.V. Mejino, A.V. Agoncillo, J.F. Brinkley, C. Rosse, and L.G. Shapiro. A relation-centric query engine for the foundational model of anatomy. In *Proceedings of MedInfo*, pages 341–345, 2004.
22. L.T. Detwiler, J.L.V. Mejino, C. Rosse, and J.F. Brinkley. Efficient web-based navigation of the foundational model of anatomy. In *AMIA Symposium Proceedings*, page 829, 2003.
23. Digital anatomist project - interactive atlases.
    http://www9.biostr.washington.edu/da.html.
24. G. Distelhorst, V. Srivastava, C. Rosse, and J.F. Brinkley. A prototype natural language interface to a large complex knowledge base, the foundational model of anatomy. In *AMIA Symposium Proceedings*, pages 200–204, 2003.
25. M. Donnelly, T. Bittner, and C. Rosse. A formal theory for spatial representation and reasoning in biomedical ontologies. *Artificial Intelligence in Medicine*, 36:1–27, 2006.
26. Dorland's medical dictionary.
    http://www.dorlands.com/.
27. Edinburgh developmental anatomy.
    http://www.ana.ed.ac.uk/anatomy/database/humat/.
28. D.W. Fawcett. *Bloom and Fawcett Textbook of Histology*. Chapman & Hall, New York, 12th edition, 1994.
29. Federative Committee on Anatomical Terminology (FCAT). *Terminologia Anatomica*. Thieme, Stuttgart, 1998.
30. FMA in OWL-full format. Published on the Web.
    http://webrum.uni-mannheim.de/math/lski/release.html.
31. FMA open source.
    http://sig.biostr.washington.edu/cgi-bin/fma_register.cgi.
32. Foundational Model Explorer.
    http://fme.biostr.washington.edu:8089/FME/index.html.
33. GALEN.
    http://www.opengalen.org/.
34. Gene Ontology.
    http://www.geneontology.org.
35. C. Golbreich, S. Zhang, and O. Bodenreider. The foundational model of anatomy in OWL: Experience and perspectives. *Journal of Web Semantics*, 4:181–195, 2006.
36. P. Grenon, B. Smith, and L. Goldberg. Biodynamic ontology: applying BFO in the biomedical domain. In P.M. Pisannelli, editor, *Ontologies in Medicine. Studies in Health technology and Informatics*, volume 102, pages 20–38, Amsterdam, 2004. IOS Press.
37. M. Haendel, F. Neuhaus, J.L.E. Sutherland, J.L.V. Mejino(Jr), C. Mungall, and B. Smith. The common anatomy reference ontology. In A. Burger, D. Davidson, and R. Baldock, editors, *Anatomy Ontologies for Bioinformatics: Principles and Practice*, New York, In press. Springer.
38. T.F. Hayamizu, M. Mangan, J.P. Corradi, J.A. Kadin, and M. Ringwald. Adult mouse anatomy dictionary. *Genome Biology*, 6(R29), 2005.
39. W.H. Hollinshead. *Anatomy for surgeons*, volume 1–3. Harper and Row, Philadelphia, 3rd edition, 1982.
40. International code of zoological nomenclature online; chapter 13: The type concept in nomenclature; article 61: Principles of typification.
    http://www.iczn.org/iczn/indes.jsp.

41. IUPS Physionome Project - body systems.
http://www.bioeng.auckland.ac.nz/physiome/anatomy.php.

42. R. Jakobovits, J.F. Brinkley, C. Rosse, and E. Weinberger. Enabling clinicians, researchers and educators to build custom web-based biomedical information systems. In *AMIA Symposium Proceedings*, pages 279–283, 2001.

43. I. Johansson, B. Smith, K. Munn, N. Tsikolia, K. Elsner, D. Ernst, and D. Siebert. Functional anatomy: a taxonomic proposal. *Acta Biotheoretica*, 53(3):153–166, 2005.

44. I.J. Kalet, J. Wu, M. Lease, M.M. Austin Seymour, J.F. Brinkley, and C. Rosse. Anatomical information in radiation treatment planning. In *AMIA Symposium Proceedings*, pages 291–295, 1999.

45. R.C. Kerckhoffs, M.L. Neal, Q. Gu, J.B. Bassingthwaighte, J.H. Omens, and A.D. McCulloch. Coupling of a 3d finite element model of cardiac ventricular mechanics to lumped systems models of the systemic and pulmonic circulation. *Ann Biomed Eng*, 35(1):1–18, 2007.

46. S. Kim, J.F. Brinkley, and C. Rosse. A profile of on-line anatomy information resources: design and instructional implications. *Clin Anat.*, 16:55–71, 2003.

47. K.L. Rickard KL, J.L.V. Mejino(Jr), R.F. Martin, A.V. Agoncillo, and C. Rosse. Problems and solutions with integrating terminologies into evolving knowledge bases. In *Proceedings of MedInfo*, pages 420–424, 2004.

48. A. Kumar, Y.L. Yip, B. Smith, D. Marwede, and D. Novotny. An ontology for carcinoma classification for clinical bioinformatics. *Stud Health Technol Inform.*, 116:635–40, 2005.

49. J.H. Martin. *Neuroanatomy Text and Atlas*. Appleton & Lange, Stamford, Connecticut, 2nd edition, 1996.

50. R.F. Martin, J.L.V. Mejino, D.M. Bowden, J.F. Brinkley, and C. Rosse. Foundational model of neuroanatomy: its implications for the Human Brain Project. In *AMIA Symposium Proceedings*, pages 438–442, 2001.

51. D. Marwede. RadiO. Personal Communication.

52. Medical Entities Dictionary.
http://med.dmi.columbia.edu/.

53. J.L.V. Mejino(Jr), A.V. Agoncillo, K.L. Rickard, and C. Rosse. Representing complexity in part-whole relationships within the foundational model of anatomy. In *AMIA Symposium Proceedings*, pages 450–454, 2003.

54. J.L.V. Mejino(Jr) and C. Rosse. Interactive radiology exercises.
http://www9.biostr.washington.edu/hubio511/.

55. J.L.V. Mejino(Jr) and C. Rosse. The potential of the digital anatomist foundational model for assuring consistency in UMLS sources. In E.G. Chute, editor, *AMIA Symposium Proceedings*, pages 825–829, 1998.

56. J.L.V. Mejino(Jr) and C. Rosse. Conceptualizations of anatomical spatial entities in the digital anatomist foundational model. In *AMIA Symposium Proceedings*, pages 112–116, 1999.

57. J.L.V. Mejino(Jr) and C. Rosse. Symbolic modeling of structural relationships in the foundational model of anatomy. In *First International Workshop on Formal Biomedical Knowledge Representation (KR-MED 2004)*, pages 48–62, Bethesda MD, 2004. American Medical Informatics Association.

58. J. Michael, J.L.V. Mejino(Jr), and C. Rosse. The role of definitions in biomedical concept representation. In *AMIA Symposium Proceedings*, pages 463–467, 2001.

59. P. Mork and P.A. Bernstein. Adapting a generic match algorithm to align ontologies of human anatomy. In *ICDE*, pages 787–790, 2004.

60. P. Mork, J.F. Brinkley, and C. Rosse. OQAFMA querying agent for the foundational model of anatomy: providing flexible and efficient access to a large semantic network. *J Biomed Inform*, 36:501–517, 2003.
61. C. Mungall. Personal Communication.
62. P.J. Neal, L.G. Shapiro, and C. Rosse. The digital anatomist spatial abstraction: a scheme for the spatial description of anatomical entities. In *AMIA Symposium Proceedings*, pages 423–427, 1998.
63. S. Nelson. Personal Communication.
64. F. Neuhaus and B. Smith. Modeling principles and methodologies – relations in anatomical ontologies. In A. Burger, D. Davidson, and R. Baldock R., editors, *Anatomy Ontologies for Bioinformatics: Principles and Practice*, New York, In press. Springer.
65. N.F. Noy, J.L.V. Mejino(Jr), M.A. Musen, and C. Rosse. Pushing the envelope: challenges in frame-based representation of human anatomy. *Data & Knowledge Engineering*, 48:335–359, 2004.
66. OBO - Open Biological Ontologies. htpp://obo.sourseforge.net.
67. T.C. Rindflesch, C.A. Bean, and C.A. Sneiderman. Argument identification for arterial branching predications asserted in cardiac catheterization reports. In *AMIA Symposium Proceedings*, pages 704–8, 2000.
68. C. Rosse. Terminologia anatomica; considered from the perspective of next-generation knowledge sources. *Clin. Anat.*, 14:120–133, 2001.
69. C. Rosse and P. Gaddum-Rosse. *Hollinshead's textbook of anatomy*. Lippincott-Raven, Philadelphia, 5th edition, 1997.
70. C. Rosse, A. Kumar, J.L.V. Mejino(Jr), D.L. Cook, L.T. Detwiler, and B. Smith. A strategy for improving and integrating biomedical ontologies. In *AMIA Symposium Proceedings*, pages 639–643, 2005.
71. C. Rosse and J.L.V. Mejino(Jr). A reference ontology for biomedical informatics: the foundational model of anatomy. *J Biomed Inform.*, 36:478–500, 2003.
72. C. Rosse, J.L.V. Mejino(Jr), B.R. Modayur, R. Jakobovits, K.P. Hinshaw, and J.F. Brinkley. Motivation and organizational principles for anatomical knowledge representation: the digital anatomist symbolic knowledge base. *J. Am. Med. Informatics Assoc.*, 5:17–40, 1998.
73. M.J. Schleiden. Beiträge zur Phytogenesis 1838. In *Transactions in Sydenham Society*, volume 12, London, 1838. Müller's Archive 1838.
74. S. Schulz and U. Hahn. Toward a computational paradigm for biomedical structure. In *Proceedings of First International Workshop on Formal Biomedical Knowledge Representation (KR-MED 2004).*, pages 63–71, Bethesda MD, 2004. American Medical Informatics Association.
75. T. Schwann. Mikroskopische Untersuchungen über die Übereinstimmung in der Struktur und dem Wachsthum der Thiere und Pflanzen, pages 1845–1856. Reimer, Berlin, 1837. Microscopical Researches into the Accordance in the Structure and Growth of Animals and Plants, translated by H. Smith, Sydenham Society, London, 1847.
76. L.G. Shapiro, E. Chung, T. Detwiler, J.L.V. Mejino(Jr), A.W. Agoncillo, J.F. Brinkley, and C. Rosse. Processes and problems in the formative evaluation of an interface to the foundational model of anatomy knowledge base. *J Am Med Inform Assoc.*, 12:35–46, 2005.
77. B. Smith. Mereotopology: a theory of parts and boundaries. *Data & Knowledge Engineering*, 20:287–303, 1996.
78. B. Smith. From concepts to clinical reality: an essay on the benchmarking of biomedical terminologies. *J Biomed Inform.*, In press.

79. B. Smith, W. Ceusters, B. Klagges, J. Kohler, A. Kumar, J. Lomax, C. Mungall, F. Neuhaus, A. Rector, and C. Rosse. Relations in biomedical ontologies. *Genome Biology*, 6(R46), 2005.
80. B. Smith, J. Kohler, and A. Kumar. On the application of formal principles to life science data: a case study in the gene ontology. In *Proceedings of DILS 2004 (Data Integration in the Life Sciences)*, Lecture Notes in Bioinformatics, pages 79–94, Berlin, 2004. Springer.
81. B. Smith, A. Kumar, W. Ceusters, and C. Rosse. On carcinomas and other pathological enitities. *Comp Funct Genom*, 6:379–387, 2005.
82. B. Smith, J.L.V. Mejino(Jr), S. Schulz, A. Kumar, and C. Rosse. Anatomical information science. In A. G. Cohn and D. M. Mark, editors, *Spatial Information Theory. Proceedings of COSIT 2005*, Lecture Notes in Computer Science, pages 149–164, New York, 2005. Springer.
83. B. Smith and C. Rosse. The role of foundational relations in the alignment of biomedical terminologies. In *Proceedings of MedInfo*, pages 444–448, 2004.
84. C.A. Sneiderman, T.C. Rindflesch, and C.A. Bean. Identification of anatomical terminology in medical text. In *AMIA Symposium Proceedings*, pages 428–32, 1998.
85. SNOMED.
    http://www.snomed.org/snomedct/index.html.
86. C.C. Teng, M.M. Austin-Seymour, J. Barker, I.J. Kalet, L.G. Shapiro, and M. Whipple. Head and neck lymph node region delineation with 3-D CT image registration. In *AMIA Symposium Proceedings*, pages 767–71, 2002.
87. R.S. Travillian, K. Diatchka, T.J. Judge, K. Wilamowska, and L.G. Shapiro. A graphical user interface for a comparative anatomy information system: design, implementation and usage scenarios. In *AMIA Symposium Proceedings*, pages 774–778, 2006.
88. R.S. Travillian, C. Rosse, and L.G. Shapiro. An approach to the anatomical correlation of species through the foundational model of anatomy. In *AMIA Symposium Proceedings*, pages 669–673, 2003.
89. R. Trelease. Anatomical reasoning in the informatics age: Principles, ontologies and agendas. *Anat Rec B New Anat.*, 289:72–84, 2006.
90. M. Tringali, W.T. Hole, and S. Srinivasan. Integration of a standard gastrointestinal endoscopy terminology in the UMLS metathesaurus. In *AMIA Symposium Proceedings*, pages 801–805, 2002.
91. Unified Medical Language System.
    http://www.nlm.nih.gov/research/umls/umlsmain.html.
92. Visible Human.
    http://www.nlm.nih.gov/research/visible/visible_human.html.
93. P.L. Williams, L.H. Bannister, M.M. Berry, P. Collins, M. Dyson, J.E. Dussec, and M.W.J. Ferguson. *Gray's Anatomy*. Churchill Livingstone, New York, 38th edition, 1995.
94. WordNet.
    http://wordnet.princeton.edu/.
95. L. Zhang, Y. Perl, J. Geller, M. Halper, and J.J. Cimino. Enriching the structure of the UMLS semantic network. In *AMIA Symposium Proceedings*, pages 939–943, 2002.
96. L. Zhang, Y. Perl, M. Halper, and J. Geller. Designing metaschemas for the UMLS enriched semantic network. *J Biomed Inform*, 36:433–449, 2003.
97. S. Zhang and O. Bodenreider. Aligning representations of anatomy using lexical and structural methods. In *AMIA Symposium Proceedings*, pages 753–757, 2003.
98. S. Zhang and O. Bodenreider. Alignment of multiple ontologies of anatomy: Deriving indirect mappings from direct mappings to a reference. In *AMIA Symposium Proceedings*, pages 864–868, 2005.

99. S. Zhang and O. Bodenreider. Law and order: Assessing and enforcing compliance with ontological modeling principles. *Computers in Biology and Medicine*, 36:674–693, 2006.

100. S. Zhang, O. Bodenreider, P. Mork, and P.A. Bernstein. Comparing two approaches for aligning representations of anatomy. *Artificial Intelligence in Medicine*, In press.

Appendices

Table 4.1. Definitions of types of high-level anatomical entities

Anatomical entity	Organismal continuant entity which is enclosed by the bona fide boundary of an organism or is an attribute of its structural organization.
Physical anatomical entity	Anatomical entity which has three or fewer spatial dimensions.
Non-physical anatomical entity	Anatomical entity which has no spatial dimension.
Material anatomical entity	Physical anatomical entity which has mass.
Immaterial anatomical entity	Physical anatomical entity which is a three-dimensional space, surface, line or point associated with a material anatomical entity.
Anatomical structure	Material anatomical entity which is generated by co-ordinated expression of the organism's own genes that guide its morphogenesis; has inherent 3D shape; its parts are connected and spatially related to one another in patterns determined by coordinated gene expression.
Portion of body substance	Material anatomical entity in a gaseous, liquid, semisolid or solid state, with or without the admixture of cells and biological macromolecules; produced by anatomical structures or derived from inhaled and ingested substances that have been modified by anatomical structures.
Anatomical set	Material anatomical entity which consists of the maximum number of members of the same class which are not directly continuous with one another. Examples: set of cranial nerves, ventral branches of aorta, set of mammary arteries, thoracic viscera, dental arcade.

Table 4.2. Definitions of salient types of anatomical structures

Biological macromolecule	Anatomical structure which has as its parts one or more ordered aggregates of nucleotide, amino acid, fatty acid or sugar molecules bonded to one another. Examples: collagen, DNA, neurotransmitter, troponin.
Cell	Anatomical structure which has as its boundary the external surface of a maximally connected plasma membrane.
Cardinal cell part	Anatomical structure which is demarcated by bona fide or fiat boundaries within a cell. Examples: plasma membrane, mitochondrion, cell nucleus, axon, apical part of columnar epithelial cell.
Portion of tissue	Anatomical structure which has as its parts cells of predominantly one type and intercellular matrix.
Organ	Anatomical structure which has as its direct parts portions of two or more types of tissue or two or more types of cardinal organ part which constitute a maximally connected anatomical structure demarcated predominantly by a bona fide anatomical surface.
Cardinal organ part	Anatomical structure which has as its direct parts portions of two or more types of tissue and is continuous with one or more anatomical structures likewise constituted by portions of two or more tissues distinct from those of their complement. Examples: neck of femur, bronchopulmonary segment, left lobe of liver, right atrium, head of pancreas, long head of biceps.
Organ system	Anatomical structure which has as its direct parts instances of predominantly one organ type interconnected with one another by zones of continuity. Examples: skeletal system, cardiovascular system, alimentary system.
Cardinal body part	Anatomical structure which has as its direct parts instances of anatomical sets of organs and cardinal organ parts spatially associated with either the skull, vertebral column, or the skeleton of a limb; in their aggregate are surrounded by a part of the skin. Examples: head, neck, trunk, limb.
Body	Anatomical structure which is the aggregate material substance of an individual member of a species.
Anatomical cluster	Anatomical structure which has as its parts anatomical structures which are adjacent or attached to one another and are together demarcated by a maximal boundary. Examples: joint, root of lung, renal pedicle, nerve fasciculus.

Table 4.3. Definitions of cell, cardinal cell parts and cell substance

Cell & Cardinal cell part	See Table 2
Nucleated cell	Cell which has as its direct part a maximally connected part of protoplasm. Examples: hepatocyte, erythroblast, skeletal muscle fiber, megakaryocyte.
Non-nucleated cell	Cell which has as its direct part a maximally connected part of cytoplasm. Examples: erythrocyte, reticulocyte, corneocyte, lens fiber, thrombocyte.
Cell component	Cardinal cell part which is demarcated from other cell parts predominantly by one or more bona fide anatomical surfaces. Examples: golgi complex, endosome, myofilament.
Cell region	Cardinal cell part which is demarcated from other cell parts by one or more anatomical planes. Examples: apical part of cell, endoplasm, head of spermatozoon.
Plasma membrane	Cell component which has as its parts a maximal phospholipid bilayer and two or more types of protein embedded in the bilayer. Examples: plasma membrane of hepatocyte, sarcolemma, plasma membrane of erythrocyte.
Cytoplasm	Cell component which has as its direct parts a portion of cytosol and one or more organelles. Examples: cytoplasm of hepatocyte, cytoplasm of erythrocyte, cytoplasm of thrombocyte, cytoplasm of neuron.
Protoplasm	Cell component which has as its direct parts a maximally connected part of cytoplasm and one or more cell nuclei. Examples: protoplasm of hepatocyte, sarcoplasm, protoplasm of megakaryocyte.
Organelle	Cell component which is surrounded by a portion of cytosol. Examples: endoplasmic reticulum, ribosome, cytoskeleton, nuclear envelope, nucleus, mitochondrion.
Cell nucleus	Organelle which has as its direct parts a nuclear membrane and nuclear matrix.
Portion of cell substance	Portion of body substance in liquid state contained in a cell cavity proper, cavity of cell nucleus or cavity of cytoplasmic organelle. Examples: mitochondrial matrix, vacuoplasm.
Portion of cytosol	Portion of cell substance in which organelles and intracellular biological macromolecules are suspended.

Table 4.4. Definitions of some high-level types of immaterial anatomical entities

Immaterial anatomical entity	See Table 1
Anatomical space	Immaterial anatomical entity which has three spatial dimensions.
Anatomical cavity	Anatomical space which is bounded by the internal surface of one maximally connected anatomical structure and contains portions of one or more body substances. Examples: lumen of esophagus, cavity of urinary bladder, cavity of lysosome, lumen of microtubule.
Anatomical compartment space	Anatomical space which is bound by the bona fide anatomical surface of two or more anatomical structures and contains two or more anatomical structures. Examples: space of anterior compartment of forearm, thoracic cavity, synaptic cleft.
Anatomical conduit space	Anatomical space which connects two or more compartment spaces or two or more anatomical cavities. Examples: pupil, nuclear pore aperture, urogenital hiatus.
Anatomical boundary entity	Immaterial anatomical entity of one less dimension than the anatomical entity it bounds or demarcates from another anatomical entity.
Anatomical surface	Anatomical boundary entity which has two spatial dimensions.
Bona fide anatomical surface	Anatomical surface which marks a physical discontinuity between two or more anatomical structures or is an interface between an anatomical space and one or more anatomical structures.
Anatomical plane	Anatomical surface which, as an imaginary plane, bisects an anatomical structure or an anatomical space.
Anchored anatomical plane	Anatomical plane which bisects an anatomical structure or anatomical space across two or more anatomical landmarks.
Floating anatomical plane	Anatomical plane which bisects an anatomical structure independent of anatomical landmarks.*
Anatomical line	Anatomical boundary entity which has one spatial dimension.
Bona fide anatomical line	Anatomical line which corresponds to the intersection of two bona fide anatomical surfaces.
Fiat anatomical line	Anatomical line which corresponds to the intersection of two anatomical planes.
Anchored fiat anatomical line	Fiat anatomical line which subdivides an anatomical surface across one or more anatomical landmarks.
Floating fiat anatomical line	Fiat anatomical line which subdivides an anatomical structure independent of anatomical landmarks.
Anatomical point	Anatomical boundary entity which has zero spatial dimension.

Anatomical landmark: part of an anatomical structure in an individual organism which is palpable or visible and can serve for anchoring a fiat anatomical line or a fiat anatomical plane.

Table 4.5. Definitions of anatomical relations

Anatomical relation	Non-physical anatomical entity which asserts an association between two or more physical and/or non-physical anatomical entities
Taxonomic anatomical relation	Anatomical relation which asserts the instantiation of types.
Is_a	Taxonomic anatomical relation which asserts the instantiation of a type by two or more subtypes or instances (individuals).
Sub_type of	Taxonomic anatomical relation which asserts the instantiation of a broader type by two or more narrower (more specific) types (subtypes).
Instance_of	Taxonomic anatomical relation which asserts the instantiation of a type by two or more instances (individuals).
Structural anatomical relation	Anatomical relation which asserts associations of a physical nature between two or more anatomical entities.
Has_dimension	Anatomical relation which associates an anatomical entity with the number of its spatial dimensions.
Has_shape	Structural anatomical relation which associates an anatomical entity with some geometric shape.
Has_boundary	Structural anatomical relation which holds between each anatomical entity of one to three dimensions and some immaterial anatomical entity of one lower dimension such that the latter demarcates (delimits) the former from its neighborhood.
Has_part	Structural anatomical relation which holds between each entity of type A and some anatomical entity of the same dimension of type B such that if A *has_part* B, there is a complement C which together with B accounts for the whole (100%) of A.
Has_generic_part	*Has_part* relation which generalizes to all specifications of the part relation.
Has_constitutional_part	*Has_part* relation which holds between each maximally connected anatomical structure and its compositionally distinct anatomical element demarcated from the complement by a predominantly bona fide boundary.
Has_regional_part	*Has_part* relation which holds between each maximally connected anatomical structure and its part demarcated from the complement by a predominantly fiat boundary.
Has_member	*Has_part* relation which holds between each anatomical set and any of its elements.
Connected_to	Structural anatomical relation which holds between each anatomical structure of type A and some anatomical structure of type B such that each structure shares some part of its bona fide anatomical surface with that of the other.
Continuous_with	*Connected_to* relation which holds between each anatomical entity of type A and some anatomical entity of type B such that there is no bona fide boundary between their contiguous constitutional parts.

Attached_to	*Connected_to* relation which holds between each anatomical structure of type A and some structure of type B such that some of the constitutional parts of the structure in type A are intermingled with some of the constitutional parts of the structure in type B across a fiat part of their maximal boundary which the related structures share.
Has_location	Anatomical structural relation which holds between an entity of any type or domain and some spatial region occupied by some physical anatomical entity.
Contained_in	Location relation which holds between a material anatomical entity and some anatomical space if the related entities are part of the same organism.
Adjacent_to	Location relation which holds between each physical anatomical entity in type A and some anatomical entity of the same dimension in type B such that their bona fide boundaries are spatially proximate, share no parts, and are separated by no physical anatomical entity of the same dimension.
Surrounds	Adjacency relation which holds between each physical anatomical entity of type A and some anatomical entity of the same dimension in type B such that the proximate bona fide boundaries of the related entities are adjacent for most of their extent.
Has_anatomical_coordinate	Location relation which holds between each physical anatomical entity in type A and some anatomical plane, line or point.
Has_organizational_pattern	Structural relation which holds between an anatomical structure and some organizational pattern.
Has_segmental_innervation	Structural relation which holds between an anatomical structure and some segment of the spinal cord.

5

Towards a Disease Ontology

Paul N. Schofield, Björn Rozell, and Georgios V. Gkoutos

Summary. The search for new mouse models of human disease has recently driven the funding of high throughput, large scale mutagenesis programmes throughout the world. As part of the attempt to deal with the data deluge resulting from these approaches together with existing hypothesis driven mouse genetics, there has been much discussion of the coding of mouse and human disease phenotypes in a way which lends itself to computer analysis, and the generation of new informatics tools. This chapter addresses current approaches to the development of a disease ontology or description framework, and critically assesses the requirements and potential solutions to the problems inherent in such an enterprise.

5.1 Introduction

Diseases have been variously classified since the time of Aristotle on the basis of the supposed underlying cause, diagnostic features or recommended treatments. The importance of a classification, or nosology of disease was clear in the late seventeeth century. Thomas Sydenham (1624-1689) wrote:

> *"It is necessary that all diseases be reduced to definite and certain species . . . with the same care which we see exhibited by botanists in their phytologies."*

By the mid-eighteenth a systematic classification of disease was proposed by Francois Bossier de Lacroix whose publication *Nosologia Methodica* was the first complete classification of disease designed specifically to aid diagnosis. The system contained 10 classes of disease, 44 orders, 315 genera, and more than 2,400 separate entities all based on similarities of symptoms and signs. To a large extent the modern development of a disease ontology reflects a continuation of the tradition of disease classification for practical purposes, but one in which the purposes of the classification framework have evolved to provide additional power to modern molecular approaches to the understanding of disease through the enablement of computation. The definition of a 'Disease' depends very much on the context and purpose. Literature definitions range from 'a medical concept which serves for communication between doctors' through 'an increased risk of adverse consequences' [31] to 'an impairment of the normal state of an organism that interrupts or modifies its vital functions'. Diseases and syndromes are difficult to distinguish as concepts, as a syndrome represents a set of signs and symptoms that appear together and characterize a disease or medical condition. Terms used alone to describe specific diseases may also constitute part of a syndrome but the syndrome itself is no more than a summary of aspects of the overall manifestation of the underlying lesion. This does suggest that any disease ontology should be able to express the occurrence of diseases as individual entities as well as the variability in the constituent entities within a syndrome.

Before embarking on a discussion of what a disease ontology should look like it is important to decide what a disease ontology is <u>for</u>. In the context of this article our prime concern will be the use of 'disease ontology' to describe the disease phenotypes of mutant mice and mouse models of human disease as a tool to predict gene function, and to discover relationships with human diseases. Specific uses might be: to take a phenotypic description of a mouse mutant and find related human diseases, to take a human disease and search for mouse models through phenotype or to take one mouse mutant and look for phenotypic overlap with others which may either provide an allelic series or aid identification of multiple genes whose proteins are involved in the same pathway. The ability to make these comparisons requires the development of either species specific disease ontologies or a common cross-species ontology capable of expressing the variation of disease manifestation in different species from the data collected.

This raises the question of the definition of a phenotype and how this fits into a definition of disease. The phenotype of a mouse may be operationally defined as the observable physical or biochemical characteristics of an organism resulting from the interaction between genotype and environment, or the expression of a specific trait, such as size or coat colour depending again on genetic and environmental influences. Such a definition is clearly compatible with a 'disease ontology' as it refers to observables in the real world. However many aspects of phenotype may not be expressed in all individual animals, yet that class of mice which is the <u>strain</u> as defined by the genotype, may be predisposed to development of the pathology. This is a class property which predicts the probability of occurrence of the disease in an individual instance, but is this predisposition itself a 'disease' or a phenotypic trait that renders a disease more likely to occur? How do we describe the predisposition to a disease where the disease is not manifested? The problem is even more acute in the case of human genetic disease. Currently the Mammalian Phenotype Ontology (MP) [26] provides an excellent high level set of phenotype terms which readily lend themselves to such phenotype descriptions. We will return to the potential relationship between MP and a disease ontology below.

In the case of man we have an enormous 'experiment of nature' where there are now more than 2000 monogenic traits alone which have been 'phenotyped' in extreme detail at population and individual levels through clinical medicine, and are the focus of much current attention as therapeutic targets [13, 21]. The systematization of phenotyping for mouse mutants, particularly in relation to large phenotype driven [1] mutagenesis screens, has become an essential development if we are at any point to use interspecific phenotype comparisons for gene and pathway discovery [5].

5.2 What is a Disease?

A 'disease' may be described as the pathological manifestation in the organism of the tissue response to an underlying lesion or set of lesions. It is a whole organism property and represents the sum total of pathological processes evolving in time. In the case of a genetic lesion this may either be somatic or inherited, in the former case there may be organism wide mosaicism or it may have been acquired only in a single cell. In the latter it will be present but possibly without consequences in all tissues. Lesions may also be induced through specific external stimuli such as infection, or factors in the environment, and the pathological response to these stimuli will interplay with the induction of new somatic lesions. Both will be subject to 'genetic background' which by and large determines the repertoire of pathological responses of the individual organism to the underlying lesion [19].

It is important to realize that what we call a disease is actually the response of normal tissue or cells to the underlying stimulus. 'Normal' is in itself a complex and shifting concept in that 'normal' might change with age and strain, and therefore must be defined in terms of deviance from a matched control group. It is an assumption in the use of animal models of human disease that phenotypic similarity, i.e. pathological similarity, can be used as a proxy for the underlying lesion(s) and therefore that the response of normal mouse and human tissue to the same lesion will be close or identical. The great success of mouse models for human disease indicates strongly that this assumption is largely justified. Recently several 'humanised' diseases have been generated through genetic manipulation of mice to demonstrate pathology not seen spontaneously in rodents, for example in mouse models of lobular carcinoma and small cell lung carcinoma [7, 18].

A description of pathology is therefore one of the most important features of any disease ontology. Pathology may be anatomically manifested – i.e. identifiable through classical anatomical pathology or histopathology, or now in conjunction with the identification of alterations in gene expression at the level of RNA and protein [2, 3]. It may also be physiologically manifest, for example in the elevation of metabolites or the failure or modification of a physiological process. The latter must however always ultimately result in cell or tissue pathology before it can be manifested as a 'disease' or a clinical disease state.

We therefore need an anatomical element of the description which will allow the tissue and cellular component affected to be described. However some diseases are not easily localized anatomically because they reflect the dysregulation of a whole organism process, for example diabetes is not localizable simply to the pancreas and the disease state encompasses pathological changes in many tissues. The problem of detailed description of pathophysiology has not yet been adequately addressed.

Diseases are dependent continuants. During the course of the disease, such as type 1 diabetes however, there will be a series of events over time and the accumulation of successive pathophysiological and pathoanatomical changes; early ones such as pre-autoimmune changes [16], initiation of autoimmune destruction of β cells, and initiation of endothelial cell death which may be classed together as dependent occurrents. Diabetes presents additional interesting problems in that the origin of the disease may be different in different individuals yet the manifestations may converge as it progresses. Thus several dependent occurrents may be common for several different disease subtypes. Specific tumour types are independent continuants, yet during the disease process - neoplasia - there will be specific dependent occurrents which might range from breach of the basement membrane, accumulation of a specific nuclear morphology or at a molecular level the acquisition of specific characteristic genetic lesions. The time course; i.e. the rate of succession of occurrents therefore also needs to be included in any disease description.

Severity is a qualifier which may reflect the genetic background, i.e. the pathological response, or the intrinsic lesion where allelic variants in the lesion might generate more or less severe consequences, for example where one allele might be hypomorphic and another allele might be a complete null. Of all the parameters we need to describe this is probably most problematical because of the highly evolved severity coding devised for many diseases which are specific to those diseases and already accepted by the community.

5.3 Terminologies and Current Approaches

Terminologies may be used for classification and for information retrieval. In themselves they are extremely useful within a defined context and examples might be the PCT codes of the American National Toxicology Programme or MeSH where the former provides a controlled vocabulary of defined terms for coding and qualifying data in a specific knowledge domain and the latter for biomedical data retrieval. Both are representations of information yet there are no meaningful relationships between the terms which allow the resulting classification to be used for logical reasoning or inference. The power of a disease description framework must lie in its ability to encapsulate complex information in defined terms (i.e. it must be highly expressive, yet unambiguous) and its ability to support logical reasoning. The latter requires that the terms and the relationships between terms are defined and meaningful allowing the use of inference to predict properties shared between members of the same class and their parent or offspring classes.

Pathological independent continuants, (tumours etc) endure through time and can undergo changes (i.e. their properties can be different at different times) whilst maintaining their identity. On the other hand, pathological processes (metaplasia etc.) do not as they may have limited duration. Hence, continuants can change their properties without changing their identity. For example the properties of an adenocarcinoma may change through time. It may change its metastatic potential or its pattern of gene expression as it progresses. This may be directly related to its degree of differentiation. Such changes will leave the individual tumour still recognizable as an instance of the class adenocarcinoma but it is important to qualify the description of the instance with a term for morphological variant or behaviour variant. Such terms should be regarded as qualifiers of the primary class and as many terms with the same meaning may be appended to different classes we see immediately that there is a need for some way of describing variants through qualifiers.

It is important that we are able to describe these properties. For this we propose the implementation of a methodology based on an ontology of qualities, termed PATO [10, 11]. This methodology, termed the EQ model, is based on a formal analysis of qualities [20]. According to this model qualities are related to bearers by virtue of inherence. For example, in order to describe an hepatocelluar carcinoma that has increased in size, we employ the following expression: increased size (PATO:0000586) *inheres_in* (MPATH:357) hepatocellular carcinoma. In human cancer, the TNM system (Sobin, 2002) provides a systematic set of stage and grade qualifiers now in common use. One challenge will be how to apply a similar system to the mouse where staging and variant description are currently rather specifically defined from tumour type to tumour type. The extension of such a system to non-neoplastic pathology would be a similar challenge and would benefit from our proposed methodology.

There has been considerable effort expended in developing systematic, and more recently computer implementable classifications of disease, since the International list for Causes of Death in 1853 which evolved into the International Classification of Diseases, now in its tenth revision [23]. ICD10 and SNOMED [22], a system, developed by the American College of Pathologists are now extensively used for clinical diagnostic recording, medical billing and insurance, yet are structured in a way which does not lend itself to logical reasoning. These approaches, which often include many terms that are etiologically or anatomically predicated, demonstrate how difficult it is to describe lesions and diseases in mice and man in a complete and applicable way.

The *Disease Ontology (DO:* http://diseaseontology.sourceforge. net/) is an ontology expressed as a directed acyclic graph with 90,000 nodes describing human disease concepts. As yet the structure of the ontology does not support reasoning and contains many terms that are either inappropriate to describe lesions in mutant mice or are redundant for rodents. However, mapping of DO onto other description frameworks shows promise of utility. There are plans to increase the semantic richness of the ontology and to increase granularity. Together with the proposed mapping of the *DO* onto the *Systematized Nomenclature of Medicine*

(SNOMED) and *Current Procedural Terminology* (CPT) codes, these will increase the application of this useful development.

An alternative approach, eVOC, has been developed by the South African National Bioinformatics Institute (http://www.sanbi.ac.za/evoc, [15]) and comprises a set of orthogonal controlled vocabularies, including a pathology ontology designed for use in humans. This was developed as a tool for the indexing of cDNA and SAGE libraries and contains 174 pathology terms organized at the higher levels as congenital anomalies, genetic, infectious, inflammatory, neoplastic, metabolic, degenerative, and other disorders. More than 8000 cDNA libraries have so far been annotated using eVOC, but there is currently insufficient coverage of pathology to use the eVOC ontology generally for the description of mutant mice.

More recently the Mouse Phenotype Analysis System (MPHASYS, http://mphasys.info/), which integrates phenotypic, pathological, and anatomical data from experiments on mutant mice, developed a rich pathology ontology derived from the National Institutes of Environmental Health Sciences (NIEHS) National Toxicology Program's (NTP) Pathology Code Tables (PCT, http://ntp-server.niehs.nih.gov/Main_Pages/NTP_PATH_TBL_PG.html).

Based on the concept that all disease has a location in the real world Rosse and co-workers have proposed a disease ontology built onto the framework of the Foundational model of Anatomy [25]. They recognise two types of pathological entity; pathological formation and pathological anatomical structure where instances of the latter are bearers of the former. However, there is no expression of pathological response which limits its use for the discovery of gene function and disease process. Interestingly they acknowledge that as long as a pathological continuant does not interfere with normal physiological processes there is pathology but no disease. It is possible therefore in this framework to have an independent pathological continuant even in a healthy organism which as discussed above may be interpreted as an abnormal phenotype (for example a predisposition or hypersensitivity) but without the presence of actual disease.

Currently the most readily applicable approach to describing disease as an aspect of phenotype is the mammalian phenotype ontology, MP [26]. The top level terms of the MP Ontology include physiological systems, behaviour, developmental phenotypes and aging and, below, physiological systems are divided into morphological and physiological phenotypes. Much manifest disease can be coded readily by MP and currently there are 88,600 annotations of approximately 21,000 genotypes. An additional resource on MGI is the human disease browser. This is based on human gene annotation to disease in OMIM [13] and provides a useful entry into the high level disease and syndrome data available from this resource. The matching is however done at the level of genes rather than phenotype/disease descriptions and it is currently not possible to automatically match phenotype with phenotype between mouse and man. At the moment MP is not structured in such a way as to allow inference and reasoning but is under continuous development and improvement. A key feature of MP is its facility of use by curators. It is structured intuitively and its use for annotation is relatively easy. The deficiencies currently are patchiness in the granularity of different segments and the inclusion of single and compound terms. The problem of wrapping up a range of aspects of a disease in a compound term become significant when considering how one matches a mouse with a human disease.

5.4 Comparing Mouse and Man; The Granularity Problem

Disease description in Man, as exemplified by ICD10 or OMIM, is tailored for clinical use, whereas those frameworks designed with mice in mind take a different approach with different ends in mind. We are left therefore with phenotypic description frameworks of rather different kinds in man and mouse, which somehow have to be related to each other. In the case of the human each 'disease' or descriptive term for a pathophysiological or pathoanatomical state is defined by a complex series of measurements and observations accumulated over years of study, using different techniques, and often differentiated by aetiology. There is a clear definition as to what is meant when we use the term 'Type 2 diabetes' or 'Beckwith-Wiedemann syndrome' in humans. However, it is highly unlikely that a human disease or syndrome will manifest itself precisely in the mouse as it does in man (for discussion of this see below) and what we are likely to be looking for in a human/mouse phenotypic comparison is an overlap of aspects of the phenotype which suggest a close relationship between the underlying lesions through the response of normal tissues which constitutes the pathology. Diabetes and its sequelae are an excellent example of complex overlap, but not identity, between what are clearly closely related diseases in the mouse and human [4, 14] as are models for autoimmune disease [16, 17]. Similar problems are well exemplified by attempts to generate mouse models of Beckwith-Wiedemann syndrome [8, 9, 28, 29, 30]. It is therefore inaccurate, misleading and limiting for hypothesis generation to give a mouse disease the same name as closely related human disorder as, though they might approximate each other, the logical implication of coding them in the same way is that indeed they are identical. Now in many cases this first approximation may be useful but when dealing with novel phenotypes it is unusable.

Another example is the relationship between the mutation carried by the "bald men of Sind' described by Charles Darwin in 1875 [6] and the mouse *Tabby* mutation [27]. Both share lesions in the same X linked gene which is manifested as anhidrotic ectodermal dysplasia. Both lose teeth, have alopecia and lack sweat eccrine glands. However, the consequences are very different for man and mouse. Mice only have eccrine sweat glands in their feet, but these are likely used for scent marking and lack of them may potentially interfere with behaviour and communication. On the other hand the lesion in man disrupts thermoregulation. Thus disaggregation of elements of the phenotype would in this case be necessary to match the diseased human with the diseased mouse in order to find the partial identity. The consequence of this is that it is problematical to code mouse diseases by using high level human terms such as those for ICD10 without a significant degree of caution.

One approach to dealing with this problem is to increase the granularity of disease description to allow for identification of overlapping human and mouse diseases. In an ideal world the granularity of the description of human diseases would be at a level similar to that in the mouse. Such descriptions should also be independent of aetiology as a similar 'disease' in human and mouse may be caused by multiple genetic lesions, environmental influences, genetic background etc. We have to decide what level of granularity and what measurements need to be recorded as part of the abnormal phenotype (disease) description. This must have enough resolution to identify partially overlapping diseases in the mouse and humans, but sufficiently practical to be applied to both organisms. To approach this we need to understand the nature of a 'disease' and the nature of 'pathology' and then to find a way of expressing the observations relating to these elements in a way which can be matched between different species. We have in this paper restricted ourselves to mammals, but if the granularity within the disease description is sufficiently fine such approaches might be used for other non-mammals - non-mammalian systems are increasingly being used as aids to understanding human disease.

5.5 A Mouse Pathology Ontology

In order to address the issue of coding pathological lesions in the mouse MPATH was developed by a group of veterinary and medical pathologists and anatomists who work extensively with laboratory mice as a description ontology for histological images of tissue lesions generated in response to underlying genetic or extrinsic damage for the Pathbase database (it was not originally intended to be used as a complete phenotypic disease description)[24]. The most recent release of MPATH contains full definitions and contains terms covering all the major classes (580 to date) of pathological lesions, with specific reference to the mouse. These classes are arranged as a hierarchy within a Directed Acyclic Graph (DAG), 6 levels deep using the *is-a* relationship with each item having an MPATH ID that can be used for database interoperability and analysis (Figure 5.1). (http://eulep.anat.cam.ac.uk/ Pathology_Ontology/index.php)

The top level of the hierarchy is arranged in categories of general pathology and covers cell and tissue damage, circulatory disorders, developmental and structural defects, growth and differentiation defects, healing and repair, immunopathology, inflammation, and neoplasia. Within these classes, each of the subsidiary terms represents an instance of the parent class and the broad arrangement is designed to be familiar to trained pathologists so as to make its use as intuitive as possible. Many tissue responses are common to multiple anatomical sites and as far as possible the redundancy of specifying a particular response in multiple tissues has been avoided, the additional topographical or anatomical information for each image coming from

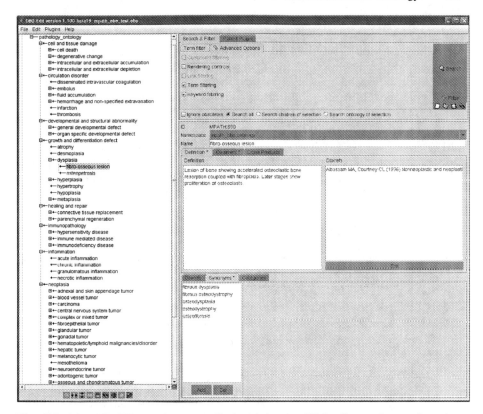

Fig. 5.1. The MPATH ontology, as displayed in the OBO-edit ontology editor (see http://www.geneontology.org, currently contains more than 570 pathology terms arranged in a simple tree format that extends to a depth of 6 levels. Each class can be viewed as a leaf attached to a higher-level node by being "an instance of" that higher level. The top levels of the ontology are arranged as general pathology and terms can be searched by ID or text using an ontology browser on the Pathbase site (http://eulep.anat.cam.ac.uk/Pathology_Ontology/index.php) and from the OBO foundry (Open Biological Ontologies) site (http://obo.sourceforge.net/) where bio-ontologies are archived.

other orthogonal ontologies and the coding for each image is therefore combinatorial. In other instances, however, there is either an intrinsic anatomical element embedded in the term or traditional pathology includes information about the cell type or tissue of origin. This is most frequent with the neoplasias and we felt that such terms were best included in their familiar form, making annotations easier for pathologists and other curators.

5.6 The Way Forward

We are currently in the position of having a group of relevant tools for the description of disease already in place but each designed for a particular purpose. The task we now have is to bring together the strengths of these tools in order to create a disease description framework that can be used as a tool for discovery.

It is clear that the measurements or observations which signify the presence of, and characterise a disease state are made quite differently in humans and mice. Indeed the reproducibility and standardisation of phenotyping in high throughput mutagenesis screens has recently been the subject of intense discussion and investigation as part of the EUMORPHIA and now the EUMODIC projects [5]. Attributing a set of symptoms or observations to the presence of a particular disease must rely on qualitative or quantitative observables and pathognomonic features. The most accurate and hypothesis neutral way to record a phenotype is through a series of statements derived from measurements or observations of a set of individuals of the same genotype. MPATH, EUROPHENOME (www.europhenome.org/) and EMPReSS [12], used in conjunction with PATO and the EQ model, are three of several tools which may be used at this level. The divergence from the prototype with regard to this set of observations, the average or normal for that strain of mouse for example, then in itself defines the disease class into which this strain should be placed. This level of description should be broadly the same as that used in the Mammalian phenotype ontology for example. It may thus be possible to generate a disease description framework that may be entered and used at different levels depending on what query is made. Comparison of two strains of mice for common disease features might be done at the level of individual phenotype observations, whilst a comparison of mouse and human diseases might be done at the higher level of what may be regarded as pre-coordinated terms.

We are at the beginning of what is likely to be an iterative activity, where adoption of a disease description framework will be dependent on its proven utility. This will depend on ease of use as much as reasoning power, and will be entirely reliant on the willingness of the community to adopt common semantic and data structure standards. The success of a disease ontology will therefore be as much a sociological challenge as a scientific one, but the rewards for cooperation are likely to be great.

Acknowledgments

The authors would like to thank Professor Jonathan Bard and Dr. John Hancock for helpful discussions on the manuscript.

References

1. J. Auwerx, P. Avner, R. Baldock, A. Ballabio, R. Balling, M. Barbacid, A. Berns, A. Bradley, S. Brown, P. Carmeliet, P. Chambon, R. Cox, D.Davidson, K. Davies, D. Duboule, J. Forejt, F. Granucci, N. Hastie, M. H. de Angelis, I. Jackson, D. Kioussis, G. Kollias, M. Lathrop, U. Lendahl, M. Malumbres, H. von Melchner, W. Muller, J. Partanen, P. Ricciardi-Castagnoli, P. Rigby, B. Rosen, N. Rosenthal, B. Skarnes, A.F. Stewart, J. Thornton, G. Tocchini-Valentini, E. Wagner, W. Wahli, and W. Wurst. The european dimension for the mouse genome mutagenesis program. *Nat Genet*, 36:925–7, 2004.

2. K. Baird, S. Davis, C.R. Antonescu, U.L. Harper, R.L. Walker, Y. Chen, A.A. Glatfelter, P.H. Duray P.S., and Meltzer. Gene expression profiling of human sarcomas: insights into sarcoma biology. *Cancer Res*, 65:9226–35, 2005.

3. J. Berman. Modern classification of neoplasms: reconciling differences between morphologic and molecular approaches. *BMC Cancer*, 5:100, 2005.

4. M.D. Breyer, E. Bottinger, F.C. Brosius, T.M. Coffman, A. Fogo, R.C. Harris, C.W. Heilig, and K. Sharma. Diabetic nephropathy: of mice and men. *Adv Chronic Kidney Dis*, 12:128–45, 2005.

5. S.D. Brown, P. Chambon, and M.H. de Angelis. EMPReSS: standardized phenotype screens for functional annotation of the mouse genome. *Nat Genet*, 37:1155, 2005.

6. C. Darwin. *The variation of animals and plants under domestication.*, volume 2. James Murray, London, 1875.

7. P.W. Derksen, X. Liu, F. Saridin, H. van der Gulden, J. Zevenhoven, B. Evers, J.R. van Beijnum, A.W. Griffioen, J. Vink, P. Krimpenfort, J.L. Peterse, R.D. Cardiff, A. Berns, and J. Jonkers. Somatic inactivation of E-cadherin and p53 in mice leads to metastatic lobular mammary carcinoma through induction of anoikis resistance and angiogenesis. *Cancer Cell*, 10:437–49, 2006.

8. J. Eggenschwiler, T. Ludwig, P. Fisher, P.A. Leighton, S.M. Tilghman, and A. Efstratiadis. Mouse mutant embryos overexpressing IGF-II exhibit phenotypic features of the Beckwith-Wiedemann and Simpson-Golabi-Behmel syndromes. *Genes Dev*, 11:3128–42, 1997.

9. A.P. Feinberg. A genetic approach to cancer epigenetics. *Cold Spring Harb Symp Quant Biol*, 70:335–41, 2005.

10. G.V. Gkoutos, E.C. Green, A.M. Mallon, J.M. Hancock, and D. Davidson. Building mouse phenotype ontologies. *Pac Symp Biocomput*, pages 178–89, 2004.

11. G.V. Gkoutos, E.C. Green, A.M. Mallon, J.M. Hancock, and D. Davidson. Using ontologies to describe mouse phenotypes. *Genome Biol*, 6(R8), 2005.

12. E.C. Green, G.V. Gkoutos, A.M. Mallon, and J.M. Hancock. EMPReSS: European mouse phenotyping resource for standardized screens. *Bioinformatics*, 21(12):2930–1, Apr 12 2005. PMID: 15827082.

13. A. Hamosh, A.F. Scott, J.S. Amberger, C.A. Bocchini, and V.A. McKusick. Online Mendelian Inheritance in Man (OMIM), a knowledgebase of human genes and genetic disorders. *Nucleic Acids Res*, 33:D514–7, 2005.

14. A.T. Hattersley. Unlocking the secrets of the pancreatic beta cell: man and mouse provide the key. *J Clin Invest*, 114:314–6, 2004.

15. J. Kelso, J. Visagie, G. Theiler, A. Christoffels, S. Bardien, D. Smedley, D. Otgaar, G. Greyling, C.V. Jongeneel, M.I. McCarthy, T. Hide, and W. Hide. eVOC: a controlled vocabulary for unifying gene expression data. *Genome Res*, 13:1222–30, 2003.

16. W.K. Lam-Tse, A. Lernmark, and H.A. Drexhage. Animal models of endocrine/organ-specific autoimmune diseases: do they really help us to understand human autoimmunity? *Springer Semin Immunopathol*, 24:297–321, 2002.

17. E. Melanitou. The autoimmune contrivance: genetics in the mouse model. *Clin Immunol*, 117:195–206, 2005.

18. R. Meuwissen, S.C. Linn, R.I. Linnoila, J. Zevenhoven, W.J. Mooi, and A. Berns. Induction of small cell lung cancer by somatic inactivation of both Trp53 and Rb1 in a conditional mouse model. *Cancer Cell*, 4:181–9, 2003.

19. J.H. Nadeau. Modifier genes in mice and humans. *Nat Rev Genet*, 2:165–74, 2001.

20. F. Neuhaus, P. Grenon, and B. Smith. A formal theory of substances, qualities, and universals. In A. Varzi and L. Vieu, editors, *Formal Ontology in Information Systems (FOIS04)*, page 4959. IOS Press, 2004.

21. T.P. O'Connor and R.G. Crystal. Genetic medicines: treatment strategies for hereditary disorders. *Nat Rev Genet*, 7:261–76, 2006.

22. College of American Pathologists. Systematized nomenclature of medicine (SNOMED), 1976.

23. World Health Organization. *Manual for international Classification of diseases and health related Problems*. Geneva, Switzerland, 10th edition, 1992.

24. P.N. Schofield, J.B. Bard, C. Booth, J. Boniver, V. Covelli, P. Delvenne, M. Ellender, W. Engstrom, W. Goessner, M. Gruenberger, H. Hoefler, J. Hopewell, M. Mancuso, C. Mothersill, C.S. Potten, L. Quintanilla-Fend, B. Rozell, H. Sariola, J.P. Sundberg, and A. Ward. Pathbase: a database of mutant mouse pathology. *Nucleic Acids Res*, 32:D512–5, 2004.

25. B. Smith, A. Kumar, W. Ceusters, and C. Rosse. On carcinomas and other pathological entities. *Comparative and Functional Genomics*, 6:379–387, 2005.

26. C.L. Smith, C.A. Goldsmith, and J.T. Eppig. The mammalian phenotype ontology as a tool for annotating, analyzing and comparing phenotypic information. *Genome Biol*, 6(R7), 2005.

27. A.K. Srivastava, M.C. Durmowicz, A.J. Hartung, J. Hudson, L.V. Ouzts, D.M. Donovan, C.Y. Cui, and D. Schlessinger. Ectodysplasin-A1 is sufficient to rescue both hair growth and sweat glands in tabby mice. *Hum Mol Genet*, 10:2973–81, 2001.

28. F.L. Sun, W.L. Dean, G. Kelsey, N.D. Allen, and W. Reik. Transactivation of Igf2 in a mouse model of Beckwith-Wiedemann syndrome. *Nature*, 389:809–15, 1997.

29. W.J. Swanger and J.M. Roberts. p57KIP2 targeted disruption and Beckwith-Wiedemann syndrome: is the inhibitor just a contributor? *Bioessays*, 19:839–42, 1997.

30. K. Takahashi, K. Nakayama, and Nakayama. Mice lacking a CDK inhibitor, p57Kip2, exhibit skeletal abnormalities and growth retardation. *J Biochem (Tokyo)*, 127:73–83, 2000.

31. L.K. Temple, R.S. McLeod, S. Gallinger, and J.G. Wright. Essays on science and society. defining disease in the genomics era. *Science*, 293:807–8, 2001.

Part II

Engineering and Linking of Anatomy Ontologies

6

Ontology Alignment and Merging

Patrick Lambrix and He Tan

Summary. In recent years many biomedical ontologies, including anatomy ontologies, have been developed. Many of these ontologies contain overlapping information and often we would want to be able to use multiple ontologies. This requires finding the relationships between terms in the different ontologies, i.e. we need to align them. Sometimes we also want to merge ontologies into a new one.

In this chapter we give an overview of current ontology alignment and merging systems. We focus on systems that compute similarities between terms in the different ontologies. We present a general framework for these kind of systems and discuss the existing strategies. We also present such a system (SAMBO) and discuss its use using anatomy ontologies. Further, we take a first step in dealing with the problem of using the best alignment algorithms for the ontologies we want to align. We present and illustrate the use of a framework and a tool (KitAMO) for comparative evaluation of ontology alignment strategies and their combinations.

6.1 Introduction

In recent years many biomedical ontologies (e.g. [18, 23]), including anatomy ontologies, have been developed. They are a key technology for the Semantic Web [19, 34]. The benefits of using ontologies include reuse, sharing and portability of knowledge across platforms, and improved documentation, maintenance, and reliability. Ontologies lead to a better understanding of a field and to more effective and efficient handling of information in that field. The work on ontologies is recognized as essential in some of the grand challenges of genomics research [4] and there is much international research cooperation for the development of ontologies (e.g. the Gene Ontology (GO) [11] and Open Biomedical Ontologies (OBO) [30] efforts) and the use of ontologies for the Semantic Web (e.g. the EU Network of Excellence REWERSE Working Group A2 [34]).

Many of the currently developed ontologies contain overlapping information. For instance, OBO lists 18 different anatomy ontologies (June 2006), some of which are deprecated (e.g. Arabidopsis anatomy and Cereal anatomy) and have been replaced

by a larger ontology (e.g. Plant anatomy) when the large amount of overlap was realized. As an example of overlapping information, in figure 6.1 we see two small
pieces from two ontologies where terms in the two ontologies are equivalent (bold
face). Often we would want to be able to use multiple ontologies. For instance, com-
panies may want to use community standard ontologies and use them together with
company-specific ontologies. Applications may need to use ontologies from different
areas or from different views on one area. Ontology builders may want to use already
existing ontologies as the basis for the creation of new ontologies by extending the
existing ontologies or by combining knowledge from different smaller ontologies.
In each of these cases it is important to know the relationships between the terms
in the different ontologies. Further, different data sources in the same domain may
have annotated their data with different but similar ontologies. Knowledge of the
inter-ontology relationships would in this case lead to improvements in search, integration and analysis of biomedical data. It has been realized that this is a major issue
and some organizations have started to deal with it. For instance, the organization
for Standards and Ontologies for Functional Genomics (SOFG) [36] developed the
SOFG Anatomy Entry List which defines cross species anatomical terms relevant
to functional genomics and which can be used as an entry point to anatomical ontologies. In the remainder of this chapter we say that we align two ontologies when
we define the relationships between terms in the different ontologies. We merge two
ontologies when we, based on the alignment relationships between the ontologies,
create a new ontology containing the knowledge included in the source ontologies.

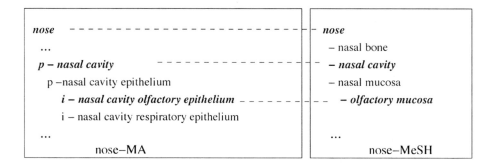

Fig. 6.1. Example of overlapping ontologies.

In this chapter we give an overview of current systems that support ontology
alignment and merging. We focus on systems that compute similarity values between terms in the different ontologies and present a general framework for this kind
of systems in section 6.2. Further, we discuss the existing alignment strategies and
give an overview of the used strategies per system. In section 6.3 we present an ontology alignment and merging system (SAMBO) and discuss its use using anatomy
ontologies. Further, we take a first step in dealing with the problem of using the best

alignment algorithms for the ontologies we want to align. We present and illustrate the use of a framework and a tool (KitAMO) for comparative evaluation of ontology alignment strategies and their combinations in section 6.4. The chapter concludes in section 6.5.

6.2 Ontology Alignment and Merging Framework

Ontology alignment and merging is recognized as an important step in ontology engineering that needs more extensive research (e.g. [31]). Currently, there exist a number of ontology alignment systems that support the user to find inter-ontology relationships. Some of these systems are also ontology merging systems. In this section we present a framework [21] for aligning and merging ontologies. The current systems that use the computation of similarity values between terms in the source ontologies[1], can be seen as instantiations of our framework.

6.2.1 Framework

The framework is shown in figure 6.2. It consists of two parts. The first part (*I* in figure 6.2) computes alignment suggestions. The second part (*II*) interacts with the user to decide on the final alignments.[2] An alignment algorithm receives as input two source ontologies. The algorithm can include several matchers. The matchers can implement strategies based on linguistic matching, structure-based strategies, constraint-based approaches, instance-based strategies, strategies that use auxiliary information or a combination of these. Each matcher utilizes knowledge from one or multiple sources. The matchers calculate similarities between the terms from the different source ontologies. Alignment suggestions are then determined by combining and filtering the results generated by one or more matchers. By using different matchers and combining and filtering the results in different ways we obtain different alignment strategies. The suggestions are then presented to the user who accepts or rejects them. The acceptance and rejection of a suggestion may influence further suggestions. Further, a conflict checker is used to avoid conflicts introduced by the alignment relationships. The output of the alignment algorithm is a set of alignment relationships between terms from the source ontologies.

Figure 6.3 shows a simple merging algorithm. A new ontology is computed from the source ontologies and their identified alignment. The checker is used to avoid conflicts as well as to detect unsatisfiable concepts and, if so desired by the user, to

---

[1] There are also some systems that use other approaches such as FCA-Merge [37], HCONE [17], IF-Map [16] and S-Match [12].

[2] Some systems are completely automatic (only part I). Other systems have a completely manual mode where a user can manually align ontologies without receiving suggestions from the system (only part II). Several systems implement the complete framework (parts I and II) and allow the user to add own alignment relationships as well.

remove redundancy.

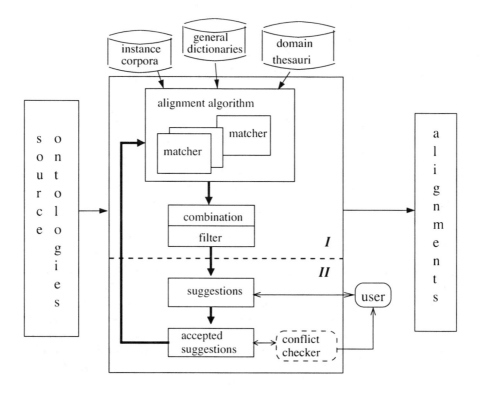

Fig. 6.2. A general alignment strategy [21].

6.2.2 Strategies

The matchers use different strategies to calculate similarities between the terms from the different source ontologies. They use different kinds of knowledge that can be exploited during the alignment process to enhance the effectiveness and efficiency. Some of the approaches use information inherent in the ontologies. Other approaches require the use of external sources. We describe the types of strategies that are used by current ontology alignment systems and in table 6.1[3] we give an overview of the used strategies per system.

- *Strategies based on linguistic matching.* These approaches make use of textual descriptions of the concepts and relations such as names, synonyms and definitions. The similarity measure between concepts is based on comparisons of the

---

[3] Also the approaches that are not based on the computation of similarity values may use these types of knowledge and are therefore included in the table.

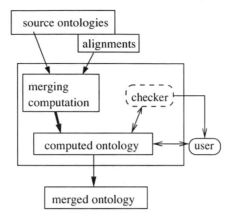

Fig. 6.3. A general merging algorithm [21].

textual descriptions. Simple string matching approaches and information retrieval approaches (e.g. based on frequency counting) may be used. Most systems use these kind of strategies.

- *Structure-based strategies.* These approaches use the structure of the ontologies to provide suggestions. Typically, a graph structure over the concepts is provided through is-a, part-of or other relations. The similarity of concepts is based on their environment. An environment can be defined in different ways. For instance, using the is-a relation an environment could be defined using the parents (or ancestors) and the children (or descendants) of a concept.
- *Constraint-based approaches.* In this case the axioms are used to provide suggestions. For instance, knowing that the range and domain of two relations are the same, may be an indication that there is a relationship between the relations. Constraint-based approaches are currently used by only a few systems.
- *Instance-based strategies.* In some cases instances are available directly or can be obtained. For instance, the entries in biological data sources that are annotated with GO terms, can be seen as instances for these GO terms. When instances are available, they may be used in defining similarities between concepts.
- *Use of auxiliary information.* Dictionaries and thesauri representing general or domain knowledge, or intermediate ontologies may be used to enhance the alignment process. They provide external resources to interpret the intended meaning of the concepts and relations in an ontology (e.g. [27]). Also information about previously aligned or merged ontologies may be used. Many systems use auxiliary information.
- *Combining different approaches.* The different approaches use different strategies to compute similarity between concepts. Therefore, a combined approach may give better results. Although most systems combine different approaches, not much research is done on the applicability and performance of these combinations.

	linguistic	structure	constraints	instances	auxiliary
ArtGen [27]	name	parents, children		domain-specific documents	WordNet
ASCO [24]	name, label, description	parents, children, siblings, path from root			WordNet
Chimaera [25]	name	parents, children			
FCA-Merge [37]	name			domain-specific documents	
FOAM [9, 5]	name, label	parents, children	equivalence		
GLUE [3]	name	neighborhood		instances	
HCONE [17]	name	parents, children			WordNet
IF-Map [16]				instances	a reference ontology
iMapper [38]		leaf, non-leaf, children, related node	domain, range	instances	WordNet
OntoMapper [33]	name	parents, children		documents	
(Anchor-) PROMPT [28, 29]	name	direct graphs			
SAMBO [21]	name, synonym	is-a and part-of, descendants and ancestors		domain-specific documents	WordNet, UMLS
S-Match [12]	label	path from root	semantic relations codified in labels		WordNet

Table 6.1. Strategies used by alignment systems [21].

6.3 An Ontology Alignment and Merging Tool

As an example of an ontology alignment and merging tool and its use, we briefly discuss SAMBO [21]. SAMBO is developed according to the framework described in section 6.2. The current implementation supports ontologies in OWL format. The system separates the process into two steps: aligning relations and aligning concepts. The second step can be started after the first step is finished. In the suggestion mode several kinds of matchers can be used and combined. The implemented matchers are a terminological matcher (TermBasic), the terminological matcher using WordNet (TermWN), a structure-based matcher (Hierarchy), a matcher (UMLSKSearch) using domain knowledge in the form of the Unified Medical Language System (UMLS) of the U.S. National Library of Medicine [39] and an instance-based matcher (BayesLearning). TermBasic contains matching algorithms based on the names and synonyms of concepts and relations. The matcher is a combination matcher based on two approximate string matching algorithms (n-gram and edit distance) and a linguistic algorithm. In TermWN a general thesaurus, WordNet [40], is used to enhance the similarity measure by using the hypernym relationships in WordNet. The structure-based algorithm requires as input a list of alignment relationships and similarity values and can therefore not be used in isolation. The intuition behind the algorithm is that if two concepts lie in similar positions with respect to is-a or part-of hierarchies relative to already aligned concepts in the two ontologies, then they are likely to be similar as well. UMLSKSearch uses the Metathesaurus in the UMLS which contains more than 100 biomedical and health-related vocabularies. The Metathesaurus is organized using concepts. The concepts may have synonyms which are the terms in the different vocabularies in the Metathesaurus that have the same intended meaning. The similarity of two terms in the source ontologies is determined by their relationship in UMLS. BayesLearning makes use of life science literature that is related to the concepts in the ontologies. It is based on the intuition that a similarity measure between concepts in different ontologies can be defined based on the probability that documents about one concept are also about the other concept and vice versa. For more detailed information about these matchers we refer to [21].

Figure 6.4 shows how different matchers can be chosen and weights can be assigned to these matchers. Filtering is performed using a threshold value. The pairs of terms with a similarity value above this value are shown to the user as alignment suggestions. An example alignment suggestion is given in figure 6.5. The system displays information (definition/identifier, synonyms, relations) about the source ontology terms in the suggestion. For each alignment suggestion the user can decide whether the terms are equivalent, whether there is an is-a relation between the terms, or whether the suggestion should be rejected. If the user decides that the terms are equivalent, a new name for the term can be given as well. Upon an action of the user, the suggestion list is updated. If the user rejects a suggestion where two different terms have the same name, she is required to rename at least one of the terms. At each point in time during the alignment process the user can view the ontologies represented in trees with the information on which actions have been performed, and she can check how many suggestions still need to be processed. Figure 6.6 shows the remaining suggestions for a particular alignment process. A similar list can be obtained to view the previously accepted alignment suggestions. In addition to the suggestion mode, the system also has a manual mode in which the user can view the ontologies and manually align terms (figure 6.7). The source ontologies are illustrated using is-a and part-of hierarchies (i and p icons, respectively). The user can choose terms from the ontologies and then specify an alignment operation. Previously aligned terms are identified by different icons. For instance, the M icons in front of 'nasal_cavity' in the two ontologies in figure 6.7 show that these were aligned using an equivalence relationship. There is also a search functionality to find specific terms more easily in the hierarchy. The suggestion and manual modes can be interleaved. The suggestion mode can also be repeated several times, and take into account the previously performed operations.

After the user accomplishes the alignment process, the system receives the final alignment list and can be asked to create the new ontology. The system merges the terms in the alignment list, computes the consequences, makes the additional changes that follow from the operations, and finally copies the other terms to the new ontology. Furthermore, SAMBO uses a DIG description logic reasoner to provide a number of reasoning services. The user can ask the system whether the new ontology is consistent and can ask for information about unsatisfiable concepts and cycles in the ontology.

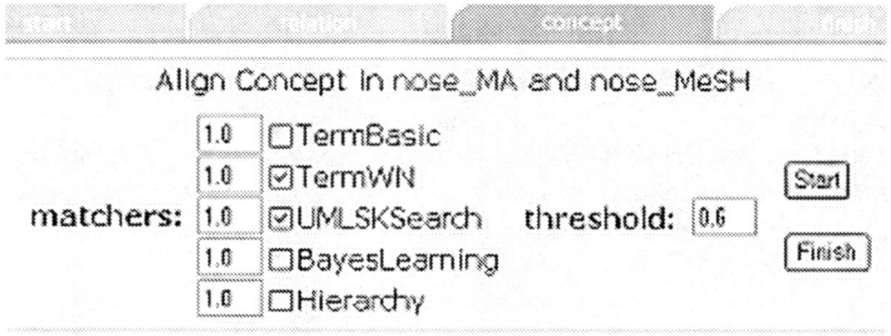

Fig. 6.4. Combination and filtering.

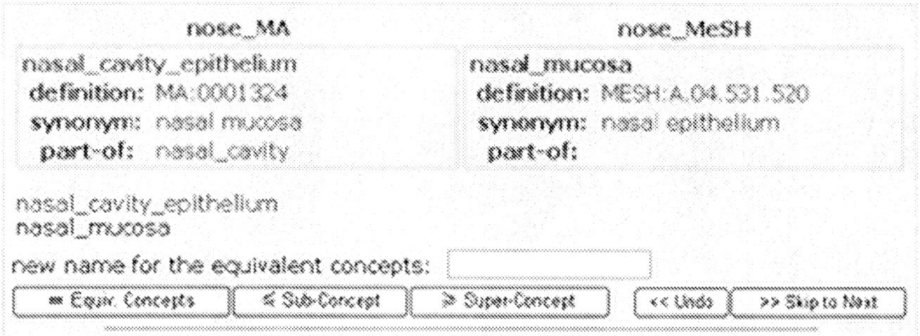

Fig. 6.5. Alignment suggestion.

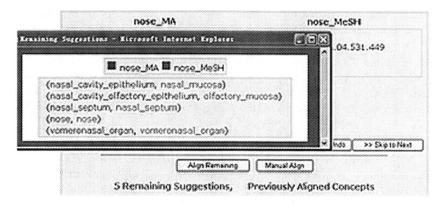

Fig. 6.6. Information about the remaining suggestions.

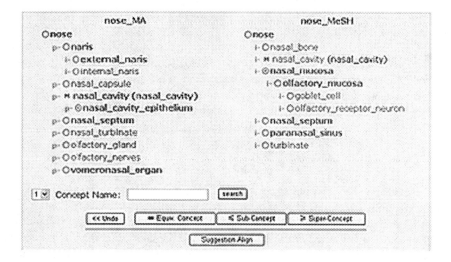

Fig. 6.7. Manual mode.

6.4 Evaluation of Ontology Alignment Strategies

To date comparative evaluations of ontology alignment and merge systems have been performed by relatively few groups [20, 21, 31] and the EON [6] and I3CON [14] contests). Among these evaluations anatomy ontologies were used in [20, 21] (Adult Mouse Anatomical Dictionary [13] and MeSH Anatomy [26]) and the 2005 EON and I3CON evaluation campaign [8] (Foundational Model of Anatomy [10] and Open-Galen[32]). The study of the properties, and the evaluation and comparison of the alignment strategies and their combinations, give us valuable insight into how the strategies could be used in the best way. To be able to do this we need tools that allow us to apply the techniques and different combinations of techniques to different types of ontologies. The tools should also support evaluation and comparison of the techniques and their combinations in terms of e.g. performance and quality of the alignment. Further, we need support to analyze the evaluation results in different ways. Although the experiments for EON and I3CON used tools for different steps in the evaluations [1, 7], there is no integrated tool that combines all the required functionalities. In this section we describe a first step towards such a tool. We first describe a framework and then a prototype implementation.

6.4.1 Framework

Figure 6.8 illustrates a framework [22] for comparative evaluation of the different alignment components (corresponding to part *I* in figure 6.2). It receives as input different alignment components that we want to evaluate, e.g. various matchers, filters and combination algorithms. It contains a database of evaluation cases which is built in advance. Each case consists of two ontologies and their expected alignments produced by experts on the topic area of the ontologies. The alignment components are evaluated using these cases.

The evaluation tool in the framework provides the wrapper which allows the alignment components to work on the ontologies in the database of evaluation cases, and provides the interface where the user can decide, e.g. which evaluation cases to use, and how the alignment components cooperate. The evaluation tool also has the responsibility to save the similarity values generated by the alignment components to the similarity database, and retrieves these similarity values from the database when required by the analysis tool.

The analysis tool receives as input data from the database of evaluation cases, similarity values retrieved by the evaluation tool from the similarity database, and possibly previously generated data from the analysis database. The analysis tool allows a user to analyze different properties of the evaluated alignment components and their combinations. For instance, it is possible to analyze such things as the similarity values between terms from different matchers, the performance of the matchers, and the quality of the alignment suggestions generated by different matchers and their combinations with different filters. Through the analysis tool the user can also save the evaluation results into the analysis database and produce an evaluation report.

6.4.2 Tool

KitAMO [22] is a prototype based on the framework introduced in section 6.4.1. It currently focuses on the evaluation of matchers and implements a weighted sum as combination strategy and filtering based on a threshold value. The matchers are added to KitAMO as plug-ins. The current database of evaluation cases consists of five small test cases including three cases based on the Adult Mouse Anatomical

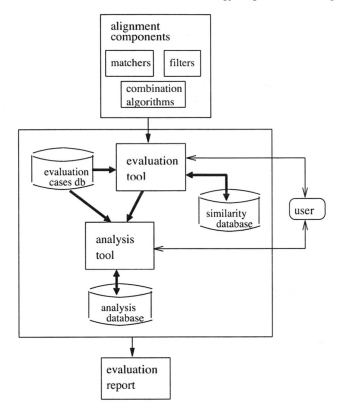

Fig. 6.8. The KitAMO framework.

Dictionary and MeSH.

The user starts the evaluation process by choosing an evaluation case. Then the user decides which matchers should be used in the evaluation from the list of matcher plug-ins configured in KitAMO. The selected matchers calculate similarity values between the terms in the chosen evaluation case, and the results are written to the similarity database. For the combination each matcher can be assigned a weight (weight in figure 6.9). The similarity values generated by the combination, i.e. the weighted sum, can also be saved to the similarity database by the user. For the filter the user can assign threshold values for individual matchers and the combination (threshold in figure 6.9).

Assuming we have chosen the *ear* case and the matchers TermWN and UMLSK Search, we receive the results as in figure 6.10. It shows the number of expected alignments (ES), the thresholds (Th), the number of correct suggestions (C), the number of wrong suggestions (W) and the number of redundant (or inferred) suggestions (I). We can save the analysis results and then experiment with other combinations and thresholds. For instance, after experimenting with thresholds 0.4, 0.5, 0.6, 0.7 and 0.8 for the two individual matchers, and different weights for the combination for the threshold 0.5, we get the results shown in figure 6.11. We have sorted the results according to the matchers. This allows us to analyze the influence of the thresholds for the matchers. For TermWN we see that the quality of the results differs significantly for the different thresholds. Although the number of correct suggestions is almost the same (25 or 26), the number of wrong suggestions goes from 3 to 8, 19, 65 and 110 when the threshold decreases. Also the number of inferred suggestions increases when the threshold decreases. This would suggest to use a high threshold for TermWN for this case. For UMLSKSearch the quality of results stays similar when the threshold changes.

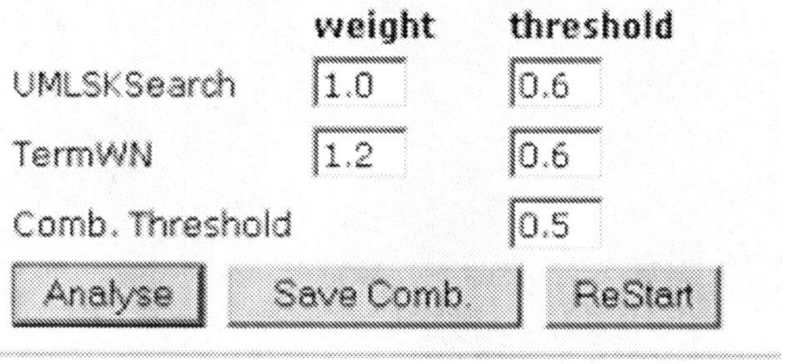

	weight	threshold
UMLSKSearch	1.0	0.6
TermWN	1.2	0.6
Comb. Threshold		0.5

| Analyse | Save Comb. | ReStart |

Fig. 6.9. The weights and thresholds assignment.

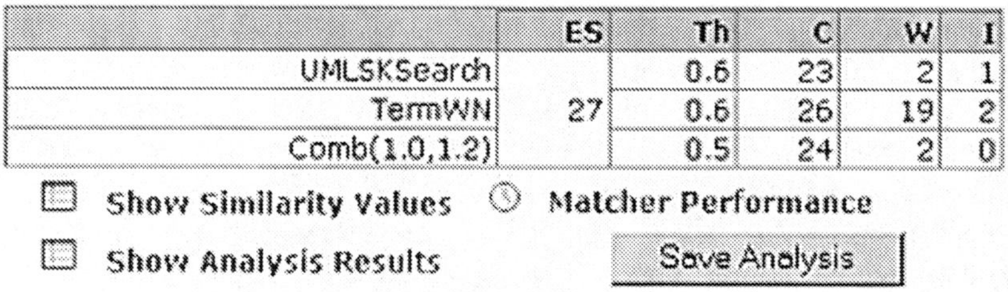

	ES	Th	C	W	I
UMLSKSearch		0.6	23	2	1
TermWN	27	0.6	26	19	2
Comb(1.0,1.2)		0.5	24	2	0

☐ Show Similarity Values ◌ Matcher Performance

☐ Show Analysis Results | Save Analysis |

Fig. 6.10. The analysis result.

For the combination the threshold is the same, but we have varied the weights for the matchers in the combination. In addition to comparing the different combinations to each other (e.g. the combinations with weights (1,1.4) and (1,1.6) give good results), we can also compare the combinations with the individual matchers. We note, for instance, that TermWN finds the correct suggestions that the combinations find.

matcher	Th	C	W	I
(1.0UM,1.0TW)	0.50	23	2	0
(1.0UM,1.2TW)	0.50	24	2	0
(1.0UM,1.4TW)	0.50	25	2	0
(1.0UM,1.6TW)	0.50	26	3	0
(1.0UM,1.8TW)	0.50	26	3	0
(1.0UM,2.0TW)	0.50	26	3	0
(1.0UM,3.0TW)	0.50	26	13	2
(1.0UM,5.0TW)	0.50	26	19	2
(1.2UM,2.0TW)	0.50	26	3	0
(1.2UM,3.0TW)	0.50	26	8	0
(1.2UM,5.0TW)	0.50	26	17	2
(1.4UM,2.0TW)	0.50	26	2	0
(1.4UM,3.0TW)	0.50	26	3	0
(1.4UM,5.0TW)	0.50	26	14	2
TermWN	0.40	26	110	19
TermWN	0.50	26	65	8
TermWN	0.60	26	19	2
TermWN	0.70	26	8	0
TermWN	0.80	25	3	0
UMLSKSearch	0.40	23	2	1
UMLSKSearch	0.50	23	2	1
UMLSKSearch	0.60	23	2	1
UMLSKSearch	0.70	22	2	0
UMLSKSearch	0.80	22	2	0

Fig. 6.11. The analysis results for the ear case.

MA	MeSH	UMLSKSearch	TermWN	(1.0UM,1.2TW)	Sug
basilar membrane	basilar membrane	1.0000	1.0000	1.0000	C
tectorial membrane	tectorial membrane	1.0000	1.0000	1.0000	C
stapedius	stapedius	1.0000	1.0000	1.0000	C
scala tympani	scala tympani	1.0000	1.0000	1.0000	C
vestibular aqueduct	vestibular aqueduct	1.0000	1.0000	1.0000	C
utricle	sacculo and utricle	1.0000	1.0000	1.0000	W
tensor tympani	tensor tympani	1.0000	1.0000	1.0000	C
middle ear	middle ear	1.0000	1.0000	1.0000	C
ear	ear	1.0000	1.0000	1.0000	C
spiral organ	organ of corti	1.0000	1.0000	1.0000	C
tympanic membrane	tympanic membrane	1.0000	1.0000	1.0000	C
auditory bone	ear ossicle	1.0000	1.0000	1.0000	C
cochlea	cochlea	1.0000	1.0000	1.0000	C
sacculo	sacculo and utricle	1.0000	1.0000	1.0000	W
incus	incus	1.0000	1.0000	1.0000	C

Fig. 6.12. The similarity table.

matchers	Performance (s)
TermWN	41.156
UMLSKSearch	137.798

Fig. 6.13. The performance table.

However, the combination finds fewer wrong suggestions.

We can also sort the table with respect to the threshold. This allows us to compare the influence of the threshold between the different matchers. We can also sort the table with respect to the number of correct suggestions. In the best case this gives us the best alignment situation. Otherwise, when there are also many wrong suggestions, it may give a good starting point for combining with other algorithms (as TermWN in the example) or for applying a more advanced filtering technique as in [2].

To examine the matchers in more detail we can use the similarity table as in figure 6.12. By sorting the table with respect to TermWN and looking at the pairs with similarity values above a certain threshold we can analyze the properties of TermWN. For instance, we observe that TermWN finds suggestions where the names of terms are slightly different, e.g. (stapes, stape). As the test ontologies contain a large number of synonyms, also suggestions where the names of terms are completely different can be found, e.g. (inner ear, labyrinth), where inner ear has labyrinth as synonym. By using WordNet, TermWN finds suggestions such as (perilymphatic channel, cochlear aqueduct) where cochlear aqueduct has perilymphatic duct as synonym, and duct is a synonym of channel in WordNet. On the other hand, since endothelium is a kind of epithelium in WordNet, TermWN generates a wrong suggestion (corneal endothelium, corneal epithelium).

The number of expected suggestions for the *ear* case is 27 (see figure 6.10). To find out the expected suggestion that is not found by any of the matchers we can check the similarity table as in figure 6.12. By sorting the similarity table according to the similarity values of a matcher, and looking at the values below the thresholds we will easily find that the only pair marked with 'C' in the 'Sug' column is (auricle, ear cartilage). This pair receives a very low similarity value from TermWN as the strings are very different and also the synonyms in WordNet are very different. We can also see that the terms are not synonyms in UMLS.

An advantage of using a system like KitAMO is that we can experiment with different (combinations of) strategies and different (combinations of) types of ontologies. For instance, the evaluation in our example may give an indication about what (combinations of) strategies may work well for aligning ontologies with similar properties as our test ontologies. However, when choosing a strategy other factors, such as time, may also play a role. For instance, KitAMO shows that UMLSKSearch is more time consuming than TermWN (figure 6.13).

6.5 Conclusion

In this chapter we presented a framework for ontology alignment and merging systems that compute similarities between terms in the different ontologies, gave an overview of existing strategies and presented a state-of-the-art system (SAMBO). Further, we discussed the problem of evaluating alignment strategies and presented a framework and a tool (KitAMO).

Alignment and merging of ontologies is an important research topic and new systems and strategies for ontology alignment will be developed. We will need more studies on which strategies work well for which types of ontologies and a system as KitAMO can provide a good environment to perform these studies. We will also see an increase of available alignments between ontologies. This will provide a new type of ontological information that can be used in, for instance, data integration [15]. Further, there are efforts to promote interoperability of ontologies, such as the OBO Foundry where it is required that the ontologies use relations which are unambiguously defined following the pattern of definitions defined in the OBO Relation Ontology [35]. The results of such efforts will provide information that should be taken into account during the alignment process.

Acknowledgements

We thank Vaida Jakoniene and Lena Strömbäck for comments on the paper and SAMBO and Bo Servenius for discussions and comments on SAMBO. We also acknowledge the financial support of the Center for Industrial Information Technology, the Swedish Research Council, and the EU Network of Excellence REWERSE (Sixth Framework Programme project 506779).

References

1. Ashpole B (2004) Ontology translation protocol (ontrapro). In: Proceedings of the Performance Metrics for Intelligent Systems Workshop.
2. Chen B, Tan H, Lambrix P (2006) Structure-based filtering for ontology alignment. In: Proceedings of the IEEE WETICE Workshop on Semantic Technologies in Collaborative Applications, pp 364-369.
3. Doan A, Madhavan J, Domingos P, Halevy A (2003) Ontology matching: A machine learning approach. In: Staab, Studer (eds) Handbook on Ontologies in Information Systems, pp 397-416. Springer.
4. Collins F, Green E, Guttmacher A, Guyer M (2003) A vision for the future of genomics research. Nature 422:835-847.

5. Ehrig M, Haase P, Stojanovic N, Hefke M (2005) Similarity for Ontologies - A Comprehensive Framework. In: Proceedings of the 13th European Conference on Information Systems.
6. Euzenat J (2004) Introduction to the EON ontology alignment context. In: Proceedings of the 3rd International Workshop on the Evaluation of Ontology-based Tools.
7. Euzenat J (2004) An API for ontology alignment. In: Proceedings of the 3rd International Semantic Web Conference, pp 698-712.
8. Euzenat J, Stuckenschmidt H, Yatskevich M (2005) Introduction to the Ontology Alignment Evaluation 2005. In: Proceedings of the K-CAP Workshop on Integrating Ontologies.
9. FOAM. http://www.aifb.uni-karlsruhe.de/WBS/meh/foam/
10. Foundational Model of Anatomy. http://sig.biostr.washington.edu/projects/fm/
11. The Gene Ontology Consortium (2000) Gene Ontology: tool for the unification of biology. Nature Genetics 25(1):25-29. http://www.geneontology.org/.
12. Giunchiglia F, Shvaiko P, Yatskevich M (2004) S-Match: an algorithm and an implementation of semantic matching. In: Proceedings of the European Semantic Web Symposium, LNCS 3053, pp 61-75.
13. Hayamizu TF, Mangan M, Corradi JP, Kadin JA, Ringwald M (2005) The Adult Mouse Anatomical Dictionary: a tool for annotating and integrating data. Genome Biology 6(3):R29.
14. I3CON (2004) http://www.atl.lmco.com/projects/ontology/i3con.html
15. Jakoniene V, Lambrix P (2005) Ontology-based Integration for Bioinformatics. In: Proceedings of the VLDB Workshop on Ontologies-based techniques for DataBases and Information Systems, pp 55-58.
16. Kalfoglou Y, Schorlemmer M (2003) IF-Map: an ontology mapping method based on information flow theory. Journal on Data Semantics 1:98-127.
17. Kotis K, Vouros GA (2004) The HCONE Approach to Ontology Merging. In: Proceedings of the European Semantic Web Symposium, LNCS 3053, pp 137-151.
18. Lambrix P (2004) Ontologies in Bioinformatics and Systems Biology. In: Dubitzky W, Azuaje F (eds) Artificial Intelligence Methods and Tools for Systems Biology, pp 129-146, chapter 8. Springer.
19. Lambrix P (2005) Towards a Semantic Web for Bioinformatics using Ontology-based Annotation. In: Proceedings of the 14th IEEE International Workshops on Enabling Technologies: Infrastructures for Collaborative Enterprises, pp 3-7. Invited talk.
20. Lambrix P, Edberg A (2003) Evaluation of ontology merging tools in bioinformatics. In: Proceedings of the Pacific Symposium on Biocomputing 8:589-600.
21. Lambrix P, Tan H (2006) SAMBO - A System for Aligning and Merging Biomedical Ontologies. Journal of Web Semantics, Special issue on semantic web for the life sciences, 4(3):196-206.
22. Lambrix P, Tan H (2007) A Tool for Evaluating Ontology Alignment Strategies. Journal on Data Semantics, VIII:182-202.
23. Lambrix P, Tan H, Jakoniene V, Strömbäck L (2007) Biological Ontologies. In: Baker C, Cheung K-H (eds) Semantic Web: Revolutionizing Knowledge Discovery in the Life Sciences, pp 85-99, chapter 4. Springer.
24. Le BT, Dieng-Kuntz R, Gandon F (2004) On ontology matching problem (for building a corporate semantic web in a multi-communities organization). In: Proceedings of 6th International Conference on Enterprise Information Systems.
25. McGuinness D, Fikes R, Rice J, Wilder S (2000) An Environment for Merging and Testing Large Ontologies. In: Proceedings of the 7th International Conference on Principles of Knowledge Representation and Reasoning, pp 483-493.

26. Medical Subject Headings. http://www.nlm.nih.gov/mesh/
27. Mitra P, Wiederhold G (2002) Resolving terminological heterogeneity in ontologies. In: Proceedings of the ECAI Workshop on Ontologies and Semantic Interoperability.
28. Noy NF, Musen M (2000) PROMPT: Algorithm and Tool for Automated Ontology Merging and Alignment. In: Proceedings of 17th National Conference on Artificial Intelligence, pp 450-455.
29. Noy NF, Musen M (2001) Anchor-PROMPT: Using Non-Local Context for Semantic Matching. In: Proceedings of the IJCAI Workshop on Ontologies and Information Sharing, pp 63-70.
30. OBO - Open Biomedical Ontologies. http://obo.sourceforge.net/
31. OntoWeb Consortium (2002) Deliverables 1.3 (A survey on ontology tools) and 1.4 (A survey on methodologies for developing, maintaining, evaluating and reengineering ontologies). http://www.ontoweb.org
32. OpenGalen. http://www.opengalen.org/
33. Prasad S, Peng Y, Finin T (2002) Using Explicit Information To Map Between Two Ontologies. In: Proceedings of the AAMAS Workshop on Ontologies in Agent Systems.
34. REWERSE. EU Network of Excellence on Reasoning on the Web with Rules and Semantics. Deliverables of the A2 Working Group. http://www.rewerse.net/
35. Smith B, Ceusters W, Klagges B, Köhler J, Kumar A, Lomax J, Mungall C, Neuhaus F, Rector A, Rosse C (2005) Relations in biomedical ontologies. Genome Biology 6:R46.
36. SOFG, Standards and Ontologies for Functional Genomics. http://www.sofg.org/
37. Stumme G, Mädche A (2001) FCA-Merge: Bottom-up merging of ontologies. In: Proceedings of the 17th International Joint Conference on Artificial Intelligence, pp 225-230.
38. Su XM, Hakkarainen S, Brasethvik T (2004) Semantic enrichment for improving systems interoperability. In: Proceedings of the ACM Symposium on Applied Computing, pp 1634-1641.
39. UMLS. http://www.nlm.nih.gov/research/umls/about_umls.html
40. WordNet. http://wordnet.princeton.edu/

7

COBrA and COBrA-CT: Ontology Engineering Tools

Stuart Aitken and Yin Chen

Summary. COBrA is a Java-based ontology editor for bio-ontologies and anatomies that differs from other editors by supporting the linking of concepts between two ontologies, and providing sophisticated analysis and verification functions. In addition to the Gene Ontology and Open Biology Ontologies formats, COBrA can import and export ontologies in the Semantic Web formats RDF, RDFS and OWL.

COBrA is being re-engineered as a Protégé plug-in, and complemented by an ontology server and a tool for the management of ontology versions and collaborative ontology development. We describe both the original COBrA tool and the current developments in this chapter.

Bio-ontologies play a crucial role in the indexing of experimental data - providing both unique IDs for aspects of anatomy, phenotype, process, cellular structure and molecular function [1, 2], and conceptual abstractions for aggregating results [3]. As discussed elsewhere in this volume, constructing ontologies of anatomy poses particular challenges including the choice of an appropriate level of granularity, how to represent spatial relationships (if at all) and how to represent the development of the organism over time. Many of the modelling decisions have been guided by the immediate use of the ontologies for indexing gene expression data, and the net result is a diversity of approaches and of interpretations for the basic elements in the anatomies, including the interpretation of the *part-of* relation. In many current anatomies the more pragmatic view of the ontology as a graph (where a *part-of* assertion is sufficient to define a concept) holds sway over the logic-oriented view that all concepts require an *is-a* relationship. This has implications for ontology editor design as the biologist will expect to see a graph that mixes *is-a* and *part-of*, rather than a pure *is-a* hierarchy that corresponds to the definitions that have been specified. These features of current anatomy ontologies had to be accounted for in the COBrA ontology editor, and its successor.

Over recent years, anatomies and other biological ontologies have grown in size, and their encoding languages have become more sophisticated, with the result that tools for creating, editing, verifying and maintaining them (e.g. version control, meta-data attribution, provenance, etc.) have become essential. This wider curation activity has been recognised as a priority for e-Science [4], and is an important concern for many communities and in standards initiatives. For example, there are efforts to standardise the names used for tissue samples assayed by microarray [5], as well as the metadata that describes the experimental results (MGED/MIAME). Ontologies are of central importance in curation, as only by defining the meaning of the terms used to describe a particular field can the underlying concepts be clarified and agreed upon among the research community, and used consistently for annotating data. A consistent, shared ontology is of critical importance to the sharing of knowledge, and has long-term value in supporting a systems-level approach to biology. For example, the Gene Ontology is in widespread use for data mining and data visualisation, and has great potential for further integration of data across the different levels of biological granularity. However, ontologies are not static: they must change to reflect changes in science, to adapt to new uses, to broaden their community or to remedy flaws. Ontologies have also been identified as key resources in numerous e-Science projects, including AstroGrid, MyGrid and the Advanced Knowledge Technologies IRC.

In parallel with expanding the range of domains being captured in bio-ontologies, and the number of terms in key resources such as the Gene Ontology (GO), researchers have been examining the formal and conceptual bases underlying ontology languages and modelling principles [6]. Initially constructed on an intuitive basis, many bio-ontologies are being scrutinised with regard to their underlying principles, and their support of inference - this being critical for automated verification. Ontologies of the same or similar conceptual domains are also being examined with respect to how they map to one another. The languages of the Semantic Web have a role to play as they provide standards, tools and techniques. For example, the Web Ontology Language (OWL www.w3.org/TR/owl-ref) has an XML syntax and a semantics designed for the sharing and reuse of ontologies over the Web. Utilising reasoners for OWL-Lite, OWL-DL and fragments of OWL-Full, OWL provides the mechanisms to address outstanding issues in bio-ontologies. For the ontology editor described here, OWL provides solutions to the problems of concept mapping and ontology verification.

Having chosen to work with OWL as the primary representation language, and to translate to and from the other bio-ontology languages, we are able to use XML databases for storage. XML querying tools can also be used for accessing and updating OWL ontologies providing we view them as XML documents.

The following sections introduce the COBrA ontology editor and its functions, then describe our solution to the curation and archiving problems that arise when individuals and communities develop ontologies.

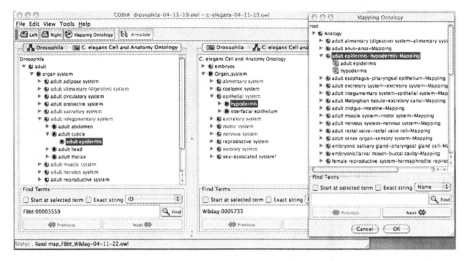

Fig. 7.1. Two anatomies displayed in COBrA, with a Mapping Ontology dialog inset right

7.1 COBrA

COBrA is an editor that allows GO and OBO ontologies to be created and explored. COBrA is also a mapping tool for ontologies that allows users to explore two ontologies simultaneously and to make links between them.

COBrA is a product of the **XSPAN** project (www.xspan.org) which uses concept mapping to express judgements of homologies and analogies between tissues across different anatomy ontologies. The resulting knowledge base will contribute to a community resource for exploring gene expression data. In XSPAN, mappings can be used to express correspondences between tissues in terms of their evolution (*Evolutionary Homology*), development (*Common Lineage Homology*) or function (*Analogy*). Creating a mapping is necessarily a human decision, made complex by the nature of the task and the size of the anatomies. Within XSPAN, COBrA supports acquisition and exploration of these human-specified mappings.

COBrA provides both a tree-based view and a node-based view of an ontology, where the latter displays the selected term's parents, children and definitional information. The tree includes all relationships used in the ontology and is not limited to only the *is-a* or only *part-of* relationships (however, the user can hide relationships if they choose to). The ontology can be edited by direct manipulation of the tree or by calling a term editor. Initial evaluation of the tool over a range of tasks, and user-types, confirms the design choices [7]. Figure 7.1 shows a mapping between *adult epidermis* (Drosophila) and *hypodermis* (C Elegans).

Concepts and relations in Semantic Web languages such as OWL require both a name and a namespace (combined into a URIRef), and COBrA provides visualisations and interfaces to these new (and potentially unfamiliar) elements of the ontology. COBrA maps OBO relationships into their OWL-Full equivalent, that is, a relationship such as *part-of* is represented as a relationship between classes (these can be formally interpreted using a translation to first-order logic as in [8])[1]. COBrA also provides a graphical interface to a number of analysis functions which we now describe.

Concept mapping, ontology merging, and verification are problems that COBrA solves through the use of OWL. A mapping is a pointer created to link a concept in one ontology to a concept in another: A mapping is a new term that relates two existing URIRefs. It can be created and saved without modifying the original ontologies. Meta-data such as authorship is associated with the mapping term, and mapping terms can be organised hierarchically, as illustrated in the right hand side of Figure 7.1. Terms with an associated mapping are shown in blue, and the user can click on such terms to automatically locate the matching term: clicking on *adult epidermis* in the Drosophila ontology causes the mapped term *hypodermis* to be found and displayed in the C. elegans ontology. The use of colour for mapped terms helps the user to locate anatomical entities that have been given a mapping. The user might critique existing mappings or seek to complete the mapping between ontologies.

Turning to ontology comparison and merger, these can be computed by finding the intersection and union, respectively, of the RDF graphs derived from the OWL representations of two ontologies. These graph-based operations improve on the equivalent operations that might be performed on textual representations of the ontologies (e.g. in CVS), but do not involve verification of the results.

For ontology verification, the semantics of the GO *is-a* and *part-of* relations must be defined, hence we use OWL *subClassOf* and define the interpretation of *partOf* [8]. These steps allow verification. An inference mechanism implements rule-based reasoning over the RDF graph, for example, to propagate properties across *partOf* links. COBrA can also perform a more complex ontology analysis that checks for cycles in the graph and in the ontology. Both graph manipulation and inference methods are provided by the Jena Semantic Web toolkit which provides Java methods to read, write and create RDF graphs (`www.hpl.hp.com/semweb`).

In addition, COBrA supports the import and export of bio-ontologies in RDF, RDFS and OWL. However, COBrA is not a generic OWL editor. The GO RDF format is that specified by the Gene Ontology Consortium, the RDFS format is a modification of that where *is-a* is replaced by *rdfs:subClassOf*. The OWL format is defined by a top-level ontology [8] which specifies a number of classes and relations

---

[1] In parallel with the development of tools for OWL-DL, a consensus on the interpretation of *part-of* in OWL-DL is emerging and so we expect to work in the OWL-DL sublanguage in future.

that are required to state GO-style ontologies in OWL.

Protégé (`protege.stanford.edu`), a generic ontology editor, and OBOEdit (`www.geneontology.org`) provide comparable editing functions to COBrA. However, neither address mapping between ontologies. Protégé would require adaptation to read GO and OBO formats, but is more fully compatible with OWL (such a plug-in tool is described below).

COBrA demonstrates the practical application of Semantic Web techniques in the Bioinformatics context by combining familiar ontology-editing functions, and compatibility with existing file formats, with additional features such as mapping, merging and verification that make use of RDF and OWL.

7.2 Ontology Curation and the COBrA Curation Tools

In common with experimental data, ontologies are created, published, and revised. Tracking and managing such changes requires new curation tools. In addition to version management, curation also includes the review of the content of the ontology, and assessment of quality. A related issue is the maintenance of ontological annotations assigned to data under a given ontology, as the ontology may change after a term has been used as an annotation and therefore one may wish for the annotation to be updated as well.

As the use of ontologies widens, the problems of tracking versions, and the changes between versions, and of reconciling differences in conceptual modelling arise. Addressing these are our main goals in the design of curation tools. Problems such as inconsistency that might arise in individual ontologies can be addressed by the graph checking that tools such as the COBrA editor can perform, or by more formal reasoning should the ontology be expressed in the description logic sub-language of OWL. We propose a server-based model for curation that allows remote users to create and submit annotated changes to ontologies and also to participate in the review process by applying some simple critiquing techniques that help identify errors. The COBrA-CT tools will support the curator by providing the appropriate management support and visualisations.

Organising the curation effort in a distributed setting, providing access to current and past versions of ontologies and providing search and related services requires an ontology management server. While it is possible (and certainly common) to simply archive different versions of an ontology, there is much to be gained from an explicit record of the changes made and their rationale. Ontology version *mappings* will be considered in the curation process, as they provide the explanation for the proposed changes. The focus on bio-ontologies is important as the problems we address are very complex in the general case. However, strategies have evolved in bioinformatics to address them, for example, concepts have IDs that are unique, and rather than being deleted, IDs persist as annotations to other concepts, or are categorised as obsolete terms.

As we are continuing to make use of the Web Ontology Language with its XML syntax as the means of data exchange, we shall be able to take advantage of both ontology-based and XML-based techniques for capturing changes. The difference between these can be illustrated as follows: from the document structure perspective, an edit to an XML-encoded ontology that asserts that class C, known to be a subclass of A, is also a subclass of B would be viewed as modifying node C in the XML document (irrespective of anything else we know about A and B). If classes A and B are known to be disjoint, then from the ontology perspective we would note a contradiction in the semantics as there can be no common subclass of disjoint classes. We propose to layer semantic checks on an XML-based ontology archiving mechanism. This approach is flexible, as XML is very widely adopted, and can exploit (but is not committed to) the logical language an ontology is expressed in.

It has been noted that changes to scientific data archives are accretive [9] - most changes are additive - although deletion and modification also occur. Scientific data is typically structured hierarchically, allowing a hierarchical key structure to be exploited in archiving changes to the data. Managing versions of a data resource can be performed on the basis of diffs (i.e. by recording the editing steps that cause the change). However, there are advantages for an approach where all objects have an associated timestamp. The central notions of hierarchical organisation, objects and timestamps [10] also apply to ontologies and ontology management, and this is the approach we plan to adopt. Given the problems noted by [11] with the simple diff approach, our approach will also be structure-based. We shall identify types of ontological changes that occur in practice, taking the procedures used in practice, e.g. by the Gene Ontology, as a starting point. As we do not assume that ontologies will make use of formalisms such as Description Logic, our approach is not reliant on the widespread uptake of this particular logic. However, we will exploit any formalism that is associated with an ontology, which may be DL or first-order logic [12].

We now present the Protégé plug-in for editing OWL bio-ontologies, named the OBO Explorer. The Ontology Version Manager is then introduced.

7.2.1 The COBrA-CT OBO Explorer

Methods for automatically converting ontologies in the Open Biological Ontologies formats into OWL have been proposed and can be utilised to create files that can be read into the Protégé ontology editor. Protégé has a large user community, and an active developer community that has created a wide range of plug-in utilities. However, Protégé is unable to display the annotations associated with OBO terms such as the database cross-references. As we aim to capture all of the content of OBO formated ontologies in OWL, both the logical structure of the ontology and the annotations, this is a significant barrier to the uptake of OWL. Therefore, there is a need for a

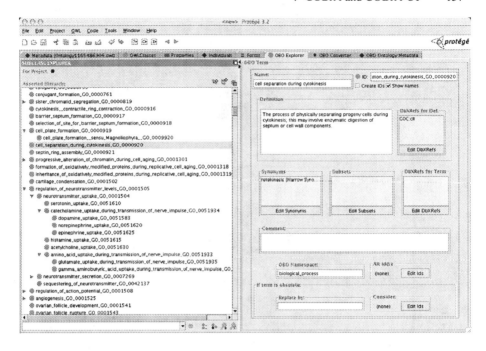

Fig. 7.2. The OBO Explorer interface

COBrA-like plug-in that will allow full visualisation and editing for OBO OWL ontologies: the OBO Explorer.

The OBO Explorer is tightly integrated to the Protégé architecture to ensure interoperability with other Protégé tools. The interface is implemented as a 'tab' that presents the term annotations on the right hand panel, with the class hierarchy on the left. The user interface components update the underlying OWL model and all changes are visualised immediately. Other Protégé GUI components viewing the OWL model behave in the same way. The user can view the term names instead of only seeing the term IDs. Protégé displays the local name of a class which is derived from the URI, however, for bio-ontologies the URI is based on the term ID as the ID (and not the name) is the primary reference. Hence, for purposes such as editing terms and navigating the ontology, the URI is temporarily modified by prefixing the term name to the ID (naturally, this change must be reversed before saving the ontology). The user can also generate new IDs for new classes, i.e. the tool finds the next ID in the series. These two features work together to allow the user to edit bio-ontologies in a familiar manner. Where the OWL ontology lacks the OWL and RDF relationships needed to represent OBO annotations, the tool creates the appropriate definitions. These features hide the underlying details of the OWL representation from the user - another contrasting feature with the built-in editor. Figure 7.2 shows the OBO Explorer tab.

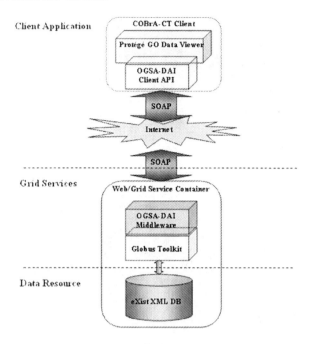

Fig. 7.3. The COBrA-CT system architecture

7.2.2 The COBrA-CT Ontology Version Manager

The COBrA-CT Ontology Version Manager allows users to access ontologies that have been published to the community and stored on the ontology server, and to store, manage and share their own ontologies. The version manager implements a simple model for assigning rights to users to allow them to download, upload, and publish ontologies. Guest users can access all public ontologies, while registered users have rights to upload and share their own ontologies. We also plan to consider explicitly representing the 'process' of curation in explicit process models, e.g. from authoring, through review, to publication and revision. The ontology curator will require a visualisation of the differences between two versions of an ontology and we can provide this through COBrA's dual-view capability. The Version Manager is implemented using Grid middleware, developed under the UK e-Science initiative, as we now describe.

Over recent years, the Grid has attracted enormous attention and gained popularity by supporting distributed resources sharing and aggregation across multiple administrative virtual organisations. Compared to the web, the Grid offers upgraded performance in terms of reliability and availability. In COBrA-CT, we developed Grid services to provide data storage and access that allow users to share their ontology information in a more scalable, secure, and dependable way. By enabling CO-

BrA-CT to operate through the Grid, the software capabilities have been enhanced greatly.

The implementation was built on top of Grid middleware, OGSA-DAI. The OGSA-DAI project (www.ogsadai.org.uk/), proposed by the University of Edinburgh, is designed to ease access to, and integration of distributed data resources via the Grid. It provides various interfaces supporting data operations, transforming and delivering with many popular (relational or XML) databases, such as Oracle, DB2, SQL Server, MySQL, Xindice, eXist etc., and file systems, such as CSV, BinX, EMBL, OMIM etc. This middleware is based on the GGF-defined OGSI specification and layered on top of the Globus Toolkit implementation. The COBrA-CT currently employs the recently-released WS-RF distribution of OGSA-DAI (OGSA-DAI WSRF 2.2), which has been designed to work with the Globus Toolkit 4 implementation of WS-RF.

The client, shown in Figure 7.3, can be implemented as part of the Protégé plug-in and uses the OGSA-DAI client libraries. Via these interfaces, the client triggers OGSA-DAI activities for uploading and downloading both ontologies and metadata. Both are passed as XML documents. XPath and XUpdate have been applied to query and modify XML database objects. XUpdate supports node-level updating in a DOM tree, which gives much more flexibility and efficiency.

The interaction between OGSA-DAI activities is illustrated in Figure 7.4. The client submits its working plan in a so-called *Perform Document*, which is a XML document consisting of a sequence of requests(*Activities*). The request is sent as encrypted SOAP message to the Grid services, which will invoke *Data Resource Accessors* (DRA) methods to connect with specific data resources. The return datasets or response message are also encrypted in a SOAP message and sent back to the client.

We use eXist (http://exist.sourceforge.net), an Open Source native XML database, to store ontology data. Compared to relational databases, the native XML database provides more powerful tools for XML processing, and so is suitable for keeping ontology and metadata information. For example, eXist supports XPath, XQuery, XUpdate, XInclude, XPointer and XSL/SXLT XML standards, and provides XML:DB API, and both DOM and SAX parsers. We also choose the eXist database because it is able to deal with large XML documents. In COBrA-CT, the ontology files sizes range from 78KB to 10,000KB. Other XML databases, e.g. Apache Xindice (xml.apache.org/xindice/) only handle documents less than 5MB, and so cannot satisfy our requirements.

In the eXist database, we store ontology files in hierarchical collections, based on user unique identifiers, ontology identifiers, and ontology version numbers. This means the physical location of a ontology OWL file is determined by these ids. To accelerate data searching, we have implemented a registry to record the ontology and

Fig. 7.4. The OGSA-DAI data flow

metadata information, and the mapping to the physical location. Current metadata information includes but not limited to:

- Ontology ownership: owner's name, id and database user roll;
- Ontology descriptions: ontology name, a text description of the version;
- Ontology file location: including the XML resource name and subcollection.
- A trace of ontology version changes, including version numbers, upload dates, and a set of previous ontologies that an ontology has been derived from. In the typical case, an ontology will simply have one previous version, but we allow for ontology merging from diverse sources, and for the concurrent editing and subsequent merging of ontology versions.
- Ontology sharing information: COBrA-CT allows a registered user to share his/her ontologies with a group of users. This is supported by associating a set of sharing users with the ontology – these users are able to download the ontology for inspection (and subsequently they may upload a modified version under their own user name). In addition to being shared with specific users, an ontology can be declared to be public, in which case it will be accessible to guest users of COBrA-CT as well as to registered users.

The client component of the Version Manager aims to provide an intuitive interface to the ontology repository. As shown in Figure 7.5, the tool shows the ontologies the user has access to and their versions, allows download and upload, and manages version numbers. User log-in using a password, however, the Grid provides other more secure methods that we shall explore in future work.

7.3 Future Work

In future work, we shall address efficiency issues in storing the OWL ontologies. Viewing the ontologies as XML data allows a range of XML techniques to be applied. We can distinguish updates to the ontology structure from updates to the annotations when analysing changes between versions. We also aim to visualise the differences between ontology versions by simultaneously displaying two versions

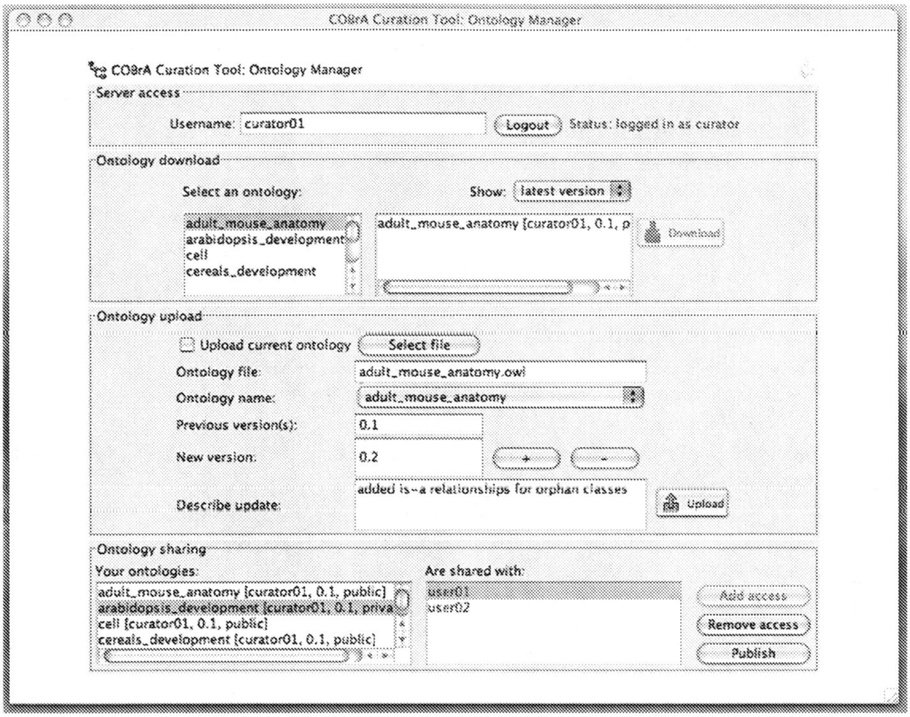

Fig. 7.5. The Version Manager client tool

and highlighting the additions and deletions graphically.

The Grid environment can provide a very high level of security covering data transmission and access to services. The Grid offers integrity (i.e. it can ensure that data has not been altered or destroyed since transmission), confidentiality, authentication, and, perhaps most importantly, availability. Currently, we have not made use of all of these features, for example, the use of certificates, and aim to explore alternative security models in future releases of the ontology tools.

Acknowledgements

This work is supported by BBSRC grants BBSRC 15/BEP 17046 and BB/D006473/1.

References

1. Gene Ontology Consortium. Gene Ontology: tool for the unification of biology. *Nature Genetics*, 25(1):25–29, 2000.

2. Open Biological Ontologies. *http://obo.sourceforge.net.*

3. J.B.L. Bard and S.Y. Rhee. Ontologies in biology: design, applications and future challenges. *Nature Review Genetics*, 5(3):213–222, 2004.

4. P. Lord and A. MacDonald. Data curation for e-science in the uk: an audit to establish requirements for future curation and provision. JISC Report.

5. H. Parkinson et al. The SOFG Anatomy Entry List (SAEL): an annotation tool for functional genomics data. *Comparative and Functional Genomics*, 5(6-7):521–527, 2004.

6. B. Smith, J. Williams, and S. Schulze-Kremer. The ontology of the Gene Ontology. Proc. AMIA 2003.

7. R. Korf. COBrA - a concept ontology browser for anatomy, 2003. Informatics Report EDI-INF-IM030022.

8. J.S. Aitken, B.L. Webber, and J.B.L. Bard. part-of relations in anatomy ontologies: A proposal for RDFS and OWL formalisations. In *Proc. PSB*, pages 166–177, 2004.

9. P. Buneman, S. Khanna, K. Tajima, and W.J.S. Tan. Archiving scientific data. In *Proc. ACM SIGMOD*, pages 1–12, 2002.

10. P. Buneman, S. Davidson, W. Fan, C. Hara, and W. Tan. Keys for xml. In *Proc. WWW 10*, pages 201–210, 2001.

11. N. Noy and M. Musen. The PROMPT suit: Interactive tools for ontology merging and mapping. *International Journal of Human-Computer Studies*, 59(6):983–1024, 2003.

12. S. Aitken. Formalising concepts of species, sex and developmental stage in anatomical ontologies. *Bioinformatics*, 21(11):2773–2779, 2005.

8

XSPAN — A Cross-Species Anatomy Network

Albert Burger and Jonathan Bard

Summary. XSPAN[1], a cross-species anatomy network, is a web-based service that aims to provide a user with information linking tissues in the anatomical ontologies of the main model species. XSPAN incorporates a database that includes the standard anatomy ontologies for mouse, *Drosophila* and human development, and adult anatomy ontologies for *Drosophila* and *C elegans*, together with mappings of their tissues to cell types from the associated cell-type ontology. It also includes further mappings that link tissues in different species on the basis of developmental lineage and functional analogy.

Mappings were either authored by domain experts or derived on the basis of other information, such as common cell-types or the lexical and structural properties of the underlying ontologies. Cross-ontology mappings were formalised in OWL files using COBrA, a newly developed bio-ontologies editor. In support of curating derived mappings, an argumentation systems approach has been explored.

The XSPAN graphical interface allows a user to select a tissue in one model species and inquire as to (1) its cell types and (2) whether there are tissues in other species to which it has a formal mapping. The system also allows the user to make a "smart" search of PubMed to find any recent papers about that tissue. Computational access to the ontologies and the mappings are provided via a web service interface.

8.1 Introduction

Two decades ago when molecular genetics was in its infancy, it was still possible for a biologist to be interested in just a single organism, but, with the appreciation of the extent to which protein sequences were conserved across the phyla, it became clear that any biologist interested in any aspect of function would need to be aware of progress across all the model organisms. With the establishment of the major gene and protein databases and the availability of sophisticated programs like BLAST, it became straightforward to compare and contrast proteins and their sequences across a wide range of organisms. Progress in doing the same thing for tissues has been

---

[1] www.xspan.org

much slower for two reasons: first, unlike proteins and genes which have defined and related sequences, there is no clear basis for comparing tissues in any analytical way; second, the grounds for establishing relationships among tissues in very different organisms are not immediately obvious.

While it might seem possible to use comparable expression patterns of homologous genes in very different organisms to identify cross-tissue relationships, this is a very ad hoc business. If one just considers the expression of the hox gene family in flies and mice, one can nod with approval about their role in establishing of the body axis for both organisms, but one has to regret that the approach fails when one considers limb development: this requires hox patterning in mice but not in flies (for review, see [4]). This example emphasises the opportunistic nature of evolution and warns us that genes can do different as well as similar things in different organisms and is of unpredictable value in comparing the genetic underpinnings of tissue relationships across organisms. If one wishes to look for other relationships, an obvious choice would be their constituent cell types. While muscle in flies and mice have their differences, they are functionally and structurally so close that they can be considered the same for this sort of comparison purposes, and it is so for most cell types. Other possible relationships include developmental lineage from similar early tissues, analogies on the basis of common function, and evolutionary homologies where they have been established. If the associated knowledge underpinning these links can be stored, it becomes possible to identify tissues in different organisms that are related on the basis of such knowledge.

This was the area that the XSPAN project set out to investigate, and its key aim was to produce an online system that would allow a user interested in a tissue in one model organism to find information about equivalent tissues in other, very different model organisms. A second aim was to enable a user to access PubMed and identify relevant publications about these tissues using smart searching. The basic approach taken was to produce a database that would hold the ontologies of developmental anatomy for the key model organisms (C elegans, Drosophila, zebrafish, mouse and humans) together with such tissue relationships that were known and appropriate data for making links across tissues and to PubMed. It soon transpired that much of the infrastructure to do this had to be produced de novo or borrowed from other resources and morphed into something that fitted within the XSPAN resource. Examples here include the cell-type ontology [2], something that turned out to be needed for a range of purposes, the COBrA tool for making links across ontologies (see Chapter 7) and the C elegans anatomical ontology, then in an early draft form.

Here, we consider what the system looks like to the user, the biological and informatics underpinnings of the XSPAN resource, how the system can be used and the wider resources available to users. The chapter ends by considering the strengths and weaknesses of XSPAN, its relationship to other resources and its possible future development.

8.2 Core Ontologies

The biological information held within the XSPAN repository includes the anatomy and cell-type ontologies together with a series of mappings linking them. In this section we discuss the ontologies, while the mappings are discussed in the next. All ontologies and mappings can be downloaded from the XSPAN site.

XSPAN currently holds ontologies of developmental anatomy for *Drosophila*, the mouse and the human (up to Carnegie stage 20, or E50) and of the adult anatomy for *C elegans* (both hermaphrodite and male). The XSPAN *Drosophila* is a slightly simplified version of the standard *Drosophila* ontology (see `obofoundry.org/browse.shtml`) as it does not include all the details of the adult tissues. The ontology includes three time stages: the embryo, the larva and the adult. This means, for example, that, although there will be separate entries for the embryonic, larval and adult guts, there is little chance of confusion by a user.

Mouse embryogenesis is far more complicated than that of *Drosophila* as it is partitioned into 26 stages with each tissue in the ontology being given a separate ID for each stage of its development [1]. We realised that using the standard staged version of the ontology would make searching unreasonably complicated for a user who was not well versed in the temporal details of mouse development (and most will not be!), and might also give rise to ambiguities. We therefore decided to use the abstract version of the mouse ontology: this includes every anatomical entity and its parts throughout development, but excludes formal stage details. In this version, a tissue has an entry for each step in its development, irrespective of the number of developmental stages in which that step is found - the abstract ontology is thus much shorter than the staged version (see Chapter 1 for details). The abstract mouse has two advantages over the normal version: first, it turns out to be useful as a high-level search index and, second, it allows a user to identify any tissue by its degree of complexity rather than by its developmental age, so simplifying the search. As the ontology of human developmental anatomy with more than 20 stages (`www.ana.ed.ac.uk/database/humat/`) is based on the ontology of mouse developmental anatomy [5], it too was handled in its abstract format.

The *C elegans* anatomy does not yet have a formal anatomy ontology, just a list of parts (`elegans.swmed.edu/parts/parts.html`). It was therefore necessary to produce one and it was partly based on the work in the worm atlas (`www.wormatlas.org`) and in discussion with David Hall) and partly formalised along the lines of the OBO anatomies (`obofoundry.org/browse.shtml`). The resulting ontology covers all the tissues in the adult hermaphrodite, and also includes male-specific tissues within alternate pathways, where appropriate. It does not cover developmental anatomy, but this is mainly laid down in the last cell division so it would not be easy to handle as a separate stage.

XSPAN also includes the cell-type ontology (see `obofoundry.org`) which includes some 600 cell types most of which can be viewed as species-independent [2]. This, unlike the anatomy ontologies which are single-parent, part-of hierarchies, has a full directed acyclic graph (DAG), multi-parent structure with cell types being organised by function, morphology, ploidy and several other high-level concepts.

8.3 The Mappings

The mapping of concepts across ontologies is based on either directly capturing expert knowledge via the COBrA tool or on computational approaches taking into consideration additional information such as common cell types across anatomical structures or a lexical and structural analysis of the underlying anatomy ontologies. Computational methods required a subsequent curation step for quality assurance. Some of the reasoning involved in this curation is modelled in the context of an argumentation system.

The expert mappings were made using the linking facility of COBrA. For tissue-cell mapping purpose, COBrA was assigned an anatomy and the cell-type ontologies and pairs of concepts were linked by hand on the basis of the relationship "has cell type", with the triad being assigned an ontology ID and the resulting set of triads being first stored in OWL, and then imported into the XSPAN database. For tissue-tissue mappings, concepts were linked on the basis of "has same lineage" or "has functional analogy" and similarly stored.

The key mappings are from each tissue in the mouse, *Drosophila* and *C. elegans* ontologies to the cell-type ontology. To do this, every terminal tissue was assigned its cell types with every higher level tissue including the cell types in its subordinate parts (human tissues were assigned the same cell types as their mouse equivalents). Assigning these cell types involved considering the histology of the tissues and it has to be said that the mappings should be considered as provisional.

Making cell-type mappings between tissues in different organisms is not exactly straightforward: if one just specifies muscles, one would obtain far too many matches as one would obtain not only every muscle, but every higher level tissue that contains a muscle. Such single-cell-type matches would only be appropriate for highly specified cell types. A better approach would be to allow the user to specify two cell types and this turns out to give a far more restricted set of matches.

There are other sorts of mappings which can be viewed as orthogonal in the sense that they are independent of one another and possibilities include developmental lineage (e.g. ectoderm gives rise to gut epithelium), functional analogy (limbs of *Drosophila* and mice) and homology (derived from a common ancestral tissue). It turns out that, in practice, there are very few evolutionary links across the vertebrates and invertebrates that are particularly useful. While there are lineage relationships between *Drosophila* and mouse, there are none between these and *C. elegans* whose ontology includes only adult tissues, and there are a few tissue analogies across the three model organisms. These mappings are included in XSPAN but the lists are far less extensive than originally expected.

In contrast to mappings derivable from common cell types, which were easily implementable as simple queries in the database, but had to be viewed with the biological caveats described above, computing potential mappings based on a lexical analysis of the terms used to identify tissues and the structural similarites in the underlying ontologies proved to be a much harder problem. Figure 8.1 depicts such a mapping between *Drosophila* and *C. elegans*.

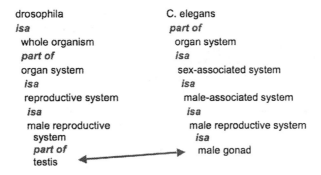

Fig. 8.1. Example of mapping from *Drosophila's* testis to *C. elegans'* male gonad, illustrating how similar terms are used to describe these tissues, but also highlighting the differences in ontology structure and names.

We assumed in proposing XSPAN that some alignment of the anatomical ontologies of the different species could be generated automatically. The two-step method we developed [6] uses both the terminology and the structure of the ontologies being aligned, to produce consistent, supported alignments. The first step finds pairs of terms from two source ontologies that are lexically similar, while the second checks pairs for structural consistency. A pair may thus have evidence of structural support, or no evidence may be found, or structural evidence may contradict the hypothesis. This method was used to suggest mapping hypotheses between human and mouse ontologies and human and *Drosophila* ontologies; these were then confirmed, denied or left "hypothetical" as the result of a manual curation process. The primary stumbling blocks preventing full automation of this process were the significant syntactic and semantic differences in the ontologies. These suggested that it would be beneficial were anatomy ontology stake holders to make a concerted effort to develop a stronger degree of standardisation than currently exists. In fact, recent efforts such as CARO (see Chapter 16) are now pursuing this issue.

In addition, we would have liked to have included cross-species mappings based on the expression of homologous genes, but it turned out that, not only were the data sets incomplete for the mouse in GXD[2] but they were mainly unobtainable on-line for the other model species. Even had more data been available, they would have been of limited use as they included homologies that were too weak to be helpful: simple cross-linkings would not distinguish between housekeeping and tissue-specific gene expression. Worse, given the large number of alternative splicings and partially truncated genes generated by BLAST, it was simply impractical to identify useful homologies other than by detailed expert analysis and this would have been of little use in an automated system of comparison.

In addition to automatically deriving anatomical mapping suggestions, as described above, some research was carried out on how to mimic in software some of the reasoning applied by a domain expert during the curation of the derived mappings. For this, an argumentation-based approach [3] has been explored. Arguments [7] are used to support the believe in some statement, e.g. that a gene is expressed in a particular anatomical structure. Arguments can also be made contrary to some statement and can attack and defeat each other. An argumentation system provides a form of non-monotonic reasoning, and is thus particularly suitable to deal with the reality of conflicting information within and across multiple bioinformatics resources.

For the purposes of XSPAN, the notion of a so-called argumentation scheme [8] was adopted. These schemes collate a series of critical questions that can be applied to argue for or against a particular statement. In the context of mapping anatomical structures across different species, the domain expert asked during curation whether the granularity of the respective structures, signified by the level of the part-of hierarchy at which they can be found, were too great. The following question was part of the XSPAN mapping argumentation scheme: Is the difference in level in the part-of hierarchy for the mapped structures greater than 3? If so, the argumentation system rejected the mapping (as did the expert).

The investigation into the use of argumentation for anatomy mapping is still in its early stages. In XSPAN, the experience was that, although some of the domain experts curation could be automated in this way, it could not entirely replace the manual curation process. A recurring problem in this context was the lack of formalisation of background knowledge required to translate questions in the argumentation scheme into an executable form.

---

[2] www.informatics.jax.org

8.4 Web-based Graphical User Interface

End-user access to XSPAN's data set is facilitated through a graphical user interface (Figure 8.2). The primary layout consists of panes presenting the two ontologies of current interest (COBrA and XSPAN use consistent visualisation paradigms). The applet has been subject to several revisions in order to accommodate feedback from end users on earlier versions of the interface. Figure 8.2 shows an example of how the GUI can be used.

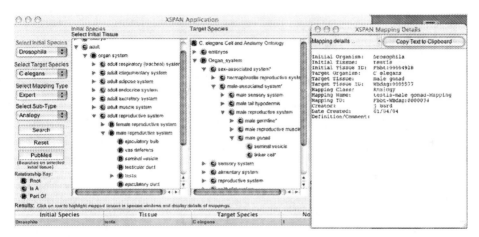

Fig. 8.2. The above screenshot shows the interface after the user selected 'testis' in *Drosophila* in the initial species panel, picked *C. elegans* as the target, and 'Expert' and 'Analogy' for the mapping type. Clicking on the 'Search' button produced the result table (with on entry) at the bottom of the window; double-clicking on that entry brings up the relevant section of the target ontology – the part of the *C. elegans* ontology containing the 'male gonads' tissue – and a small window with details on the mapping that has been found.

Integrated into the GUI is a query mechanism to find relevant PubMed[3] entries. Users can select an anatomical entity in one of the ontologies and trigger a search of PubMed for papers relating to this entity based on MESH headings, which are automatically generated based on the selected anatomy ontology term (for an example, see Figure 8.3).

8.5 XSPAN Server

This user interface and the interaction with COBrA are supported by a computational backend, the XSPAN server. Figure 8.4 shows the main components of the system: a

---

[3] www.pubmed.org

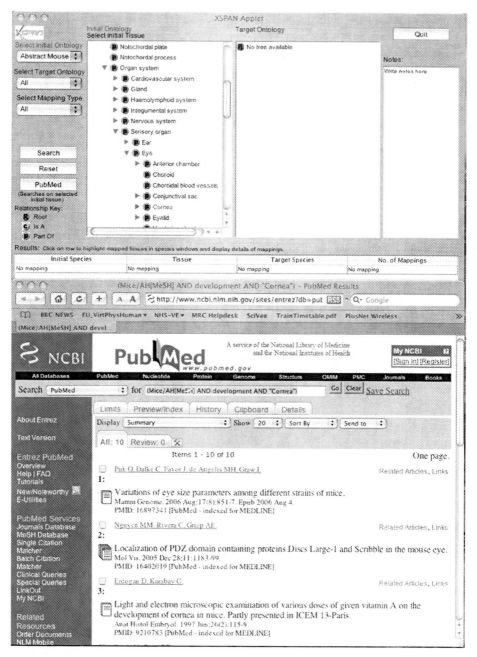

Fig. 8.3. PubMed Example: A user has selected the 'Cornea' tissue in the mouse ontology and clicked the PubMed button on the XSPAN interface. This resulted in a call to PubMed using (Mice/AH[MeSH] AND development AND "Cornea") as the search query. Relevant papers found are shown in the PubMed web page that is returned to the user.

database for the persistent storage of the ontologies and the mappings, a web service for access to this database and clients that facilitate access to the data via the web service.[4] The web service also handles the link to PubMed.

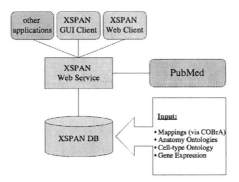

Fig. 8.4. Basic XSPAN Architecture: the data layer in terms of a database, the computational layer encoded in the XSPAN web service, and the user interface layer in terms of GUIs for end users or some other applications, such as a workflow.

XSPAN anatomy ontologies and the mappings recorded between them are persistently stored in a relational database system whose design schema is generic in order to accommodate the various underlying structures of the original anatomy ontologies. A loader program has been developed that imports RDF/OWL files generated by COBrA. Programmatic access to the database has been abstracted into a separate software layer, allowing flexibility in the choice of DBMS (we have used IBMs DB2 and mySQL).

The XSPAN Web Service allows programmatic access to the ontologies and their mappings in the database. Making such a web service publicly available allows workflow tools such as Taverna[5] to include the XSPAN functionality in more complex analysis applications. A simple cross-species gene-expression workflow including XSPAN is illustrated in Figure 8.5. This XSPAN web service implementation is based on the Apache AXIS toolkit and its interface (its WSDL description) is available from the XSPAN project web pages, as is a sample client application that demonstrates the use of the web service.

---

[4] The non-web version of the XSPAN GUI client is not available publicly. It's primary function was for internal testing and all its functions are also available via the XSPAN Web Client GUI.

[5] taverna.sourceforge.net

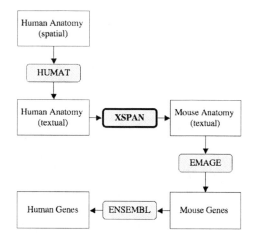

Fig. 8.5. Workflow Example using XSPAN (boxes in rounded corners represent web services, the remaining boxes represent input/output data for these services): Consider the scenario where one wishes to find possible gene expression in a human tissue by exploring gene expression data available for mouse. The workflow might use a human atlas (HUMAT) to determine the name of the tissue identified spatially. XSPAN finds the equivalent tissues for mouse, which in turn are used to retrieve the genes expressed in these tissues, for example by accessing the EMAGE gene expression database (genex.hgu.mrc.ac.uk). ENSEMBLE can then be used to determine the homologues human genes, which might be considered as likely to have expression in the original human tissue. This workflow has been implemented as a test case using the Taverna workbench.

8.6 Discussion

Over the past decade, bio-ontologies have been made for a range of knowledge domains and are now readily accessible in a variety of formats (obofoundry.org). The anatomy ontologies have, it should be said, mainly been used within their home organism databases and it is a little disappointing that they have not been more generally exploited for annotation purposes in other contexts, as the ontology IDs are key to interoperability across resources. Two important exceptions here are the Gene Ontology (www.geneontology.org) which is used to annotate Uniprot (www.uniprot.org), the protein resource, and is the basis of many other applications and resources (www.geneontology.org/GO.tools.shtml) and the mouse developmental anatomy ontology which is used to annotate GXD, the mouse gene expression database (www.informatics.jax.org).

XSPAN sets out to exploit the anatomy and cell-type ontology resources, not in the context of annotation, but to expand the ontological knowledge that can be associated with tissues from adult *C. elegans*, all stages of *Drosophila* development, and for mouse and human embryos. In a sense, it aims to provide a rich and integrated ontological resource by linking the existing anatomical ontologies with the cell-type ontology through adding additional knowledge. This is done partly through expert mappings and partly through mappings derived computationally. At its most basic, XSPAN offers the user information on links between tissues and cell types, and across tissues that has not hitherto been available computationally. These links are available as downloadable OWL files to be used with whatever computational resources a user wishes. More importantly, XSPAN also offers an interface that allows users to select a tissue or a cell type and look for mapped information on the basis of the information in the XSPAN database. It also allows a user to conduct searching of Pubmed to identify papers associated with tissues.

We hope that XSPAN will be helpful to biologists, whether or not they have a strong informatics bent, and to facilitate this, the graphical user interface (GUI) to XSPAN has been designed with the intention of making it easy to use. Nevertheless, it is worth pointing out that the current version of XSPAN has several limitations. The most obvious one is that, while every effort has been made to ensure that the tissue-to-cell-type and tissue-to-tissue mappings are correct, mistakes may have been made. Such mistakes are of two sorts: first, the exact cell type for every developing tissue has not been precisely identified in the literature and there are times when guesses, albeit sensible, have been made; second, it has, of necessity, to be assumed that, if the cells of a developing tissue look alike, they are alike, unless there are good reasons to suppose otherwise (e.g. many early mouse tissues have mesenchyme that includes neural crest cells); where the mappings are oversimple, they will need to be corrected in the light of future knowledge. The mappings also assume that the developmental anatomies are both full and accurate: while there is no reason to doubt this in most cases, the later stages of the mouse developmental anatomy may well be a little oversimplified as it hard to capture the full richness of mammalian anatomy. This is just to say that the XSPAN knowledge base, like any other ontology or information resource will need regular curation.

A second problem lies with the derived mappings, particularly those based on cell types. At the moment, XSPAN allows a user to select a cell type from the cell-type ontology and ask the system for all tissues in a choice organism that include that cell type. It is also possible to choose any tissue and request its constituent cell types. What the system does not yet allow is a user to choose a tissue and ask for all tissues that have the same cell types, or to specify more than one cell type for the analysis. In practice, this means that a query may well generate more tissues than a user might find helpful (in most cases, there are one-to-many mappings both within and across species). We actually put some effort into trying to produce algorithms that gave short and accurate responses, but found that there were usually too many false negatives for accuracy and, in the end decided that it would be more sensible to generate responses that included unnecessary positives that a user could eliminate by hand.

As to the anatomy ontologies, those for *C elegans* and *Drosophila* are straightforward to use, but those for the mouse and human are not simple. The obvious reason here is that mammalian development is not only complicated but is multi-staged. While *C elegans* only covers adult tissues and *Drosophila* has three easily distinguished stages (embryo, larva, adult) which are not further temporally subdivided in the ontology, the mouse has 26 and the human 23 stages. We felt that including all these stages with their many tissues would be too hard to handle for anyone without a fair amount of knowledge on mouse and human anatomy. In addition, when matching tissues across stages, it is not sensible to equate, say, the gut of *Drosophila* embryo with one or another developmental stage for the mouse or human gut.

XSPAN therefore handles the mouse and human developmental ontologies in their abstract (stage-independent) forms which are still quite complicated but includes each tissue (name) path only once (e.g. the early mouse heart without, say, an atrial septum has a separate entry from a heart that has developed an atria septum, but each has a single entry, irrespective of in how many stages it is present). Inexperienced users may require a few minutes of experimentation to navigate the system. The human ontology of developmental anatomy has been similarly formulated, but it works at a slightly coarser granularity than that for the mouse which was designed to handle high-resolution gene-expression submissions in the GXD database. It should also be pointed out that, where appropriate (and this means for the great majority of the tissues), the mouse-cell type mappings have been transferred directly to the human embryo.

A further improvement that we hope to introduce is to link the tissues to their corresponding CARO concepts so as to be able to compare tissues on the basis of their anatomical type. The current version of CARO (see Chapter 16) is based on the tissues of the adult human that are included in the FMA. The next version of CARO is expected to be adapted to incorporate annotations for developmental tissues which are often relatively simple, but in a state of flux. Once this is done, it should be possible to extend XSPAN's mapping facility on the basis of the nature of a tissue (e.g. its geometry, its complexity and whether or not it is acellular).

A further enhancement of the XSPAN mapping would be to relax the constraint of one-to-one mappings, i.e. to allow one tissue in one organism to be mapped to a collection of tissues in a different organism, thus partly addressing issues of different granularities across ontologies.

8.7 Conclusion

The prime aim of XSPAN was to provide biologists interested in one model species with information on equivalent tissues in other model species, with equivalence being defined as sharing some important property such as cell types, developmental lineage or functional analogy. Although XSPAN does not attempt to integrate the underlying anatomy and cell-type ontologies into a *single* ontology – the semantics of the concepts and relations in the underlying ontologies are too heterogeneous – it does provide a single environment for searching anatomical knowledge across multiple species.

Among the secondary aims were the wish to be able to search Pubmed for literature associated with these tissues in a way that was more sophisticated than usual as XSPAN employed some of the hidden functionalities of that database. These aims were met by producing a searchable user interface for biologists, underpinned by a database containing ontologies, linking files and search engines; the net result is an on-line tool that we hope will be useful to the community.

Acknowledgements

XSPAN was funded by UK's Biotechnology and Biological Sciences Research Council (BBSRC 15/BEP/17045 and 15/BEP/17046). Funding support by the European Union – projects Sealife (FP6-2006-IST-027269) and REWERSE (FP6-2006-IST-506779) – for the work on argumentation systems is also kindly acknowledged.

We would like to thank our XSPAN colleagues at the University of Edinburgh (Stuart Aitken, Sarah Luger and Bonnie Webber) and at Heriot-Watt University (Kenneth McLeod, Gus Ferguson, Patrik Holt and Werner Nutt), as well as colleagues from other institutes who have collaborated with us on this project, in particular the Mouse Atlas group at the MRC's Human Genetics Unit in Edinburgh, who are now hosting the XSPAN system.

References

1. J. Bard, M.H. Kaufman, C. Dubreuil, R.M. Brune, A. Burger, R.A. Baldock, and D.R. Davidson. An internet-accessible database of mouse developmental anatomy based on a systematic nomenclature. *Mech Dev*, 74:111–120, 1998.
2. J. Bard, S.Y. Rhee, and M. Ashburner. An ontology for cell types. *Genome Biology*, 6(2):R21, 2005.
3. D.V. Carbogim et al. Argument-based application to knowledge engineering. *Knowledge Engineering Review*, 15(2):119–149, 2000.
4. S.F. Gilbert. *Developmental Biology*. Sinauer Press, 8th edition, 2006.
5. A. Hunter, M.H. Kaufman, A. McKay, R. Baldock, M.W. Simmen, and J. Bard. An ontology of human developmental anatomy. *Journal of Anatomy*, 203:347–355, 2003.
6. S. Luger, S. Aitken, and B. Webber. Cross-species mapping between anatomical ontologies: Terminological and structural support. In *Proc. of Bio-Ontology Workshop, ISMB*, July 2004.
7. J. Pollock. Defeasible reasoning with variable degrees of justification. *Artificial Intelligence*, 13(1-2):233–282, 2003.
8. D. Walton. *Appeal to expert opinion: arguments from authority*. Penn State Press, PA, USA, 1997.

9

Searching Biomedical Literature with Anatomy Ontologies

Thomas Wächter, Dimitra Alexopoulou, Heiko Dietze, Jörg Hakenberg, and Michael Schroeder

Summary. Many ontologies and vocabularies have been designed to annotate genes and gene products based on evidence from literature. They are also useful to search literature systematically. GoPubMed is such an ontology-based literature search engine. It allows users to explore PubMed search results with hierarchical vocabularies such as the Gene Ontology or MeSH. We demonstrate the use of GoPubMed and MeshPubMed to answer questions relating to anatomy. Then, we discuss MousePubMed, the adaption of GoPubMed to vocabularies used in the Edinburgh Mouse Atlas with genes, tissues, and developmental stages. We develop a specific text mining algorithm for MousePubMed and demonstrate its usefulness by evaluating it on the Mouse Atlas. For nearly 1500 genes and over 10.000 triples of gene, tissue and stage, we are able to reconstruct with MousePubMed 37% of genes, 31% of gene-tissue associations and 13% of gene-tissue-stage associations from PubMed abstracts. These figures are encouraging as only abstracts are used.

9.1 Introduction

Ontologies and vocabularies such as the Gene Ontology [2], UMLS [8], Mesh[1], OBO[2], Snomed[3], and GALEN[4] are widely used for annotating biomedical data. They typically contain thousands of terms and cover broad subject areas of biomedical research. Additionally, many species-specific vocabularies for anatomy have been designed covering, among others, plant [18], C. elegans [1], drosophila [16], mouse [4, 5], and human [26] anatomy. These vocabularies are used to facilitate communication between scientists in different communities and inter-operability between databases. Annotators, who are usually human, assign terms from such terminologies for example to genes. These assignments are ideally based on direct evidence from literature. Therefore, it is an important problem to automatically identify terms from ontologies in literature to support and even partly automate the annotation process.

---

[1] nlm.nih.gov/mesh
[2] obo.sourceforge.net
[3] snomed.org
[4] opengalen.org

However, if terms from ontologies can be found in text, then ontologies can serve directly in literature search. Recently, a number of such knowledge-based search engines were published; for instance, Textpresso [23], XplorMed [25], and GoPubMed [12]. The ontological background knowledge can serve to answer questions with such tools. Consider for example a researcher interested in the Pax6 gene. He/she might have the following questions:

- Which processes is Pax6 involved in?
- Which diseases is Pax6 involved in?
- At which developmental stages is Pax6 active in mice?

Literature holds answers to these questions, but a classical literature search cannot answer the questions directly, as articles will not mention gene, disease or process, but rather specific instances such as Pax6, Aniridia, or eye development. Since ontologies contain knowledge that Pax6 is a gene, Aniridia is a disease, and eye development is a process, they can help to answer such questions.

In this chapter, we will show how ontology-based literature search with GoPubMed can answer questions as the ones above. To accommodate the specifics of anatomy we will also discuss the use of specialised background knowledge. In particular, we will devise an algorithm and a system, called MousePubMed, to work with genes, tissues, and developmental stages as used in the Edinburgh Mouse Atlas [4]. We evaluate MousePubMed's automated annotation of PubMed abstracts with the handcurated annotations of the Edinburgh Mouse Atlas. Before we go into the details of ontology-based literature search, we will discuss the general problem of identifying ontology terms in text with a specific emphasis on anatomical and developmental terminology.

9.2 Databases and Text Mining

Curating Databases

The large amount of species-specific databases today helps researchers to easily access various kinds of information on many organisms. Most such databases are manually curated by domain experts and constantly improved in terms of quantity and quality with input from the respective research communities. This manual curation process guarantees high quality and degree of reliability of the data. Annotations, for instance of genes and gene products, are stored in structured manners (associated functions, phenotypes, etc.), so that they can easily be queried by a researcher. Controlled vocabularies and ontologies designed for specific types of annotations reduce the amount of ambiguity for both curation and later access.

Database curators constantly scan the relevant literature to find evidence for new annotations related to their domain. These annotations are standardised terms from controlled vocabularies, often referred to as ontologies. For genes and gene products, annotations reflecting functions, locations, and processes are sought [2]. For drugs, it is interesting to find known digestive pathways and respective (desired and undesired) targets. Such facts often are reported in the literature, spread over a large variety of journals and other publication formats.

Ontologies as Semantic Frameworks for Cross-Database Queries

Efforts are under way to design ontologies suited not only for a single species, but rather a range of organisms. Some of these ontologies have already reached advanced stages and are widely used for annotations by many databases. One example is the Gene Ontology (GO), a hierarchy of concepts related to biological processes, molecular functions, and cellular components of genes and gene products. Many of the databases curating data on genes and proteins use GO for their annotations such as UniProt and EntrezGene. Another example is the Plant Ontology, a controlled vocabulary reflecting plant structures and developmental stages [18]. It is used by TAIR, Gramene, MaizeGDB, and other databases [6, 19, 29]. The use of such common ontologies that are applicable to disparate databases, which may be species-centred like SGD or gene-centred like EntrezGene, alleviates cross-database queries. An example is a query across multiple species to find similarly annotated genes, possibly restricted to a common type of tissue. The proper design of exhaustive ontologies and/or controlled vocabularies to annotate, for instance, genes and gene products with structures, functions, processes, stages, or phenotypes, and their installment in relevant databases present major tasks towards facilitating comprehensive annotations and queries.

Databases vs. Literature

Queries across disparate databases are required to exploit available data. However, a lot of data are not yet stored in such a structured form. This is due to two main reasons. For one, there is no immediate interest for researchers to submit their findings to (one or more) relevant databases, as scientific publications function as the main instrument for making information accessible and gaining reputation. The second reason comes with the necessary process of manual curation of database entries and annotation to maintain a certain quality standard. Another resource of data are aforementioned scientific publications themselves. Fairly often, these provide insight into more recent findings than databases. In addition, more information can be found in texts, such as, background knowledge, descriptions of experimental settings, etc., showing broader context as well as in-depth details. Natural language often is more suitable to express facts than the structured form of any database. Moreover, many annotations in databases come in the form of free text, for instance functions and diseases in UniProt, or phenotypes in MGI. This shows that scientific publications and other textual descriptions present important resources to be considered when searching for certain information. In the following sections we will describe, how ontological terms can be found in text.

Text Mining

In biomedical text mining, researchers use techniques from natural language processing, information retrieval, and machine learning to extract desired information from text [20]. Even when the concepts to extract are available in a structured form, such as a controlled vocabulary or ontology, finding them in free text is not always an easy task. For instance, a recent assessment for extracting Gene Ontology terms revealed performances around 20% success rate only [14]. The difficulty of automating manual annotation is evident from the fact that only as few as 15% of manually annotated terms appear literally in the associated abstracts.

Ad-hoc Variations of Names To begin with, terms in vocabularies and labels of concepts in ontologies appear in many, slight or severe, variations in natural language texts.

- orthographic: IFN gamma, Ifn-γ
- morphological: Fas ligand, Fas ligands
- lexical: hepatitic leukaemia, liver leukemia
- structural: cancer in humans, human cancers
- acronyms/abbreviations: MS, Nf2
- synonyms: neoplasm, tumor, cancer, carcinoma
- paragrammatical phenomena/typographical errors: cerevisae, nucleotid

Some of the terms encountered in texts are rather ad-hoc creations, which cannot be found in any term lists.

Synonymity of Ontological Terms As mentioned before, terms in a vocabulary or ontology might not appear literally in a text, but authors rather use synonyms for the same concept. First of all, this complicates proper searches: When searching for "digestive vacuole", results should also contain texts that mention "phagolysosome"; mentionings of "ligand" refer to the concept "binding"; an "entry into host" might occur as an "invasion of host". In the Plant ontology for example, many synonyms exist for the same structure in different species. "Inflorescence" is referred to as "panicle" in rice, and as "cob" in sorghum, and "spike" in wheat, for instance. We note that there are also intra-ontology synonymities: "eye" in AnoBase can refer to the eye spot or the adult compound eye. In a similar manner, the Edinburgh Mouse Atlas contains unspecific mentions such as "cavity" or "body" for the mouse.

Ambiguity of Ontological Terms Terms can have a very specific meaning in biomedical research, but mean other things in other contexts. Examples are "development", "envelope", "spindle", "transport", and "host". Protein names such as "Ken and Barbie", "multiple sclerosis" or "the" that resemble common names, diseases, or common English words are especially hard to disambiguate. The same problems arise from drug names like "Trial" or "Act". Table 9.1 lists some anatomical terms that have other meanings in different domains. Especially where cross-ontology or cross-database queries are needed, one has to consider ambiguity, for instance when applied to different organisms: "gametogenesis" (sexual reproduction) in plants is different from "gametogenesis" in metazoans.

Table 9.1. Some anatomical terms that have other meanings in different domains. Some misinterpretations occur only when certain spelling variations are allowed, for instance, ignored capitalisation or plural forms.

Term	Other meaning
rod	common English
iris	species: plant; common English
axis	species: deer; common English
chin	common English
beak	common English
pons	protein: Serum paraoxonase/arylesterase 1 (PON)
penis	protein: Penicillinase repressor (penI)
sigma	common English/Greek
patella	species: limpet
cicatrix	disease: scar
nephrons	drug: bronchodilator (Nephron)
hemocytes	drug: iron supplement (Hemocyte)
chondrocytes	drug: cartilage cells for implantation
hippocampus	species: seahorse

Stemming and Missing Words Some aspects for finding terms in text refer to the actual processing of natural language and appear rather technical. Very often, words will appear in different forms, such as "binding" and "binds". These refer to the same concept, which can be solved by resolving words to their stem ("bind"). However, the analogous reduction of "dimerisation" to "dimer" is more questionable. The former talks about the process, the latter about the result. A similar example is "organisation", where a transformation into "organ" is invalid.

Texts contain additional words that are missing in the ontological term. This happens, for instance, when a text contains further explanations that describe findings in more detail. An example is "tyrosine phosphorylation of a recently identified STAT family member" that should match the ontology term "tyrosine phosphorylation of STAT protein." In general, matching is allowed to ignore words such as "of", "a", "that", "activity", but obviously not "STAT". Additional background information on term variations is needed to know that a "family member" can refer to a protein.

Formatting of terms represents another source for potential matching errors. Terms in an ontology contain commas, dashes, brackets, etc., which require special treatment. For "thioredoxin-disulfide" the dash can be dropped, for "hydrolase activity, acting on ester bonds" the clause after the comma is important, but unlikely to appear as such in text. Terms containing additions such as "(sensu Insecta)" contain important contextual information, but are also less likely to appear in text.

9.3 Ontologies and Text Mining

Three main key dimensions of ontologies have been defined by Uschold: formality, purpose, and subject matter [28]. The degree of formality by which a vocabulary is created and meaning is specified varies among different ontologies. The purpose refers to the intended use of an ontology. Domain ontologies (such as medicical or anatomical), problem solving ontologies, and representation ontologies comprise examples for different subject matters an ontology is characterising.

In contrast to ontologies designed primarily for annotating biological objects, there is a clear distinction to ontologies designed for text mining. We will describe this distinction and its impact on text mining strategies as well as on the redesign of dedicated ontologies. In the case of a text mining ontology, compromises must be made on the relationships and on the labels used. Labels need to be descriptive and they or associated synonyms must be used in text. The ontology does not need to be very formal in terms of containing many different relationships between terms (such as 'derives from', 'causes', 'part of', etc.) or of distinguishing between 'classes' and 'instances'. It should be constructed in a way, that it is possible to obtain a structured vocabulary with only one type of directed relationship defining a hierarchy, i.e. 'is_a' relationships or simply parent child relationships. In general, there has to be a compromise to obtain a correct ontology with valid relations and still get the best possible results from text mining. The most prominent topics considering ontology design for text mining are the following.

- *Term overlaps* — some concepts can overlap in their labels or synonyms: in many cases there is a difference between what authors write and what they actually mean to express. Unfortunately, researchers do not have strict and formal ontologies or nomenclatures in their minds when composing a scientific article; in most of the cases they might use parent terms to refer to a child term, or vice-versa. For example, many people are treating the MeSH terms 'cardiovascular disease' and 'coronary artery disease (CHD, CAD)' the same, although the latter is a child of the first.
- *Descriptive labels* — in most of the cases, the labels in an annotation ontology cannot be used for text mining, usually due to their explanatory nature. For example, it is unlikely that the Gene Ontology term "cell wall (sensu Gram-negative bacteria)" will appear as such in text. Terms like "positive regulation of nucleobase, nucleoside, nucleotide and nucleic acid metabolism" and "dosage compensation, by inactivation of X chromosome" are almost complete sentences and are also unlikely to be found as such in text.
- *Ambiguity* — results either from identical abbreviations for different terms, or, in general, tokens that can refer to terms that might or may not be of our interest. An example of an ambiguous *abbreviation* is "CAM" that can stand for "constitutively active mutants" , "cell adhesion molecule" , or "complementary alternative medicine" . The second category of ambiguities — and the most difficult to handle — is that of terms that (in the context of anatomy) can refer to

different species. An example of such ambiguities is "embryo", which can be a chicken, mouse, human, or even zebrafish embryo. Therefore, if we are interested in the different developmental stages of the mouse embryo nervous system, we need to retrieve articles focusing on studies on mouse embryos only. If the term "embryo" is inserted in the Mouse Anatomy ontology as such, then the search engine will return articles on all kinds of embryos. If the term "mouse embryo" is inserted in the ontology, the number of articles retrieved will not be the real number of articles mentioning the term "mouse embryo", since not all of them will mention the term as such. A similar example is that of organs/tissues common to different species, such as "eye" or "lens".

- *Generic and specific labels* — when using the ontology for text mining in a specific biomedical sub-domain (anatomy, disease, glucose metabolism, etc.), the ontological concepts must be specific for that domain. The articles retrieved must be anatomy-specific or disease-specific or glucose-metabolism-specific. Therefore, we need a vocabulary specific enough to distinguish between relevant and irrelevant articles, but general enough to not exclude potentially relevant articles. If the concepts are too generic, they could be referring to many other domains. For example, during the design of a glucose-metabolism ontology, we might need to include information on kinetics. "Kinetics" as such is too generic to be used as a term, as it can refer to different kinds of kinetics (kinetics of phase transition, hydrolysis kinetics, kinetics of equilibrium reactions). On the other hand, the term "glucose kinetics" might be too specific, as it might seldom appear as such in a text. The decision on which terms should be used in the ontology ideally should only be made after exhaustive searches with different variations of terms.

We can derive some simple rules from all these observations, which can be used for (re-)design of ontologies when they should serve as resources for text mining applications.

- Avoid descriptive labels and synonyms: they should be likely to appear in texts as such – avoid "and", "of" and the like;
- Avoid improper spelling variations: capitalisation, noun plural forms, verb flexions;
- Use common names as labels or include them as synonyms;
- Add structural and lexical variations wherever possible;
- Keep the nomenclature consistent, precede terms with superstructure name;
- Use different representations of a concept in the ontology.

For a proper extraction of terms and subsequent term disambiguation in case of homonyms, the occurrence of parents helps to decide on the exact term. As, especially in anatomical ontologies, terms can have multiple representations, such multiple hierarchies should also reflected by the ontology. Examples are spatial and systemic representations of a tissue — "lung" is a "body part", and also a specific "organ system". Depending on the context in which "brain" is found, parent terms below "head" might not be found in the text at all, but rather terms related to "organ system." An ontology should therefore cover at least the most likely paths to subsume a tissue.

All of the above problems mean that extracting terms from literature will not be error-free. However, despite all of these problems, ontology-based literature with text mining can answer questions as posed in the introduction. Next, we introduce three such engines, GoPubMed, MeshPubMed, and MousePubMed and illustrate how they help to answer questions.

9.4 GoPubMed and MeshPubMed

GoPubMed [12], MeshPubMed and MousePubMed, which is discussed in the next section, index articles provided by PubMed with ontology terms from GO, Mesh, and Mouse anatomy/development, respectively. As an example consider Fig. 9.1, which shows a screenshot of MeshPubMed when queried for Pax6. The key difference to a classical search is that all the documents are annotated with terms from the domain specific ontology. Therefore, the user interface shows ontological information on the left and the documents on the right side. Beside the complete hierarchy of relevant terms found in documents mentioning the given keywords, a list of frequently occurring terms is placed above. Clicking on any of these terms reduces the result set and allows users to quickly filter large result sets to the necessary documents needed to answer their question.

Let us consider the three questions about Pax6 from the introduction:

- Which processes is Pax6 involved in? A query in GoPubMed for Pax6 shows that the most frequent process mentioned is development. Opening the development branch reveals the processes of brain and eye development as well as organ morphogenesis including pancreas development. Indeed the corresponding articles support this essential role of Pax6 as transcription factor and master control gene in development of eye, brain and pancreas [21].
- Which diseases is Pax6 involved in? A query in MeshPubMed for Pax6 shows that the most frequent disease mentioned is aniridia. Hovering the mouse over the term gives an explanation that it is "a congenital abnormality in which there is only a rudimentary iris. This is due to the failure of the optic cup to grow. Aniridia also occurs in a hereditary form, usually autosomal dominant." A click on aniridia shows articles mentioning both the disease and the gene such as for example [9], which confirm the answer.
- At which developmental stages is Pax6 active in mice? A query in MousePubMed for Pax6 shows that Theiler stages up to 14 (9 dpc, days post conception) are frequently mentioned supporting Pax6's role in early development. Clicking on a stage reveals e.g. the statement "In the early development of the vertebrate eye, Pax6 is required for..." in [3]

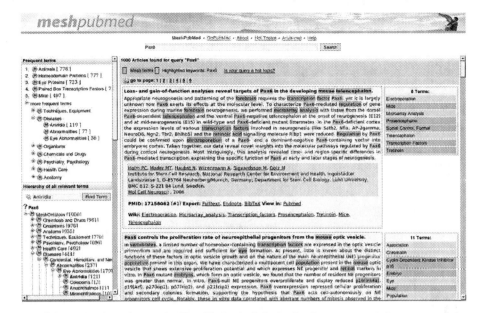

Fig. 9.1. MeshPubMed query for "Pax6". On the left, the five frequent terms, frequent terms by category and all relevant terms are shown. The most frequently mentioned disease is aniridia. Clicking the term and retrieving the articles mentioning aniridia confirms that Pax6 is involved in aniridia.

Indeed, Pax6 is the most researched gene of the family of Pax genes and appears throughout the literature as a 'master control' gene for the development of eyes and is of medical importance because heterozygous mutants produce a wide spectrum of ocular defects such as aniridia in humans. We can now further check in Mesh-PubMed whether aniridia is a 'hot topic' and who the most active authors publishing on aniridia are. Consider Fig. 9.2. It turns out that V. van Heyningen is the number one published author having the most collaborations, especially together with A. Seawright, as shown on the co-authorship network in Fig. 9.2.

9.5 MousePubMed

To use ontology-based literature search for developmental biology, we built Mouse-PubMed using vocabularies for mouse anatomy (EMAP), human anatomy (EHDA), mouse genes (from EMAGE), and mouse developmental stages (Theiler) as resources.[5] To demonstrate MousePubMed's usefulness, we evaluate it against tissue and developmental stage annotations in the Edinburgh Mouse Atlas. Before we discuss this evaluation, we introduce the matching algorithm developed.

---

[5] Further details on the Edinburgh Mouse Atlas (EMAP and EMAGE) can be found in Chapter 12 of this book.

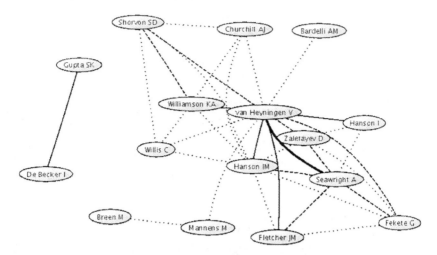

Fig. 9.2. Part of the co-authorship network for "aniridia" in MeshPubMed showing V. van Heyningen and A. Seawright as the authors most active in this area.

9.5.1 Extracting Gene Names, Anatomy Terms and Developmental Stages

Ontology based text mining is not restricted to finding words or word groups in texts. The structure of the ontology can be used to state the relation between a term and a document by finding the children of the term. This task is reasonably well solvable for the Gene Ontology where its term labels are self-descriptive. Many terms in GO are contained in their child terms [24]. As an example, the term "envelope" is refined into "organelle envelope" and further to "organelle envelope lumen". The ontology for the Abstract Mouse contains anatomical concepts in the mouse embryo at different embryonic developmental stages. The vocabulary is used to annotate images of mouse embryos. It unifies the vocabulary needed to describe the different parts throughout 26 Theiler stages. Concepts like organs or body parts are further refined into tissue types, unspecific loci such as "cavities", "left", "upper", as well as general terms such as "node" or "skin". Considering only the textual labels, one cannot distinguish between the different ontological concepts. For example, "chorion" has the children "mesoderm", "ectoderm" and "mesenchyme". "Amnion" and "yolk sac" have children sharing the same labels. Searching for documents related to "chorion" will retrieve very similar document sets to searching for "amnion", only because the documents mention "mesoderm", in this case with meaning "mesoderm specific to amnion". Different anatomical concepts share the same term label. For instance, there exist 171 individuals with label "epithelium". These all refer to different body parts at a specific stage in development.

Ontology-based text mining relies on the assumption that unique or similar types of directed non-cyclic relationships exist, which can be unified in the hierarchical relationships creating a taxonomy. This assumption does not hold for the Abstract

Mouse ontology. There does not always exist a path to the common root supported by only one type of hierarchical relationships. Therefore, in our analysis, a document is annotated with a term from the Abstract Mouse ontology taking the term label and its synonymous labels into account. In the Abstract Mouse Ontology the term labels follow various creation patterns. Sometimes a child term contains information of the parent term (for example, "cavities" has the child "amniotic cavity"). In other cases a term like "umbilical vein" has the children "left" and "right", rather than "left umbilical vein" and "right umbilical vein", respectively. These short and common sense labels make the text annotations arbitrary.

For our experiments we slightly adapted the ontology. For the terms "left", "right", "upper", "lower", "common", "anterior" and "posterior" we expanded the term labels with its parents labels. "Eyelids" thus became "upper eyelids" and "lower eyelids", for instance, and we removed the children terms "upper" and "lower" accordingly. To distinguish between common terms such as "skin" occurring — for instance, for different organs — the matching algorithm took text annotations for ancestor terms into account. Terms with the same label were grouped according to the number of text annotations for their ancestors in the same document. Only annotations of the top ranked group were confirmed. Figure 9.3 shows an example for the term "skin". There were multiple possibilities to resolve this term to a specific tissue. Only when a parental term (shoulder, upper arm, etc.) was found, the text was annotated with the specific skin.

Finding gene names in documents is done using exact matching against gene names contained in EMAGE. We enriched this set using additional names and synonyms for each gene taken from the MGI database[6]. We tested all 1437 genes mentioned in EMAGE for their annotations with tissues and Theiler stages in PubMed.

We analysed 123,074 abstracts retrieved from PubMed with the query "mouse AND development". This amounted to approximately 0.7% of all documents listed in PubMed. Based on the document annotations with ontology terms, we issued in total 36,358 statements on relations between genes, tissue and developmental stages, which we extracted from EMAP/EMAGE. Cases with multiple Theiler stages from EMAP were split into separate statements. We evaluated the tissues mentioned using EMAP's Abstract Mouse ontology and the anatomy part or MeSH. For path descriptions like "embryo.ectoderm" in EMAP we required the matching document to be annotated with the terms "embryo" and "ectoderm". For MeSH, as in MeshPubMed, we also included descending terms. A document was annotated with the term "embryo" if annotations for its descendants, for example, "germ layers" or its children "ectoderm", "endoderm" or "mesoderm", were found.

To find mentions of Theiler stages in texts, it was not enough to search for them directly, as they seldom occur as such in abstracts ("Theiler stage 12", "TS12", etc.).

---

[6] See http://www.informatics.jax.org.

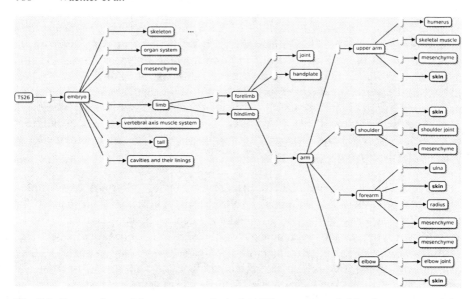

Fig. 9.3. Excerpt from the anatomy ontology, for different types of skin. Occurrences of the term "skin" (yellow concept nodes) in a text were resolved using the hierarchical dependencies. Only when a parental node was also found, for instance, "shoulder", we annotated the text with "skin."

We therefore compiled a set of regular expressions based on two main notions, the mentioning of embryonic days (E) and of days post coitum (dpc). These expression had to capture occurrences like:

- "embryonic day 10.5",
- "day 9 mouse embryos",
- "between E3.5 (E = embryonic day) and E8.5",
- "12.5 days post coitum", and also
- "7.5-13.5 days post-conception".

As mentionings of Theiler stages do not often occur, but rather general time spans are given ("early embryonic development"), we decided to assign Theiler stages one to 14 to "early development", and stages 20 to 27 to "late development," respectively. Every mention of an "early developmental stage" thus was treated as a match for stages one through 14. Both assignment were based on statements found in PubMed relating days to general time spans.

9.5.2 Experiment Designs

To assess the potential of ontology-based literature searches, we designed two experimental scenarios. For the first, we manually collected two sets of queries and detailed answers. For the second scenario, we evaluated the complete EMAP/EMAGE data.

Using the methodology described in the previous section, we tried to find textual evidences for all sets in PubMed. This means that we searched PubMed for abstracts that shared annotations for each collected triple consisting of a gene, tissue, and Theiler stage.

Manually Curated Test Set

We first selected set of questions manually to study results in detail. The idea was to send simple keyword queries to MousePubMed, asking for mouse abstracts that discuss a certain tissue and embryonic day. MousePubMed should then identify all genes mentioned in the top-ranking abstracts. Questions and retrieved answers were as follows:

- Which genes play a role in the development of the nervous system in Theiler stage 14? A query for "mouse development nervous system 9 dpc" finds the genes Adamts9, Hoxb4, Otx3, and EphA4 within the first eight abstracts[7]. In addition, the genes EphA2, A3, A7, B1, B2, and B4 are found, which are not yet annotated in the EMAGE database.
- Which genes play a role in sex differentiation during murine embryo development? A corresponding query for "mouse sex 10 dpc" results in a set of eight genes within the first fifteen abstracts: Fgf9, Asx11, Sry, Sox9, Usp9x, Maestro/Mro, Wt1, Amh1 and Fra1[8]. Only half of the genes can be found in EMAGE so far.
- Which genes play a role in the development of the murine embryonic liver? A query for "mouse 'liver development'" results in a set of several genes, most of which can be found in EMAGE as well: Shc, Pxn, Grb2, PEST/Pcnp, GATA6, HNF4a, Foxa1/2, Zhx2, HNF6, Mtf1, SEK1, Nfkb1, c-Jun, Itih-4, and Hex. To answer this question exactly, however, too few abstracts mention particular Theiler stages or days post congestion. They rather refer to "early stages of development", and the exact time span might be presented in the full text article only.

All the results, in particular where genes and exact Theiler stages are concerned, are highly dependent on the ordering of abstracts as provided by PubMed. Whenever a new publication appears containing the same search keywords, it will displace abstracts potentially more informative regarding the original question. Abstracts answering the original question might not appear among the first few and be immediately present to the user. However, text mining methods will still extract all the data, even from older publications, and still the right set of articles can easily be found.

---

[7] Important for answering this query are returned PubMedIDs 12736215, 12055180, 11403717.

[8] Important are PubMedIDs 16540514, 16412590, 14978045, 14684990, 14516667, 12889070, 9879712, 9115712.

The abstracts resulting from a keyword search occur in the same ordering as provided by PubMed. That is, in general, the most recent articles occur first. However, querying for species, tissues, and stages still returns the abstracts that discuss the interesting genes. Although corresponding expression patterns might first have been described in older publications, even in recent publications the desired genes reappear quite often.

Reconstructing Outcomes of Large-scale Screening

Thut et al. provided a list of 62 genes found expressed during eye development in mice, together with developmental stage and substructure [27]. Of the 62 genes, 26 were not previously reported (as of 2001); to 16 genes, novel valuable information could be added; 20 genes were fully reported before. Expression patterns were summarised for E12.5, E13.5, E14.5, E16.5, E18.5 and P2. Using MousePubMed, we tried to reconstruct the result of this large-scale screen of 1000 genes.

As Table 9.2 shows, nine PubMed abstracts contained the full information as stated by Thut et al., mentioning gene, tissue, and specific stages (days). For most cases, however, not all data were contained in one single abstract. In three cases, we were not able to automatically spot the gene name (left column), in all cases this was due to synonyms lacking in EMAP and MGI. Note that the assessment of recognising genes was based only on genes mentioned in EMAGE. The tissue could be found in almost all of the cases; from most abstracts, even the specific part of the eye could be extracted.

Complete EMAP Test Set

To evaluate capabilities of automated searches against the complete EMAGE data, the experimental setting was as follows. Genes in EMAGE have annotated tissues, in which they were detected at various stages of embryo development. Thus, we queried MousePubMed with each gene and checked which tissues were mentioned in the resulting PubMed abstracts. This was based on co-occurrence of the gene considering, a tissue, and a Theiler stage (day) in the same abstract. Currently, there are 1437 genes in the EMAGE database annotated with (sometimes multiple) tissues and stages. All in all, we identified 18,179 such triples — gene, tissue, and stage — in EMAGE. Many of the annotations consist of general annotations for tissue, like "mouse", "embryo", "left", "female", "node". We removed such trivial instances, because they were very frequently found. 12,782 triples referred to specific tissues, and we tried to find these triples using the aforementioned term extraction (also see Table 9.3).

As Table 9.4 shows, we were able to reconstruct 31% of the gene-tissue associations in EMAGE using PubMed abstracts. Only 13% of the full information (gene, tissue, exact stage) was contained in abstracts. All in all, the data recovered from PubMed included information on about 37% of the EMAGE genes. We noted that

Table 9.2. Expression patterns identified by MousePubMed in articles derived from [27]. Often, an abstract does not mention a (specific) developmental stage; MousePubMed did not find this particular fact; otherwise: facts as identified by MousePubMed. Given are only tissues related to the murine eye.

Gene	Tissue	Stage	PubMedID
Sparc	retina, RPE, eye	E4.5, E5, E10, E14, E17	9367648
Sparc	lens	embryonic day (E)14	16303962
Stat3	retina, RPE, eye	-no specific stage-	12634107
Stat3	lens	E10.5	14978477
Pedf	RPE	-no specific stage-	7623128
Pedf	retina	E14.5, 18.5	12447163
Runx1	inner retina	embryonic day 13.5	16026391
Col15a1	conjunctiva, cornea	E10.5-18.5	14752666
Otx2	outer retina	-no specific stage-	15978261
Edn1	retina	-no stage-	11413193
IGF-II	eye, cornea, retina, scleral cells	E14	2560708
Wnt7b	anterior eye, cornea, optic cup, iris	-no specific stage-	16258938
CDH2	—	-no stage-	9210582
—	lens	-no stage-	9211469
Col9a1	eye, lens vesicle, neural retina, ciliary epithelial cells, cornea	13.5, 16.5-18.5 d.p.c.	8305707
Tgfb2	cornea, lens, stroma	-no specific stage-	11784073
Thra	retina	-no specific stage-	9412494
BMP4	retina	E5	17050724
Bmp4	optic vesicle, lens	-no specific stage-	15558471
BMP4	lens, optic vesicle	-no specific stage-	9851982
—	eyes	N/A	15902435
Sox1/2	lens	-no stage-	15902435
—	retina, eye axis	E2, E3, E5	15113840
Notch1	eye	-no specific stage-	11731257
Notch2	eye	-no specific stage-	11171333

Table 9.3. Types of information and quantity contained in EMAGE.

Type of information	Amount of data
Genes with tissues, stages	1437
Genes with at least one non-trivial tissue, stages	1346
Triples of gene, tissue, stage	18,179
Triples of gene, non-trivial tissue, stage	12,782
Tuples of gene, non-trivial tissue	8653

Table 9.4. Number of tuples/triples consisting of gene and tissue or gene, tissue and stage found in PubMed abstracts retrieved by the query "mouse AND development."

Type of information	Amount of data
Triples of gene, non-trivial tissue, stage	1637 (12.8%)
Tuples of gene, non-trivial tissue	2667 (30.8%)
Genes with at least one tissue and stage	537 (37.4%)

in many cases, abstracts do not mention specific time points during development. Sometimes, "early" and "late development" are mentioned, which we resolved as described previously in this section. On the other hand, mentions like "in early liver development" could not be resolved to specific overall-stages without background information. Cross-checks revealed that indeed much of the necessary information was only mentioned in the full text of references annotated by EMAP for a specific association.

9.6 Conclusion

Ontologies are widely used for annotation. They are also useful for literature search, but the extraction of terms from text is a difficult problem due to the complexity of natural language. Here, we demonstrated the use of the ontology-based literature engines GoPubMed, MeshPubMed, and MousePubMed to answer questions in the context of development. We discussed the specific extraction algorithms needed for MousePubMed and evaluated them small scale on examples relating to eye development and large scale on gene-tissue-stage triple from the Edinburgh Mouse Atlas. We were able to reconstruct 37% of genes, 31% of gene-tissue associations and 13% of gene-tissue-stage associations from PubMed abstracts. These figures are encouraging as only abstracts are used.

References

1. Altun ZF, Hall DH (eds) (2002-2006) WormAtlas. http://www.wormatlas.org
2. Ashburner M, Ball CA, Blake JA, Botstein D, Butler H, Cherry JM, Davis AP, Dolinski K, Dwight SS, Eppig JT, Harris MA, Hill DP, Issel-Tarver L, Kasarskis A, Lewis S, Matese JC, Richardson JE, Ringwald M, Rubin GM, Sherlock G (2000) Gene ontology: tool for the unification of biology. The Gene Ontology Consortium. Nat Genet 25(1):25–9.
3. Azuma N, Tadokoro K, Asaka A, Yamada M, Yamaguchi Y, Handa H, Matsushima S, Watanabe T, Kida Y, Ogura T, Torii M, Shimamura K, Nakafuku M (2005) Transdifferentiation of the retinal pigment epithelia to the neural retina by transfer of the Pax6 transcriptional factor. Hum Mol Genet, 14(8):1059–68.
4. Baldock RA, Bard JB, Burger A, Burton N, Christiansen J, Feng G, Hill B, Houghton D, Kaufman M, Rao J, Sharpe J, Ross A, Stevenson P, Venkataraman S, Waterhouse A, Yang

Y, Davidson DR (2003) EMAP and EMAGE: a framework for understanding spatially organized data. Neuroinformatics 1(4):309–25.

5. Bard JL, Kaufman MH, Dubreuil C, Brune RM, Burger A, Baldock RA, Davidson DR (1998) An internet-accessible database of mouse developmental anatomy based on a systematic nomenclature. Mech Dev 74(1–2):111–120.

6. Berardini TZ, Mundodi S, Reiser L, Huala E, Garcia-Hernandez M, Zhang P, Mueller LA, Yoon J, Doyle A, Lander G, Moseyko N, Yoo D, Xu I, Zoeckler B, Montoya M, Miller N, Weems D, Rhee SY (2004) Functional annotation of the Arabidopsis genome using controlled vocabularies. Plant Physiol 135(2):745–55.

7. Berneis K, Rizzo M (2005) Ldl size: does it matter? Swiss Med Wkly 134(49–50):720–4.

8. Bodenreider O (2004) The Unified Medical Language System (UMLS): integrating biomedical terminology. Nucl Acid Res 32.D267–70.

9. Brinckmann A, Rüther K, Williamson K, Lorenz B, Lucke B, Nrnberg P, Trijbels F, Janssen A, Schuelke M (2006) De novo double mutation in PAX6 and mtDNA tRNA (Lys) associated with atypical aniridia and mitochondrial disease. J Mol Med.

10. Corcho O, Fernandez-Lopez M, Gomez-Perez A (2003) Methodologies, tools and languages for building ontologies: where is their meeting point? Data Knowl Eng 46(1):41–64.

11. Couto FM, Silva MJ, Coutinho P (2005) Finding genomic ontology terms in text using evidence content. BMC Bioinformatics 6(1):S21.

12. Doms A, Schroeder M (2005) GoPubMed: Exploring PubMed with the GeneOntology. Nucl Acid Res 33(Web Server Issue):W783–6.

13. Duineveld AJ, Stoter R, Weiden MR, Kenepa B, Benjamins VR. Wondertools?: a comparative study of ontological engineering tools. Int J Hum-Comput Stud 52(6):1111–33.

14. Ehrler F, Geissbuehler A, Jimeno A, Ruch P (2005) Data-poor categorization and passage retrieval for gene ontology annotation in Swiss-Prot. BMC Bioinformatics 6(1):S23.

15. Fang Z, Cone K, Sanchez-Villeda H, Polacco M, McMullen M, Schroeder S, Gardiner J, Davis G, Havermann S, Yim Y, Vroh Bi I, Coe E (2003) iMap: a database-driven utility to integrate and access the genetic and physical maps of maize. Bioinformatics 19:2105–11.

16. Grumbling G, Strelets V (2006) FlyBase: anatomical data, images and queries. Nucl Acid Res 34(Database issue):D484–8.

17. Harbach RE, Knight KL (1980) Taxonomist's Glossary of Mosquito Anatomy. Plexus Publishing Inc, Marlton, NJ, USA.

18. Jaiswal P, Avraham A, Ilic K, Kellogg EA, McCouch S, Pujar A, Reiser L, Rhee SY, Sachs MM, Schaeffer M, Stein L, Stevens P, Vincent L, Ware D, Zapata F (2005) Plant Ontology (PO): a Controlled Vocabulary of Plant Structures and Growth Stages. Comp Funct Genom 6(7–8):388–97.

19. Jaiswal P, Ni J, Yap I, Ware D, Spooner W, Youens-Clark K, Ren L, Liang C, Zhao W, Ratnapu K, Faga B, Canaran P, Fogleman M, Hebbard C, Avraham S, Schmidt S, Casstevens TM, Buckler ES, Stein L, McCouch S (2002) Gramene: a bird's eye view of cereal genomes. Nuc Acids Rcs 34(Database issue):D717–23.

20. Jensen LJ, Saric J, Bork P (2006) Literature mining for the biologist: from information retrieval to biological discovery. Nat Rev Genet 7:119–129.

21. Kleinjan DA, Seawright A, Mella S, Carr CB, Tyas DA, Simpson TI, Mason JO, Price DJ, van Heyningen V (2006) Long-range downstream enhancers are essential for Pax6 expression. Dev Biol.

22. Lambrix P, Habbouche M, Perez M (2003) Evaluation of ontology development tools for bioinformatics. Bioinformatics 19(12):1564–71.

23. Müller H-M, Kenny EE, Sternberg PW (2003) Textpresso: an ontology-based information retrieval and extraction system for biological literature. PLoS Biol 2.

24. Ogren PV, Cohen KB, Acquaah-Mensah GK, Eberlein J, Hunter L (2004) The compositional structure of Gene Ontology terms. Pac Symp Biocomput. 214–25.
25. Perez-Iratxeta C, Perez AJ, Bork P, Andrade MA (2003). Update on XplorMed: A web server for exploring scientific literature. Nucl Acid Res 31(13):3866–8.
26. Rosse C, Mejino JLV (2003) A Reference Ontology for Bioinformatics: The Foundational Model of Anatomy. Journal of Biomedical Informatics 36:478–500.
27. Thut CJ, Rountree RB, Hwa M, Kingsley DM (2001) A large-scale in situ screen provides molecular evidence for the induction of eye anterior segment structures by the developing lens. Dev Biol 231(1):63–76.
28. Uschold M (1996) Building Ontologies: Towards a Unified Methodology. Conf Brit Comp Soc SIG Exp Sys, Cambridge, UK.
29. Vincent PL, Coe EH, Polacco ML (2003) Zea mays ontology – a database of international terms. Trends Plant Sci 8(11):517–20.

Part III

Anatomy Ontologies and Spatio-Temporal Atlases

10

Anatomical Ontologies: Linking Names to Places in Biology

Richard A. Baldock and Albert Burger

Summary. An ontology captures knowledge of a domain in a structured and well-defined way. Anatomy provides a conceptual framework for the parts and tissues of an organism including structural (parts and sub-parts), functional (system elements), developmental (change of structure during embryogenesis) and derivation. Anatomy is also used to define place or spatial location. This can be in the form of topological relationships, e.g. adjacency or connectedness, or geometric, e.g. distance and direction. Atlases provide a direct or iconic framework to describe the spatial organisation of an organism with no reference to the conceptual anatomical framework. In this chapter we discuss how anatomical and atlas frameworks can be used to provide a rich ontology encompassing both conceptual, topological and geometric spaces. We also introduce the notion of a *natural coordinate system* both as a robust tool for navigation within an organism (e.g. the mouse embryo) and as a mechanism for cross-atlas interoperability.

10.1 Introduction

It is clear from this collection of articles that the notion of ontology in the context of biological resources has a wide interpretation from the strictly formalised structures and definitions of the OBO Foundry[1] through to rather more loosely defined controlled vocabularies. Here we take a very pragmatic view, we want a greater degree of automation in accessing and analysing bioinformatics resources using ontologies, but in the context of what is achievable in terms of capturing the knowledge in a formalised way. Experience tells us that if the barrier to building the ontology is too high, then the effort from the community will not be available and a usable standard will not be achieved, furthermore, even if that were possible, the ontology will not be employed.

The purpose of an ontology is primarily one of standardisation, at the syntactic as well as the semantic level. For computational systems and/or people to interact effectively, they must agree on the representation and meaning of the concepts that

---

[1] www.obofoundry.org

are part of that interaction. For example, by consistently annotating gene products in different databases with GO terms, a level of interoperability can be achieved that allows the development of applications which are able to "integrate proteomic information from different organisms, assign functions to protein domains, find functional similaritites in genes that are overexpressed and underexpressed in diseases and as we age" (from GO web site[2]). The purpose of anatomical ontologies is to provide syntactic and semantic interoperability across a number of tissue-based information sources, e.g. *in situ* gene expression data.

Traditionally a field of study in philosophy, *ontologies* have emerged as a key topic for the development of the *Semantic Web* [4] — the next generation of the World Wide Web — as well as for the Semantic Grid [8]. The promise of these semantic-based infrastructures lies in the automation of services that far exceeds what is currently possible. For this to be achievable, however, knowledge underpinning these services must be formalised and represented in software systems. The aim of ontology research in the Artificial Intelligence community has, therefore, been to develop knowledge representations that can be shared and reused by machines as well as people; the most widely accepted definition of ontologies in the modern sense is given by Gruber [9]:

> *"An ontology is a formal, explicit specification of a shared conceptualization".*

The debate of what constitutes an ontology is of wide interest to philosophers, in terms of the general properties and constitution, and of a narrower interest in computer science, in terms of content and structure including representation and language. The debate is widespread and occasionally acrimonious. For our purposes we take the much more pragmatic view that it is a structured and clearly defined encapsulation of knowledge about a field that can be used for annotation and reasoning within that domain. This includes automated reasoning in the context of database query and database interoperabllity and computational analysis.

In this chapter we omit an overall review of other work to do with anatomy ontologies, as this is comprehensively dealt with in other chapters, but instead focus on issues related to linking anatomical names to places in geometric space. The work is rooted in the Edinburgh Mouse Atlas experience (see Chapter 12), but is intended to be more generally applicable. The chapter reflects current ideas and plans of how to develop the work relating to anatomy ontologies and biomedical atlases instead of presenting a summary of work carried out thus far (which can be found in Chapter 12).

---

[2] www.geneontology.org

10.2 Embryo Part, Place and Space

The formal study of anatomy has been declining as an academic discipline. However, with the development of atlas databases as a standard reference framework for biomedical research it is now witnessing a renaissance as attempts are made to capture the concepts of anatomy – tissues and structures – for use in database systems. Sets of anatomical terms have appeared in many vocabularies/ontologies. The purpose of these ontologies has been primarily to provide a controlled vocabulary so that annotation and referencing can be more easily checked and compared but there is also an attempt to capture anatomical relationships and knowledge. The relationships are generally "part_of" which provides a hierarchical view of the structure from large to small, and "type_of" (often termed is_a) which define taxonomies of classes. The anatomical terms in these ontologies represent biological *concepts* which can be defined by reference to other terms in the ontology or by reference to other works. The concepts by their nature represent a *class* of (putative) objects in the real world, i.e. instances of an anatomical concept are parts (or features) of individuals.

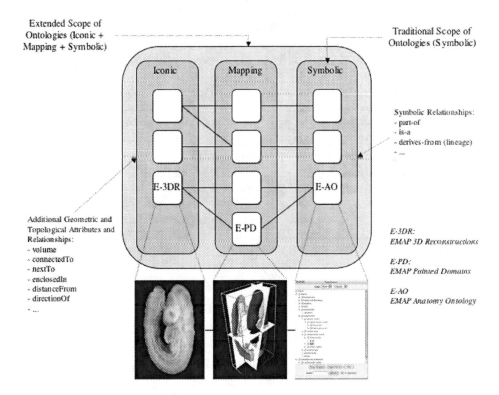

Fig. 10.1. Ontologies, atlases and mappings.

This type of relationship also holds with *representations* of individuals for example the atlas models that are being implemented [2, 11, 13]. Figure 10.1 shows these relationships. For any given domain (e.g. mouse embryo development) there could be many ontologies, capturing the anatomical concepts, perhaps in different ways and perhaps with different interpretations and definitions for the *same* concept in the sense of the *same name*. In figure 10.1 these are represented by the right-hand column with an example from the Edinburgh Mouse Atlas shown at the bottom. Each of these may have their own definitions in text or relationship terms but may also have a *graphical* representation. This graphical representation, represented by the middle column in figure 10.1, may also have a number of representations but most importantly may include alternative and inconsistent views of the underlying concepts. This brings to the fore a critical development of the notion of what constitutes an ontology. By formal definition an ontology should be consistent, but now we try to capture alternative views of the underlying terms, i.e. build in inconsistency. This raises a problem, how can we develop an ontoltogy for practical purposes and yet let it evolve as scientific understanding progresses. Of course consistency can be rescued by sub-dividing the concept into separate classes, e.g. *hindbrain-view1* and *hindbrain-view2*, or bay allowing alternative definitions for a single concept, nevertheless the idea is to capture the current state of knowledge in a domain which will evolve as understanding changes. Now the ontology is almost a database and the ontology forms part of the framework [7] such that what was experimental data at one stage will be part of the current model or theory at a later stage.

The graphical representation is in some cases the extension of the definition language to a graphical form in order to capture a definition more accurately. This definition is in some cases, however, in terms of a particular *individual*. In these cases a specific example of the concepts concerned – in the Mouse Atlas case, part or all of a mouse embryo – is selected and used as a realization of the concept in order to provide an unambiguous definition. The example may be from a single individual animal or may be synthesized and averaged from a group of individuals. Either way there is a selection of a representative model within which the ontological concepts can be interpreted. These cases are usually referred to as atlases. There may be of course many such atlases as indicated by the left hand column in figure 10.1. An atlas therefore consists of at least three parts, an ontology or controlled vocabulary (sometimes implicit, e.g. a list of countries), a representative individual to define the spatial extent and coordinates (may also include time), and a *mapping*, or the interpretation of one within the other, between the two.

A simple example of an anatomy is that developed as part of the Edinburgh Mouse Atlas[3], [1, 7, 5]. This ontology is designed to capture the structural changes that occur within the time-course of embryo development and is a set of hierarchies, one at each developmental stage [17]. The ontology can be displayed as a set of hierarchical "trees" with each anatomical term subdivided into its constituent parts.

---

[3] http://genex.hgu.mrc.ac.uk/

There is no requirement that each concept is divided into non-overlapping structures, or that each component has only one parent, and therefore in graphical terms the ontology can be represented as a *directed acyclic graph* (DAG). Each node of each graph represents the biological concept, e.g. heart, at that particular time. Many of the terms and structures are repeated at multiple stages and it is possible to collapse the set of terms onto a single large hierarchy that includes all of the terms from all stages. This tree is stage-independent and referred to as the "abstract-mouse" and the terms now represent the biological concepts for all stages. Within the EMAP database the abstract mouse and stage terms can be independently referenced although the stage terms can be considered to have a class-instance type relationship. In addition to the "part-of" and "instance-of" relations, emap includes "derived-from" as a putative lineage relationship between tissues. These link the stage-components so that it becomes possible to query the lineage derivation (and destination) of any given tissue.

Rosse [15] provides a comprehensive discussion of the ontological issues with developing an anatomical nomenclature using a set of well defined principles and a structure provided by the Protégé system[4]. These principles derive from a formal analysis of the problem and the definition of anatomy both as a specific discipline and as a set of concepts.

By this means the Foundational Model of Anatomy (FMA) has been implemented as an ontology providing the anatomical concepts (tissue names) with their associated relationships. The basic relationships are part-of and type-of as above but these are further sub-classed. The FMA provides the most complete and ontologically sound representation of human anatomy (see Chapter 4).

10.3 Extending to Embryo Space

In philosophy, since classical times, ontology has been about the "essence of being". The "part" or "part-of" relationship has a special niche in philosophy under the heading of mereology, the ontology of parts and wholes. Casati and Varzi [6] provide a full analysis of these theories, particularly the classical extensional mereology of Lesniewski [16]. This provides a constrained logical representation but does not meet the requirements of a useful theory dealing with the type of parts and wholes encountered in biological and other applications. These shortcomings are discussed by [19] and relate primarily to the observation that different types of "part" imply different reasoning requirements, particularly if the results of such inference are to agree with common language usage. Examples are the notion of part collections, connected parts, mixtures and so on. In fact in the medical ontology GALEN 23 different type of "part-of" relationship are recognized (for more details on GALEN see Chapter 3).

Some authors have attempted to extend these mereologies to include spatial reasoning. For example the convex hull has been proposed as a mechanism for defining

---

[4] http://protege.stanford.edu

enclosure in order to define a "mereo-topology". This may help in some instances but will clearly be an inpoverished topology and cannot capture the complexity of biological structures. By this argument the vascular system of arteries, veins and capillary will enclose the rest of the body! The problem arises because of an attempt to encode information about the spatial distribution and relationships in a constrained logic. This is analogous to trying to reason about uncertainty with Boolean logic. The way to solve the latter is to use probability, the way to solve the former is to use geometry (and topology).

Thus ontologies to date have concentrated on symbolic representation and spatial relationships have been introduced to enhance both the power of the spatial representation and to enable more sophisticated spatial queries. The required fidelity and complexity of potential spatial relationships will undoubtedly confound any such scheme. The problem is one of representation - words can not efficiently capture the set of spatial and geometric properties and relationships that are required, the most efficient representation is a *coordinate* framework with the query language defined using standard geometrical and topological concepts. We therefore need to extend the notion of ontology to include such explicit knowledge.

The question is how do we extend an ontology based on a simple logical representation of type and structural knowledge, even with some topological relationships (e.g. adjacency or connectedness) to a framework that can interpret descriptions such as:

1. the expression domain extends 300 μm from the notochord in all directions.
2. gene X shows expression in the dorsal half of the somite,
3. the caudal limit of expression of gene Y is defined by a transverse plane through the hind-limb axes.

These are examples that could be encountered as parts of descriptions of regions of gene-expression but of course similar statements could apply to any spatial data for which a description in terms of anatomy is required. Descriptions in the literature and communication using text must of course always be of this form, i.e. combinations of anatomical structure, anatomical directions and geometrical concepts. Although apparently simple these examples illustrate that it is not sufficient to develop an atlas coordinate system based on the rectilinear image coordinates. This is sufficient to capture the regions associated with given anatomical structures and can be used to interpret the first of the expressions above. The second two statements however both use a notion of biological direction e.g. dorsal and caudal. The last example also includes the notion of axes and a special plane (transverse) which is defined as perpendicular to a given direction.

These expressions are the way location and space are communicated in biology and any atlas framework should be able to interpret them. Extensions to the anatomy ontology will never get there, we contend that the way forward is to extend the ontological framework to include the atlas with its coordinate system. In addition we need to define within the context of the atlas coordinates the sense of direction defined by the underlying biological tissue. We coin the term *natural coordinate system* (NCS) to capture the types of spatial description presented above. Biologists are able to use these coordinates to navigate and describe spatial locations and distributions, for an anatomical framework to be useful in this context we need to define a natural coordinate system that is integrated both with the simple coordinates of the atlas models and the anatomical ontology.

10.4 Natural Coordinate Systems

The issue that we need to face is that most biological references are not co-ordinates in the numerical sense that we understand from say a set of x, y, z axes but more often a description of direction (see figure 10.2 with a location defined by some biological structure with associated directions and route description for defining the location uniquely. Distance is then typically a relative value - distal-proximal and a precise measure is not necessary and maybe not useful. One area where a more typical co-ordinate frame has been established is for the brain, stereotaxic co-ordinates are defined with respect to particular locations, e.g. the *bregma* which is a point defined by the intersection of two suture lines in the skull plus orientations providing the standard anterior-posterior, dorsal-ventral and left-right. The co-ordinate values are typically just the distance in millimeters from the bregma in the appropriate direction. This however is not the typical case. Within the context of the brain it may be possible to define the biological coordinates to be a simple mapping of a simple rectilinear coordinate, in most cases however this will not work because the biological axes are curved and depend on the posture of the organism and in addition many sub-structures (e.g. a limb) will have an independent coordinate set, also depending on posture and flexion. This is illustrated in figure 10.2 which compare the adult organism with the embryonic state.

These natural co-ordinates can be considered as a generalised version of the stereotaxic co-ordinates as defined for the brain [14] and may be useful for visualisation and comparison of data. In addition this new frame may be a simpler way to define lineage and growth relationships between embryo stages, however, that is not the subject of this chapter and can only be demonstrated by experiment. Here we discuss what is needed within a natural coordinate system and discuss how these might be implemented. In the first instance we must capture the standard terms for direction and measure that are commonly employed.

10.4.1 Biologically Defined Directions and Measures

In this section we consider terms used for navigation and viewing that relate to the underlying biology, i.e. from structures recognised in the organism. They come from many sources and for different purposes. The first set of biological directions define both directions within the body at any point and a standard set of planar sections and

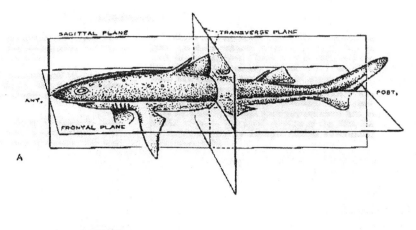

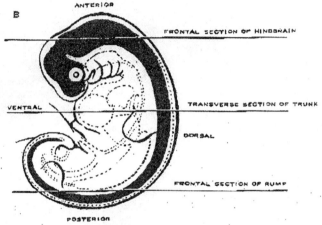

Fig. 10.2. Co-ordinate directions as defined by Ballard [3]. Note the three section planes, frontal, sagittal and transverse are nominally orthogonal and planar for the adult organism (A). For the embryo (B) this is not true.

viewing directions. There is a complication arising from the the fact that the head orientation for upright vertebrates is roughly perpendicular to that for "horizontal" animals - animals, human or otherwise look forwards. anterior (forwards) - posterior (backwards) in a mouse is nose to tail and pretty well always in the same direction in "real"-space, i.e. an external cartesian co-ordinate frame, but in primates the anterior-posterior axis turns through $90°$ somewhere near the neck.

Traditional Terms

By traditional terms we mean all the standard terms used in describing anatomy for vertebrates including man. We need to collect these terms and definitions for the purposes of interoperability even though some may be deprecated. In most cases we will use definitions from Ballard [3] (figure 10.2 supplemented by those in Grays [18], established thesauri and ontologies such as UMLS and the FAM and finally Internet sources such as online dictionaries[5]. These terms typically denote a direction or a relative placement. Ballard points out that many of the terms from the medical literature relate to posture, e.g. superior-inferior, which is not very useful for other animals which do not walk upright. However, we need to understand these terms in order to establish synonyms for purposes of interoperability and moreover note that whilst the names are inspired from the posture their usage is strictly in terms of the body axes and well understood. Table 10.1 provides a short list with definitions (with inconsistencies noted) of the most used terms.

There is clearly a significant scope for confusion and an ontology that attempts to capture these definitions will need to cope with use and re-use of the same text symbol in different contexts - positional, directional and view. In addition there are no "co-ordinates" defined in the sense of a continuous variable defining a position in space. The terms can be used to define relative positions but without some sort of metric it is not possible to establish if something is more or less proximal unless it is explicitly stated. From this we see that full use of the NCS may well required the capability of natural language processing (NLP) for correct interpretation of expressions. Many of these terms are already in the ontology associated with the digital anatomist (see figure 10.3) but unfortunately without definitions, although with a more exhaustive list of terms and defined synonyms. It is clear from the above discussion, however, that the synonyms are context dependent.

10.4.2 Definition and Implementation

How can we define a coordinate frame which captures the directions and positional terms described above? The solution most likely to be useful, although technically difficult, is to accept that the biology will not be easily constrained and therefore impose co-ordinates by the process of non-linear spatial mapping. If we can define the biological or "natural" coodinates in the context of the atlas then by establishing the mapping between the atlas and an experimental individual we thereby establish the standard co-ordinates in the context of the experimental animal. How are such "natural" co-ordinates to be defined? It is clear that such coordinates will have to be defined by expert input, somehow the three directions need to be defined throughout the extend of the organism. In many cases it may be sufficient to define a rule and apply a simple rectilinear coordinate frame as is the case with the stereotaxic coordinates in the brain. In general however a curvilinear body axis will need to be defined with the orthogonal axes additionally defined to enable propagation through the volume of the organism. In addition many sub-structures, e.g. limb components, will need internal coordinates to capture these directions independently of articulation or posture. This will require significant expert input and the protocol by which these coordinate are defined will have to be adopted as part of the ontological standard.

---

[5] e.g. http://www.wordreference.com/ or http://www.m-w.com/

Table 10.1. A selection of the directional and location terms commonly used in biological descriptions of place and view. The full list is significantly longer.

Term	Description
Dorsal-Ventral	Dorsal refers to the backbone side, ventral refers to the belly side. This defines an axis through the body. Note in the head this gets confused because of body posture and Gray's Anatomy seems to be at variance with other authors.
Cranial-Caudal	Cranial is head end, caudal the tail end.
anterior-Posterior	In non-human anatomy texts these are synonymous with cranial-caudal respectively. In Grays for example they are synonymous with dorsal-ventral.
Superior-Inferior	In Grays synonymous to cranial-caudal.
Left-Right	Left-right as defined looking from dorsal to ventral.
Medial and Lateral	Relative positional terms, medial structures are relatively close to the mid-line or primary body axis, lateral are further to the side. Note lateral implies a relative distance and a direction, i.e. left or right.
Proximal and Distal	Relative positional terms, proximal is closer to the centre of the body or the point of attachment in the case of e.g. a limb, and distal is further away.
Transverse	Any plane dividing the individual into a cranial and a caudal part. Can be oblique up to a point but is often additionally constrained to be perpendicular to the body longitudinal axis. (Note:Ballard says transverse, frontal and sagittal may be oblique. In general any oblique section may a bit transverse and bit frontal and indeed partly sagittal. Presumably how a section is described is rather subjective, furthermore any section has a real spatial extent and it is unlikely that if it is "perfectly" transverse over the whole section because all individuals will exhibit some degree of curvature and twist with respect to the nominal cartesian co-ordinate system.)
Frontal	A plane roughly perpendicular to the dorsal-ventral direction.
Sagittal	A plane roughly perpendicular to the left-right direction. The *midsagittal* divides the individual into mirror image parts (apart from non-symmetric organs) and *parasagittal* planes are to one side.
Coronal	Typically refers only to sections through the head and is perpendicular to the dorsal-ventral axis (Grays) or caudal-rostal (other neuroscience).
Axial	Seemingly only used for the head, synonymous to transverse or frontal depending on the Grays view or Ballard view of directions in the head.
Longitudinal	synonymous to frontal.

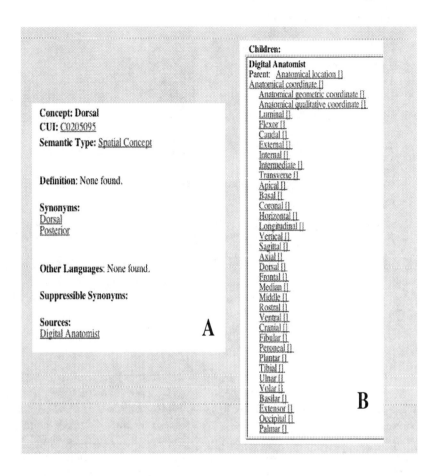

Fig. 10.3. Example spatial terms from the Digital Anatomist as provided from UMLS Metathesaurus (http://umlsks.nlm.nih.gov).

In terms of implementation we envisage a hierarchical approach with more detailed coordinate frames associate with selected sub-structures. This is now illustrated in the context of the mouse embryo with the spatial extent of the embryo divided into however many parts are required to capture the required biological description. Each part will have a corresponding domain defined in the space of the atlas model and each region or *domain* will be a subset of the parent part. The top-level part will include the domains of all sub-parts as sub-domains therefore each part will have a hierarchical set of co-ordinate frames defined by the parent parts. How is this represented? Clearly we will need to make compromises. The ideal might be a fully dynamic model, e.g. based on the articulations and joints model defined for the adult human [12, 10]. We would then need to define how the tissues moved as the joints moved and flexed so that the segmentation would remain valid i.e. no overlaps or gaps appearing. This is too hard for this purpose. At the moment we need a static description so that for mapped data we can query and analyse in terms of biology and if required map the co-ordinates back to the source.

So with this compromise, what do we have? The assumed basic input is:

1. 3D voxel-based models and domains defining the embryo and embryo component extents.
2. A set of voxel dimensions to define the sampling frequency in external units e.g. microns.
3. A mechanism or rule for defining biological direction at each point.

In addition we propose to add:

4. A defined origin and primary axis through the embryo or component, presumably manually defined.

We also assume that, with the origin, axis and definition of biological direction at each voxel location (in reality of course it will not be at each voxel and we will be doing some interpolation) then we can define a "stereotaxic" set of co-ordinates with a non-linear mapping onto the domain/voxel co-ordinates. The actual co-ordinate values or distances within this space of course have a complex relation to true distance in the original embryo.

The embryo extent within the voxel model is defined by a set of domain for each of the regions for which we define a set of natural co-ordinates. The number and complexity of these domains will vary through embryo development and will be a compromise between complexity and effort. For example at early stages there is likely to be only one domain whereas for later stages there may be many, especially for the articulating limbs. We will organise these domains hierarchically with each part being a sub-part of an enclosing domain, except of course for the top-level which is the domain of the whole embryo. Each domain is defined in terms of the voxel co-ordinates of the embryo voxel model and has a mapping onto natural co-ordinates represented by a non-linear transformation. Because each domain has a transform to a natural co-ordinate frame, each sub-part has therefore multiple natural co-ordinate frames at different resolutions. This is useful because a spatial reference may be be in the context of the primary body axis or say the limb axis, both may be needed. This organisation is illustrated in figure 10.4. Note the organisation of domains for the co-ordinate frames does not have to follow the tissue organisation of the anatomy ontology. This may well follow an ordering defined by connectivity with a view to the nested frames associated with articulation.

10.5 Discussion

In this chapter we have discussed the extended use of anatomical ontologies for describing spatial location and spatial distributions or patterns. It is clear that to at-

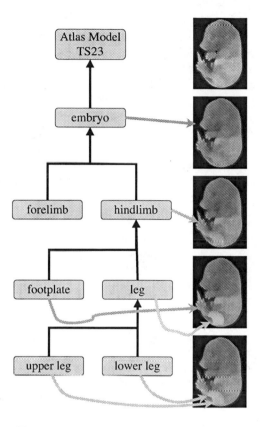

Fig. 10.4. An example domain hierarchy for defining a series of nested natural co-ordinate systems. Each sub-part domain is a proper subset of the domain of the parent part. The domain links are depicted by the arrowed lines. Note here the representation is 2D, in practice the domains will be defined within the context of the 3D atlas models. Also please note these domains are merely illustrative and bear little relation to the true limb domains.

tempt to include all the topological and geometric relationships needed in this context within a "standard" ontology mechanism, i.e. predicate or description logics is a hopeless task and a better solution is to extend the notion of ontology to include a more complete geometric framework. We have introduced the notion of "natural coordinates" with an example of how they might be implemented. In the Edinburgh Mouse Atlas we are implementing natural coordinates in order to extend the query capability of any databases using the atlas as a spatio-temporal framework. We expect that the benefits of this approach will arise in a number of areas. The first is usability for the querying the database with more natural descriptions of location, the second is a more meaningful mechanism for analysing the data, particularly in the context of anatomical development. The third is to provide interoperability between atlas and anatomy-based databases avoiding the need for explicit coordinate mappings, this extends to comparison and query of data between developmental stages, i.e. temporal queries. In particular it is likely that descriptions using NCS will be more robust and reliable in terms of capturing the true biological meaning.

In practical terms, as well as establishing the mapping between the NCS and the low-level voxel coordinates (or what ever has been adopted for a given reference framework) we also need to develop mechanisms for inference. Some of this will come directly from the underlying geometry and topology. In addition it may be possible to describe location by a "descriptor set" by which we mean a set of NCS expressions that when taken together specify, perhaps redundantly the required place. An interesting question for query and interoperabilty is how can such descriptor sets be compared? Therefore, as well as the need for mechanisms of inference we may also need pattern recognition.

References

1. R.A. Baldock, J. Bard, M.H. Kaufman, and D Davidson. A real mouse for your computer. *BioEssays*, 14:501–502, 1992.
2. Richard Baldock, Jonathan Bard, Albert Burger, Nicolas Burton, Jeff Christiansen, Guangjie Feng, Bill Hill, Derek Houghton, Mathew Kaufman, Jianguo Rao, James Sharpe, Allyson Ross, Peter Stevenson, Shanmugasundaram Venkataraman, Andrew Waterhouse, Yiya Yang, and Duncan Davidson. Emap and emage: A framework for understanding spatially organised data. *Neuroinformatics*, pages 309–325, 2003.
3. W. W. Ballard. *Comparative Anatomy and Embryology*. Ronald Press, USA, 1964.
4. Tim Berners-Lee, James Hendler, and Ora Lassila. The semantic web. *Scientific American*, May 2002.
5. Albert Burger, Duncan Davidson, and Richard Baldock. Formalization of mouse embryo anatomy. *Bioinformatics*, 20:259–267, 2004.
6. R. Casati and A.C. Varzi. *Parts and Places, The Structures of Spatial Representation*. The MIT Press, Cambridge, Massachusetts, 1999.
7. Duncan Davidson and Richard Baldock. Bioinformatics beyond sequence: Mapping gene function in the embryo. *Nature Reviews Genetics*, 2:409–418, 2001.
8. D. de Roure, N.R. Jennings, and N. Shadbolt. The semantic grid: A future e-science infrastructure. In F. Berman, G. Fox, and A.J.G. Hey, editors, *Grid Computing - Making the Global Infrastructure a Reality*, pages 437–470. John Wiley and Sons Ltd., 2003.
9. T.R. Gruber. A translation approach to portable ontology specifications. *Knowledge Acquisition*, 5:199–220, 1993.
10. Isam Hilal, Serge van Sint Jan, Alberto Leardini, and Ugo Della Croce. D3.2 final report on data collection procedure - annex 1. Technical report, University of Brussels, 2002. Report from EU project 10954 Virtual Animation of the Kinematics of the Human for Industrial Educational and Research Purposes.
11. Janet Kerwin, Mark Scott, James Sharpe, Luis Puelles, Stephen C Robson, Margaret Martinez de-la Torre, Jose Luis Ferran, Guangjie Feng, Richard Baldock, Tom Strachan, Duncan Davidson, and Susan Lindsay. 3 dimensional modelling of early human brain development using optical projection tomography. *BMC Neuroscience*, 5:27, 2004.

12. Anderson Maciel, Luciana Porcher Nedel, and Carla Maria Dal Sasso Freitas. Anatomy-based joint models for virtual human skeletons. In *Computer Animation 2002 (CA 2002), 19-21 June 2002, Geneva, Switzerland*, pages 220–224. IEEE Computer Society, 2002.

13. Allan MacKenzie-Graham, Erh-Fang Lee, Ivo D Dinov, Mihail Bota, David W Shattuck, Seth Ruffins, Heng Yuan, Fotios Konstantinidis, Alain Pitiot, Yi Ding, Guogang Hu, Russell E Jacobs, and Arthur W Toga. A multimodal, multidimensional atlas of the c57bl/6j mouse brain. *J Anat*, 204(2):93–102, 2004.

14. George Paxinos and Keith B. J. Franklin. *The Mouse Brain in Stereotaxic Coordinates*. Academic Press, San Diego, 2nd edition, 2001.

15. C. Rosse and J.L.V. Mejino. A reference ontology for biomedical informatics: the foundational model of anatomy. *Biomedical Informatics*, 36:478–500, 2003.

16. S.J. Surma, J.T. Srzednicki, D.I. Barnett, and F.V. Rickey, editors. *Stanisław Leśniewski: Collected Works*. Kluwer Academic Press, 1992.

17. Karl Theiler. *The House Mouse*. Springer-Verlag, New York, 1989.

18. P. L. Williams, R. Warwick, M. Dyson, and L. H. Bannister, editors. *Gray's Anatomy*. Churchill Livingstone, London, 37 edition, 1995.

19. Morton E. Winston, Roger Chaffin, and Douglas Herrmann. A taxonomy of part-whole relations. *Cognitive Science*, 11:417–444, 1987.

11

Time in Anatomy

Duncan Davidson

"So anatomy is not merely the separation of parts, the accurate description of bones, ligaments muscles, vessels, nerves and so forth, but an attempt to grasp the totality of body structure, engaging many disciplines, constantly searching for underlying principles and viewing the living frame as an extraordinarily complex, labile entity with a temporal dimension..." (page 2 of [20]).

11.1 Introduction

Time is important in anatomy. One anatomical structure develops into another, each changes in size, shape and composition during the period of its existence and different structures interact over particular time windows when they are competent to signal and respond. Indeed this temporal co-ordination of parallel developmental trajectories is essential for the integration of anatomical structure on a large scale. Closely timed reciprocal tissue interactions co-ordinate development in all organ systems, as for example the 'inductive' interactions between the lens placode and optic vesicle drive the development of the lens and retina of the vertebrate eye (for a review, see [13]). The importance of temporal co-ordination is not confined to development. Anatomical structures mature, are renewed by molecular turnover, undergo progressive changes in disease and healing and ultimately degenerate. The dynamics of these processes is central to the study of anatomy. Take for example, the continuous remodelling of the bony skeleton in vertebrates. Competing bone resorption by osteoclasts [49] and deposition by osteoblasts [19] not only determine the shape of the bone during development, but continuously renew the adult skeleton. According to some estimates the human skeleton is renewed every 10 years [39]. The rates of resorption and deposition change through life and, in man, the relative decrease in the rate of deposition in later life can result in osteoporosis [39].

Anatomy ontologies must include temporal concepts. One important purpose of anatomy ontologies is simply to represent traditional anatomy. Such a description needs to be simple enough to support machine-readable annotation of, for example, gene expression patterns and mutant phenotypes on a large scale and rich enough to accommodate new kinds of data as techniques in biology evolve. Thus ontologies need concepts to indicate that structure A is part of B, and a member of the class C and temporal concepts to indicate those structures that are precursors of A, those that A develops into and those that are present at the same stage of development as A, perhaps in the same vicinity. Therefore, we need to represent the period of existence of an anatomical structure, developmental relationships between structures in the same developmental trajectory and temporal relations between structures in different trajectories. A typical practical aim here is to use large-scale data to correlate mutant phenotype or gene expression in one structure with, say, gene expression in its precursors or descendents in order to explore causal relationships. In addition to direct questions such as 'which genes are expressed, or have mutant phenotypes, in the precursors of structure X?', one would like to apply temporal reasoning to annotated data, for example 'from the available expression and phenotype data, is it plausible that the expression of gene X is necessary for the expression of gene Y?'. Present biological methods are producing information that will challenge the power of anatomy ontologies to represent time. Current ontologies deal with discrete structures – heart, lung, etc. But much of anatomy deals with continuous change. Live imaging using genetically marked cells is being used increasingly to describe detailed temporal relations between events during anatomical development (see for example, [24, 43, 23]). Thus, anatomy ontologies need concepts to represent, for example, cell migration or epithelial folding during morphogenesis. Cell-lineage tracing using single-cell injection or genetic markers induced at specific times and locations will yield information about the clonal origin of the cells that make up visible structures. This description may even engender a clonal view of post-gastrulation anatomy in vertebrates quite distinct from the traditional one (see, for example, [22, 3, 29]). To accommodate this data, ontologies must relate the concepts of discrete anatomy to descriptions of the growth and segregation of cell clones. Lastly, and importantly, computational models are exploring the development and stability of anatomical structure in terms of the behaviour of complex dynamical systems. The challenge here is to articulate a set of temporal concepts that will integrate existing anatomical ontologies with models that use continuous, time-dependent variables. The practical aim is to ensure that annotated data on a large scale provides, on one hand, a substrate for modelling and, on the other, a test-bed for hypotheses.

Despite its importance, a good deal of work remains to be done to represent time in anatomy ontologies. Most ontologies incorporate temporal concepts implicitly in the names of structures, for example, anlage in statu nascendi, anlage, primordium (Flybase, flybase.org; see also [51]), presumptive notochord (ZFIN, zfin.org), future brain (EMAP, genex.hgu.mrc.ac.uk). Most include developmental stages (for example, larval stages in Flybase) and some give stage ranges to represent the period of existence of an anatomical structure (e.g. ZFIN and EMAP). Some anatomy ontologies include relations between two or more structures that can capture sequential relations within developmental series (for example, 'develops_from' in Flybase and ZFIN). Formal definitions of these relations have been proposed ([47], Haendel et al

Chapter 16 of this volume). But neither axioms of time nor any temporal calculus have been incorporated into anatomy. Temporal relationships between structures in different developmental pathways cannot be formally represented. Anatomical ontologies do not accommodate continuous change and there is no formal process ontology for anatomy that incorporates the time dimension.

This chapter explores the formal aspects of time in anatomy. In Section 11.2, I discuss briefly the basic concepts needed to represent temporal relations since these are well developed in other fields but have not been applied to anatomy. Section 11.3 discusses the temporal aspects of the concept of an anatomical structure, particularly from the perspective of continuity and change. Section 11.4 reviews temporal relations in anatomy ontologies and Section 11.5 discusses the concepts needed to articulate processes and dynamical systems models.

11.2 Representing Time

Anatomy naturally entails the intuitive, classical-mechanical view of time as linear, continuous, directional and frame-independent. The classical view entails concepts of time-point, period and order. A thorough formalisation of these concepts, initially in the context of natural language semantics, artificial intelligence and planning, has been made in terms of an interval calculus by J.F. Allen and his associates [2]. J.R. Hobbs and J. Pustejovsky have described a temporal ontology for the semantic web [30] that is based on Allen's interval calculus and has been articulated in DAML-OWL (http://www.cs.rochester.edu/ ferguson/daml/). The main features of interval calculus can be summarised, after Allen and Ferguson, as follows. A time-point is a location in time. In contrast, a period is defined as starting at one time-point and ending at another. Allen and Ferguson distinguish two kinds of period, the interval which is a period that has sub-periods and the moment which has none[1]. Thus the temporal entities of interval calculus are *time point, period, moment,* and *interval*; moment is_a period; interval is_a period. The primitive relation between periods is '**meets**'. This relation captures temporal continuity: two periods **meet** if one **precedes** the other, there is no time between them and they do not **overlap**. The direction of time is captured by the ordering relation '**precedes**' which applies to time points and periods. There is no beginning or ending of time and no infinite periods. Every period has a period that **meets** it and another that it **meets**. Periods that **meet** can be concatenated. Each period uniquely defines an equivalence class of periods that **meet** it. Thus, representing periods by the letters i,j, k and l, if i **meets** j and i **meets** k then any period l that **meets** j must also **meet** k (note that the linearity and direction of time are implicit in this assertion). From this formal basis, the interval relations shown in Table

---

[1] Note that the relation of discrete points and intervals to a continuum raises well-known fundamental problems that are common to other domains (space, set theory). Their resolution in temporal ontologies, for example the idea of a moment, a period that is indivisible, but still of time, reflects these difficulties.

11.1 and Figure 11.1 can be articulated. For example, the relation '**Before**' is defined as follows.

i **before** j, there exists an interval m such that i **meets** m and m **meets**j.

Before and *meets* are subclasses of *precedes*. *After* and *met_by* are subclasses of *preceded_by*.

Unless explicitly stated, we will assume homogeneity of all predicates over time. That is, if predicate P holds over interval t, then P holds over all of interval t; i.e. there is no subinterval of t during which P does not hold. We will also assume that negation takes the strong form: if ¬P(t) then P does not hold over any subinterval of t.

In this chapter I will use the following conventions for the sake of brevity. A time-point is written 't' and t_s and t_e denote the start time and end time of an interval. Instances, for example instances of anatomical structures or instances of intervals, will be denoted by lower-case letters, classes by upper-case.

Table 11.1. Interval Relations

Relation	Inverse	Applies to
Precedes(i,j)	Preceded_by (j,i)	time_points and periods
Before(i,j)	After(j,i)	time_points and periods
Meets(i,j)	Met_by(j,i)	time_points and periods
Overlaps(i,j)	Overlapped_by(j,i)	periods
Starts(i,j)	Started_by(j,i)	periods
During(i,j)	Contains(j,i)	periods
Finishes(i,j)	Finished_by(j,i)	periods
Equals(i,j)		periods

Since an interval is defined by an ordered pair of time points, relations between intervals can be described as relations between their respective t_s and t_e. An interval calculus can thus be reduced to formalisms that deal with the two binary relations; **before** and **simultaneous**. 'Point algebras' thus provide an alternative formal system to calculate temporal relationships. Where the beginning and end points of intervals are known, interval calculus allows temporal relations to be calculated. But for reasoning over intervals with undefined end times, interval calculations can be essentially unsolvable. For certain domains of this problem, point algebra can provide a solution [52]. A formal general study of approximate qualitative temporal reasoning has also been made by Bittner [7].

An alternative way to represent time in anatomy is to use a measure of real time relative to some reference time point in the life of the organism. For example, time

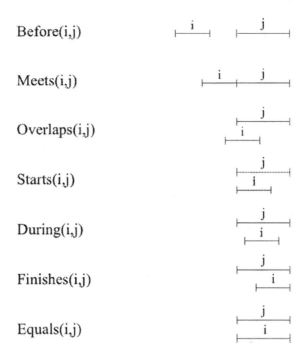

Fig. 11.1. A graphical representation of Allen's temporal intervals. See Table 11.1 for details.

frames that refer to the whole organism use fertilization or hatching or birth as reference points (for example, hours post fertilization in zebrafish and medaka, or years after birth for people). Other time points can also be used, for example the beginning of a particular developmental or disease process where this is the focus of interest. We will refer to this as the 'time post-reference' (TPR) representation of time. Clearly, intervals can be mapped to a TPR time frame and point algebras can be applied to TPR data. At the instance level, **tpr** is simply time (for example, in hours) after the time of a reference event. At the class level for any event E that occurs in all organisms of a given species; then 'E occurs at time T post-reference' implies that each instance e of E in a given organism, occurs at time T after the reference time in that organism. Of course, variation between individuals means that this assertion is an approximation.

Temporal concepts for anatomy. In summary, the key elements of an ontology of time for anatomy are therefore as follows: the axioms that time is linear, continuous and ordered and that an entity can exist over a period of time, the concepts of time point, period and order, and the formal relations between time points and between intervals described above. In addition, it is necessary to be able to relate time to the life of an individual organism. These concepts can be used to capture the temporal relations between anatomical structures. When we consider processes in anatomy we also need the usual concepts of rate and frequency.

11.3 Representing Anatomy

11.3.1 Anatomical Structure

Anatomy is founded on dissection [20]. The concept of an 'anatomical structure' as a part of the body that can be recognised and named in a dissected specimen has, of course, been extended by microscopy and, more recently by advanced imaging techniques, to include cellular and sub-cellular structures that can be distinguished visually. Modern anatomy has extended this concept logically to include macromolecules and molecular complexes that can be considered as 'structural'. In this broad sense, then, anatomical structures range in scale from gross anatomy to cells and molecules.

The view that the body is an assemblage of spatially and temporally discrete parts is the basis of all anatomical ontologies. Smith et al [47] have expressed this view from a philosophical perspective, noting that an anatomical structure is a continuant in the same sense as are other familiar objects in the world; it exists as a whole at every moment when it exists at all.

The immediate temporal implications of this view are simple. First, each structure exists continuously over some interval of time. Formally, at the instance level, each instance c of a type of structure C exists continuously during an interval i_c. i_c can be defined as follows:

i_c *interval of existence of c = Let t_s and t_e be the start and end times respectively of i_c; then for all times t such that t_s **precedes** t AND t **precedes** t_e, c exists at t AND c does not exist at any time t_2, where (t_2 precedes t_s OR t_2 **preceded** by t_e). Note that the duration of i_c is $t_e - t_s$.*

At the class level, I_c is the class of i_c of all c instance of C. C denotes the universal class of all instances of c (see below).

I_c is the interval used to represent temporal relations between anatomical structures at the class level, for example in developmental series. At the class level, intervals of existence can be related using the relations described in Section 11.2. For example,

I_c *meets I_d = for all instances d of D there exists a corresponding instance c of C such that i_c meets i_d.*

More than one instance of a structure may be present in the same individual (for example, cells of any particular type, or buds in a branching epithelial system, or nephrons, etc.). It is therefore useful to have a concept to represent the interval during which some instance of c exists in an individual organism. When we speak of a class of structures, we often refer to the universal class that includes all instances in all individuals of that 'species' of organism. We will use C to denote this universal class. But it is not useful to refer to the interval of time when there exists some instance of a structure across all individuals of the species. We need an intermediate entity, referring to the class of structure of which c is an instance in an individual organism. We also need to distinguish between an instance of this class (i.e. the class of all c in a particular individual) which we will call C' and a class of all C' across all organisms which we will call C. There are two alternative definitions of the interval of existence of C' at the instance level. One demands that at all moments **during** this interval there is some instance c of C in the given organism. The other simply defines the interval as the time from formation of the first c to the end of existence of the last c in the given organism. We will use the latter definition here. Thus,

At the instance level, that is in a particular individual organism, iC' is the interval of existence of C' = iC' is the interval starting at the same time as the start t_s of i_c for the earliest-formed instance c of C' and ending at the same time as the end t_e of i_c for the latest-formed instance c of C' in the same organism.

Notice that iC' is continuous and that this definition allows that there is some moment **during** iC' when no c exists.

At the class level, IC is the class of iC' of all C' instance of C.

The set of discrete anatomical structures that comprise an organism thus defines an equivalent set of intervals, i, that can, in principle, be related to one another as described in the preceding section. At the instance level, where one structure is **part_of** another (as the left ventricle is **part_of** the heart), the interval of existence of the part (left ventricle) clearly must have a **starts, during, finishes** or **equals** relation to the interval of existence of the spatial parent structure (heart). For is_a relations between anatomical structures (as the left ventricle **is_a** ventricle), if A **is_a** B then, I_A *starts, during, finishes* or *equals* I_B bearing in mind that A and B refer to the same individual organism. The relations between intervals of existence are discussed further in the context of 'developmental stages' in Section 11.4.3.

The second implication is that each structure is formed from one or more previously existing structures. Such a developmental series of structures, c1, c2, ..., cn defines an equivalent series of intervals of existence i_{c1}, i_{c2}, ..., i_{cn}. The first interval in the series begins at conception of the organism and the last ends before or at the onset of post-mortem disintegration. ('Development' is used here in the broad sense to include not only prenatal changes, but also changes during ageing, disease and healing.) At the class level, for all structures D there is some structure C which is also a member of the series such that each instance d of D was a corresponding instance c of C AND i_c meets i_d (except where C is the last member of the series).

Note that, under this view, each named structure changes abruptly into the next in the series.

By considering the identity of each structure as discrete in time, anatomical ontologies gain the immediate advantage of allowing simple reference to structures and thus simple annotation of, say, gene expression patterns or mutant phenotypes. This kind of annotation is well suited to binning data in order to support queries such as 'which structures express gene X?' or 'which structures are affected in mutant Y?'. However, this view of anatomy raises the difficulty of defining the interval of existence of each structure. In the great majority of cases information about interval boundaries is both incomplete and imprecise. This is necessarily so. In most cases the change from one structure to another is gradual. An example is the change from an otic placode to an otic pit during vertebrate ear development. Like their spatial counterparts (for example, the boundary between limb and trunk) these are 'fiat' boundaries with a degree of arbitrariness that is usually undefined. In a few cases, one boundary can be quite precisely defined, for example the closure of an epithelial vesicle (the lens vesicle or otic vesicle, for example). But even in cases where there is a clear and rapid change in topology, the correlation between the time of the local event and the transition to a new anatomical structure can be complicated. For example, in higher vertebrates the neural plate folds and fuses dorsally to form the neural tube, but this event spreads along the antero-posterior axis over a period of hours or days (about 24 hours in the mouse). The closure of the neural tube can be represented with a simple, instantaneous temporal boundary at the time when the posterior neuropore closes, but this fails to represent the local transition from plate to tube that moves antero-posteriorly with time. The time of events in anatomy is often a function of their position in the organism. We will return to this point in Section 11.5.

11.3.2 Continuity and change

Much of development is a process of progressive differentiation. The many developmental series that represent the anatomy of an organism during its life can, in principle, be combined in a single directed graph in which vertices represent anatomical structures and edges transitions from one structure to another. A small section of such a graph is illustrated in Figure 11.2. Notice that the graph in Figure 11.2 is based not on times of existence like an interval representation, but on continuity, representing transfer of material from one structure to another. The key distinction to be made is between vertices of degree 2 (with one in-degree and one out-degree) and those of degree >2; that is, between structures that form by internal change and those that form by gain or loss. Most vertices of degree >2 are degree 3 with either 2 in-degrees (fusion) or 2 out-degrees (fission). Most higher order vertices can be reduced to sequences of vertices of degree 3 except where the same graph is applied to mixed levels of granularity; for example, to transitions between dispersed cells and an intact gross anatomical structure (aggregation and dissociation). Smith et al [47] have emphasised the importance of distinguishing those anatomical structures in a series that are different objects (material continuants) from those that are merely different states of the same object. For example, the limb bud mesenchyme and the ulna bone are different objects, whereas the ulna can pass through successive states each with a different shape, size and composition while still being the same bone. While this distinction is important, it can be arbitrary in some situations. For example, when one structure buds off from another, the newly formed, free structure is clearly a new object, but is the parent structure the same object as before? The formalisation represented by a directed graph avoids this issue. The formal representation of temporal relations in developmental series is discussed in Section 11.4.1. Here, we discuss the biological meaning of the graph.

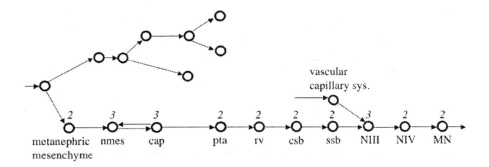

Fig. 11.2. Developmental series represented as a directed graph. The graph shows part of the development of a mouse embryo including an hypothetical version of nephron development for the purpose of illustration. Vertex degrees are indicated on the graph. The direction of development is indicated on the arcs. See text for abbreviations of the names of anatomical structures in the nephron series.

Consider the formation of a new anatomical structure. Each mature structure has both form and substance. For example, the Foundational Model of Anatomy (FMA) defines 'an anatomical structure' as a material entity which has its own inherent 3D shape and which has been generated by the coordinated expression of the organism's own structural genes [45, 41]. When representing the formation of a new structure it is useful to distinguish the two different kinds of temporal discontinuity that mark, respectively, the creation of a new anatomical region and the creation of a new subdivision of material (see also [41]). These may occur in sequence or, as is more common, concurrently.

Take, for example the initial development of a structure by a fission process. Suppose that an undifferentiated, 'starting' population of cells responds to a pattern of diffusible, cell-bound and matrix-bound signals. This response establishes, and may continue to refine, the region where the structure will develop. The cellular response depends on complex interactions between signalling and transcriptional regulation circuits so the shape of the differentiated region is usually not related in any simple way to the original pattern of signals. Anatomically speaking, this is a discontinuity in development: a new region has been defined *de novo* that establishes the location and initial morphology of the incipient structure. But at this stage not all cells in the region, or their descendents, will necessarily become incorporated into the final structure; cells may even continue to move in and out of the molecularly-defined region, reversibly expressing early regional markers in a transient response to local signals.

Complementary, but distinct, processes determine the material composition of the new structure. Of particular importance is the closure of the population of cells (the precursor population[2]) whose descendents will become part of the anatomical structure, or die. In cellular terms this may be passive, simply the consequence of the formation of a physical barrier to dispersal, or active, the result of one, or a series, of binary cell fate switches. There are important questions here about the temporal aspects of cell fate in anatomical development. Is fate determined by a switch that is intrinsic to the cell? When do the first cells switch? What is the origin and fate of cells that enter the precursor population at different times? Does the precursor population every really become closed; if so, when? The closure of a precursor population is a discontinuity in anatomical development: from this time onward, the structure as a whole has material continuity over time in the sense that it is composed mainly, or entirely, of descendents of this population of cells. We can define a precursor cell as follows.

*Q is a precursor cell of structure C = for any cell q instance of Q at t there exists some cell r that is **part_of** c instance of C at t and that is q or is a descendent of q.*

Transfer of cells between structures is one of the most important kinds of change that occurs during the formation and continued existence of an anatomical structure. Apart from non-cellular structures, such as exoskeleton, continuity of cell lineage is the main sense in which temporal continuity of material applies to anatomy. Any more strict interpretation of continuity is confounded by continual turnover of both the cellular and molecular constituents of gross anatomical structure. Cell lineage is important because different lineages potentially contribute differently filtered genetic input into a structure by virtue of their progressively restricted potential for co-ordinated gene expression. It is clearly important to be able to represent the relation between cell lineage and anatomical structure in an anatomy ontology. This is so from a practical as well as formal point of view. We want to be able to address such questions as: could changes in gene expression within a structure at a particular

---

[2] The term 'precursor population' is used here in the broadest sense to denote the population of cells from which the structure forms.

time in development correlate with the recent addition of new cells? Can a mutation in genes expressed in one structure affect phenotype in another through cell-lineage-dependent mechanisms? Do different sub-structures develop from different lineages?

In practice, there are many known – and presumably many unknown – cases where cells move between structures in the course of development and during health and disease. The number of cells may be large or small, their effects significant or insignificant and the periods over which transfer occurs may be short or prolonged. Classic examples in development are the migration of myoblasts from the trunk into the developing limb where they form the entire limb musculature, and the contribution of neural-crest-derived cells to many organs, for example, melanocytes in the skin. Examples in the adult are metastatic invasion by malignant cells, and the migration of stem cells into mature structures, for example, osteoclast precursors and osteoblast progenitor cells to sites of bone remodelling. The fact that gain and loss of cells is so widespread creates problems for any definition of anatomical structure that is based strictly on cell-lineage continuity. However, it is interesting to seek a defining moment for the formation of a new anatomical structure in cell-lineage terms. One interesting possibility is the first moment when all cells in the region that has spatio-temporal continuity with the mature structure have descendents in the mature structure.

More important is to have a formal expression to represent cell transfer. As a step towards this goal we can attempt to define temporal cell-lineage continuity. This can be used as a null condition against which cell-transfer events can be represented.

*Cell-lineage continuity of an anatomical structure C during its interval of existence = let t_s and t_e, respectively, be the start and end points of the interval i_c and let t_1 and t_2 be any two time points such that t_s **precedes** t_1 AND t_1 **precedes** t_2 AND t_2 **precedes** t_e; then for all c instance of C, every cell nucleus that is **part_of** c at t_2 is the same cell nucleus as, or has descended from, some cell nucleus that is **part_of** c at t_1 AND every cell nucleus that is **part_of** c at t_1 is the same cell nucleus as, or is the ancestor of, a cell nucleus that is **part_of** c at t_2 or degenerates during the interval t_1 to t_2.*

This definition implies that when cells are added from, or lost to, another structure, an anatomical structure no longer has cell-lineage continuity. To avoid exceptions in the case of multinucleate cells arising from cell fusion, the definition of lineage relation relates nuclei rather than cells (see [36]).

11.4 Temporal Relations in Anatomy

11.4.1 Relations between Structures in the Same Developmental Series

Two formal relations between structures in a developmental series have been proposed, *transformation_of* and *derived_from* [47]. These relations are defined by Smith et al in a context that emphasises the view that an anatomical structure is a 'material continuant'. At the instance level, **transformation_of** is the relation between two states of the same continuant at different times, for example, between the child and adult states of the same individual. *Transformation_of* is defined as follows:

transformation_of $C_1 = C$ and C_1 for all c, t, if c is an instance of C at t, then there is some t_1 such that c is an instance of C_1 at t_1, and t_1 precedes t and there is no t_2 such that c is an instance of C at t_2 and c is an instance of C_1 at t_2 [47].

Derives_from is a relation between two material continuants. It applies in three situations: 1) where the whole of one continuant c becomes the whole of another continuant d, 2) where c undergoes fission the relation holds between c and each of its fission products, and 3) where c fuses with another continuant the relation holds between c and the fusion product. At the instance level, the relation d **derives_from** c is immediate in the sense that 'the spatial region occupied by d as it begins to exist at t overlaps with the spatial region occupied by c as it ceases to exist in the same instant' (from [47]). At the class level, *derives_from* is defined in two parts [47]. First, C *derives_immediately_from* C_1 is defined as follows: for all c, t, if c is an instance of C at time t, then there is some $c-1, t_1$, such that: t_1 is earlier than t and c_1 is an instance of C_1 at t_1 and c **derives_from** c_1. This concept of immediate derivation is then used to define *derives_from*: C *derives_from* C_1 if and only if there is some sequence $C = C_k, C_{k-1}, ..., C_2, C_1$, such that for each C_i ($1 \leq i < k$), C_{i+1} *derives_immediately_from* C_i.

The *derived_from* relation between two whole continuants (case 1) above) is distinguished from the *transformation_of* relation between two states of the same continuant by the assertion that the *derived_from* relation entails a change of identity, that is a change of one continuant into another. Brochhausen [9] pointed to the arbitrariness of the attribution of identity and suggested that *transformation_of* should represent all cases where one anatomical structure gives rise to another single structure and that '*derives_from*' should apply only in fission and fusion. *Derived_from* has been renamed '*arises_from*' which is defined using a *part_of* relation without specifying quantity of material (Haendel et al, Chapter 16 of this volume). Haendel et al also suggest a third relation, '*develops_from*' under which *transformation_of* and *arises_from* are subsumed (for definitions see Chapter 16).

The *transformation_of* and *derives_from* / *arises_from* relations entail assertions about the transition process in terms of the temporal discreteness of the structures concerned and their material relationships. In this sense, these are special cases of more general temporal relations between anatomical structures. From this they gain directness of application to the situations they describe, but they lack more general power. A more general approach would be to disentangle temporal relations from the transition process and thus from the concepts it entails. A full description of this approach is beyond the scope of this chapter, but the principles are described briefly below.

The most direct representation of a temporal relation is the assertion d **was** c (and its counterpart c **will_be** d). The **was** relation entails temporal sequence without implications about the nature of the temporal continuity between the structures, for example whether morphological or cellular. The relation does not entail assertions about whether the transition was gradual or discontinuous, or about temporal relations between the intervals of existence of c and d except only that i_c **precedes** i_d. At the instance level, **was** is a bare expression of the axioms of time and existence.

d was c = c at t_1 is d at t_2 AND t_1 precedes t_2

At the class level,

*D was C = for all d instance of D at t_2 there is some c instance of C at t_1 such that d **was** c AND t_1 **precedes** t_2*

Note that the assertions D *was* C and C *will_be* D are not equivalent. Examples of the use of this relation are, s-shaped body **was** comma-shaped body and adult mouse *was* Theiler Stage 23 mouse embryo. The assertion e **was** c does not imply that c is the immediate precursor of e in a developmental series. The **was** relation is transitive and the series c,d,e implies that e **was** d, e **was** c and d **was** c. The *was* relation can apply to any anatomical structure in the ontology and also to states of a named structure within its interval of existence.

'*Was*' can be combined with the '*part_of*' relation common to all anatomy ontologies using the form: d *was* x AND x *part_of* c. Providing that the interpolation of the structure x is understood, this relation can be abbreviated for convenience to d *was part_of* c; similarly, **part_of** d **was** c.

At the Class level,

*D was part_of C = for all d instance of D at t_2, d **was** x at t_1 AND x at t_1 **part_of** c instance of C at t_1.*

*Part_of D was C = for all d instance of D at t_2, x **part_of** d at t_2 AND x **was** c instance of C at t_1.*

*Part_of D was part_of C = for all d instance of D at t_2, x **part_of** d at t_2 AND x at t_2 **was** y at t_1 AND y at t_1 **part_of** c instance of C at t_1.*

These relations are illustrated in Figure 11.3. Examples are, renal vesicle *was part_of* nephrogenic mesenchyme and *part_of* vertebra T2 *was part_of* somite T1. This representation can, of course, explicitly use the different kinds of *part_of* relations appropriate to anatomical structures [41]. In particular, the *part_of* relation can refer explicitly to regions or material or both or to the *member_part_of* relation. Notice that cells, cell populations and clones can be *part_of* a structure and that a cell can be *member_part_of* a cell population or clone. We can recognise three special kinds of cell population e that have a *part_of* relation to a structure c: e might be all the cells of which c is comprised, all the cells of a particular type in c or all the cells belonging to a particular clone in c.

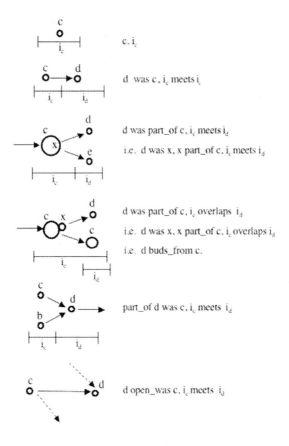

Fig. 11.3. Temporal relations based on the was relation represented as vertices of a directed graph. The relations discussed in the text are shown.

Besides these relations it is useful to have to hand a single relation to express the weaker assertion that *all or part_of* D *was all or part_of* C. This requirement is similar to that met by the *develops_from* relation proposed by Haendel et al (Chapter 16 of this volume). This allows one to represent a temporal relation in situations where one needs a general assertion that encompasses instances with different *part_of* relations or an assertion where the detailed relationships are unknown as is common in developmental anatomy. To avoid the implication of any developmental mechanism, the term '*open_was*' ('*OWR*') might be used for this relation. For example, renal vesicle *open_was* cap mesenchyme. This is broadly equivalent to *develops_from* which is already in use in the zebrafish and flybase ontologies.

The *was* relation can be combined with time points, for example d at t_2 **was part_of** c **at** t_1 (i.e. d at t_2 **was** x at t_1 AND x **part_of** c at t_1) and with time intervals, for example, d **was** c AND i_c **meets** i_d. The latter specifies an immediate transition of c to d. This might be abbreviated to d **was_directly** c. Similarly, at the instance level:

*d **buds_from** c = d **was part_of** c AND (i_c **overlaps** i_d, OR i_c **contains** i_d, OR i_c **finished_by** i_d).*

At the class level:

D buds_from *C = for all d instance of D there is some corresponding c instance of C such that d* buds_from *c.*

Importantly, these relations allow us to represent transitions that take time without giving names to intermediate anatomical structures, d **was** c AND i_c **meets** i_t AND i_t **meets** i_d where i_t is the interval of time occupied by the transition between c and d: : i.e. i_t is the transition interval for c to d.

An important property of any ontology is completeness of coverage in its domain. This is distinct from resolution. For example, an anatomical ontology based on *part_of* relations should refer to the complete set of substructures of any structure A such that all the material of A is accounted for (see also [41]). This is crucial if the ontology is to be used for annotation and reasoning. For example, structure A has substructures B and C: if gene X is expressed in A but not in B, then gene X is expressed in C. This requirement extends to temporal completeness when one considers the *was* relation and its derivatives, except the *open_was* relation which is too indeterminate to support this constraint. It is important to know, for example, if structures E and F are the only time-children of D and that D is the only time-parent of E and of F. Notice the asymmetry of these relations. Temporal completeness is important in order to support reasoning across temporal relations, for example, to follow cell-autonomous effects of the expression of a given transcription factor earlier in development or to trace an observed mutant phenotype back to the primary lesion.

How can these relations be used in practice? To illustrate this, let us look at development of the nephron, the filtration unit of the kidney. The most detailed developmental anatomy ontology published to date represents the mouse genitourinary system from the 10 days post-coitus (dpc) to adulthood [37] (see also metanephros in the Foundational Model of Anatomy, [45] and see [35] for a detailed nomenclature of the adult kidney that is not based on ontological principles). The kidney of higher vertebrates has long been used as a model system to study general principles of development (for a review, see [17]). In man, the kidney is affected in a number of important diseases and offers possibilities for tissue engineering and its development is also the focus of intense research for this reason. Many aspects of nephrogenesis have therefore been described in detail. Nevertheless, there are significant gaps in our understanding and current research will certainly produce large amounts of gene expression, mutant phenotype and cell lineage data that will require annotation and integration using the ontology of Little et al and its subsequent adaptations (see for example, http://www.gudmap.org and http://www.euregene.org). This ontology, though representing a large number of anatomical structures and their parthood relationships, does not include developmental or temporal relations so it is interesting to ask if the concepts discussed above can represent those aspects of nephrogenesis that are important for current and future research.

Before we look at how these formal relationships can be applied we need an outline picture of nephron development. Thousands of primary nephrons form in each mouse kidney. While we can generalise to give an outline description of a notional instance of a developmental series, it is important to realise that each actual instance occurs in a particular location in the kidney and over a particular period in the overall development of the organ. Differences in location and time of development may have major effects on the nephron formed. While ontologies presently ignore this subtlety, annotations using an ontology should be used with this in mind. To address this issue, it is possible to give a two tier temporal annotation – one level referring to time within the developmental series, the other referring to time within the interval of existence of the class C' of structures in the individual organism (see Section 11.3).

The starting tissue for nephrogenesis is the *nephrogenic mesenchyme* at the periphery of the growing kidney. This mesenchyme is invaded by a branching epithelial ureteric tree that forms the urine-collecting system. Signals that initiate nephron formation in the nephrogenic mesenchyme originate at least in part from the ureteric epithelium. Mesenchyme surrounding the tip of each new ureteric branch (*cap mesenchyme*) responds to these signals by expressing particular genes. Indeed, the boundary between cap mesenchyme and the neighbouring nephrogenic mesenchyme is defined, in practice, by the expression of molecular markers, such as *Crym* [37]. The cells of the cap mesenchyme move under the ureteric tip as the ureteric branch grows. Here they begin to form a ball, the '*pretubular aggregate*', that can now be distinguished by its morphology and location. This structure, in turn, undergoes epithelialization to form a small hollow vesicle, the '*renal vesicle*'. It is not yet clear from published evidence just how much cell exchange occurs between the nephrogenic mesenchyme and cap mesenchyme, or indeed between nephrogenic mesenchyme and pretubular aggregate. But it is clear that by the renal vesicle stage, the precursor cell population of the nephron is closed. Thus, on grounds of cell-lineage continuity, the renal vesicle is an individual nephron, albeit in an early developmental state. Other cells of the nephrogenic mesenchyme that are not incorporated into the vesicle give rise, for example, to interstitial tissue. During subsequent development, the renal vesicle, already becoming differentiated into epithelial regions marked by the expression of specific genes, elongates and bends, becoming first the '*comma-shaped body*' and then the '*s-shaped body*'. This is a gradual process. There is no moment when a renal vesicle becomes a comma-shaped body, for example. The s-shaped body continues to change its shape, becoming differentiated both morphologically and cytologically into Bowman's capsule, proximal tubule, loop of Henle, distal tubule and connecting duct. These structures themselves become further differentiated into named regions. Structures that are temporally intermediate between s-shaped body and mature nephron are termed '*Stage III (capillary loop stage) nephron*' and '*Stage IV (maturing) nephron*'. Here, however, we can see the competing interests of practical use and formal consistency. From stage III onward, the structure denoted by 'nephron' in the ontology of Little et al includes an additional component that develops outwith the s-shaped body. This is the developing vascular capillary system (for example, the *presumptive mesangium* at stage III and *maturing glomerular tuft* at stage IV) that eventually forms the '*glomerulus*'. Because of its intimate structural association with Bowman's capsule, the glomerulus is very reasonably represented as part of the nephron. This is a classic fusion event in development. In contrast, the collecting duct, despite being physically fused with the nephron, is traditionally not represented as part of it. These kinds of issues are common in building anatomy ontologies, at least the current ones. They arise from the need to accommodate important biological relationships, common usage and convenience, for example in applications such as manual annotation of gene expression patterns.

The development of a nephron is shown as a directed graph in Figure 11.2. The graph and the text below use the following abbreviations: nmes = nephrogenic mesenchyme, cap = cap mesenchyme, pta = pretubular aggregate; rv = renal vesicle; csb = comma-shaped body; ssb = s-shaped body; stage III nephron = NIII; stage IV nephron = NIV; MN = mature nephron. We wish to represent the temporal relationships between individual structures in development. For clarity, we will illustrate temporal relations on the instance level. The simplest way to represent the relations between the main structures in this developmental series is to use the **open_was** relation, applied to each arc in the graph, in conjunction with the **meets** relation between the intervals of existence of each consecutive structure. For example, stage III nephron **OWR** s-shaped body AND i_{ssb} **meets** i_{NIII}; stage III nephron **OWR** nmes AND i_{nmes} **contains** i_{NIII}; comma-shaped body **OWR** renal vesicle AND i_{rv} **meets** i_{csb}; etc. Alternatively, the equivalent '**develops_from**' relation can be used in place of **OWR** (see above). This representation shows the advantage of the **OWR** relation. There is no available ontology term for the mesangial precursor cells in the nephrogenic mesenchyme from which the presumptive mesangium is formed. The OWR relation allows us to deal with this, albeit in a rather uninformative way.

This approach is sufficient to support a good deal of broad-brush data-mining in data annotated with these terms. But many of the biologically interesting pieces of information in our outline description are lost in this representation. Take the simple relationship between renal vseicle, comma-shaped body and s-shaped body, for example. These structures share the same cell lineages and each is represented by a vertex of degree 2: they are simply successive developmental states of the nephron. These relations can be captured simply as:

s-shaped body **was** comma-shaped body; AND i_{csb} **meets** i_{ssb} ;

comma-shaped body **was** renal vesicle AND i_{rv} **meets** i_{csb}.

This expression is more informative, but no more complex, than the **OWR** relations above. One can easily imagine additional temporal subdivisions of the renal vesicle stage, for example, to segregate successive steps in the progressive differentiation of the epithelium or the introduction of a transition interval between pretubular aggregate and renal vesicle in order to annotate data specifically relating to the epithelialisation process.

The relations between the nephrogenic mesenchyme, cap mesenchyme and pretubular aggregate are more challenging, as is typical of the initial stages of structural development. Bearing in mind that the true cell lineage relations between these structures are as yet uncertain, let us imagine for the sake of illustration, the following hypothetical situation. The cap mesenchyme is regionally defined by response to epithelial and mesenchymal signals; some cells in the cap mesenchyme have a nephron fate, but others are exchanged with the nephrogenic mesenchyme; the nephron precursor population is closed at the time when the pretubular aggregate forms and occupies the whole of the pretubular aggregate. This can be represented as follows:

Precursor cell is defined in Section 11.3.2. Here we give a corresponding definition for a cell that is not a precursor cell.

N is a non-precursor cell of C where C is a type of anatomical structure = for all n instance of N it is not the case that there exists some cell in c instance of C at t that is n or is a descendent of n.

Let t_s and t_e respectively be the start and end of i_{cap}, then t_s **precedes** t_1 , t_1 **precedes** t_2 , t_2 **precedes** t_e; let nmes(p) be the population of cells that comprise the nmes; let cap(p) be the population of cells that comprise the cap mesenchyme; let cap(py) be the population of cells that are not nephron-precursor cells in the cap mesenchyme; let cap(px) be the population of nephron-precursor cells in the cap mesenchyme. Let pta(p) be the population of cells that comprise the pta.

1. cap region **was part_of** nmes region; i_{cap} **during** i_{nmes}, **(regional part_of)**
2. cap region has regional continuity from t_s to t_e
3. cap region r_1 at t_1 is the region occupied by cap(p) at t_1; cap region r_2 at t_2 is the region occupied by cap(p) at t_2; there is not cell-lineage continuity between cap(p) at t_1 and cap(p) at t_2.
4. **part_of** cap(py) at t_2 **was part_of** nmes(p) at t_1
5. **part_of** nmes(p) at t_2 **was part_of** cap(py) at t_1
6. cap(px) **part_of** cap(p) at t where t_s **precedes** t and t **precedes** t_e;
7. cap(px) = cap(p) at t_e.
8. pta **was** cap at t_e; i_{cap} **meets** i_{pta}, (full regional and material relation)
9. rv **was** pta; i_{pta} **meets** i_{rv}, (full regional and material relation)

1) and 2) formation of cap region; 3) absence of cell-lineage continuity in cap between t_1 and t_2; 4) and 5) exchange of non-precursor cells between cap and nmes; 6) and 7) differentiation of cells in the cap region into precursor cells; 7) closure of the nephron precursor population at the time when the cap becomes the pta; 8) formation of the pta from cap; 9) formation of rv from pta.

This example shows how we can use temporal relations to begin to build a dynamic picture of development. At the simplest level, this representation allows us to ask questions about the temporal relationships between gene expression patterns annotated with the anatomical terms from the ontology. For example, which genes are expressed in the s-shaped body, but not in its predecessors? Using the kinds of relations illustrated in the representation of development from cap mesenchyme to renal vesicle, where cell lineage information is available one can begin to annotate gene expression in genetically marked cell lineages, map clonal contributions to anatomical structures and update this information as new evidence becomes available. Data annotated in this way can be used to explore the cellular dynamics of gene expression.

11.4.2 Relations between Structures in Different Developmental Series

Why compare the times of development of structures in different developmental series? There are at least three reasons, each on a different scale. The first is to divide the development of the whole organism, or of individual organs, into standard stages for practical purposes. This is discussed in Section 11.4.3 below. The second is to uncover interactions between tissues within a single structure, for example the eye or the nephron, by analysing data that is annotated with terms from an anatomical ontology. The issues here are similar to those encountered in relating intervals of existence of anatomical structures to developmental stage boundaries and are therefore discussed at the end of Section 11.4.3. A third, and important, motivation is to carry out detailed investigations of such interactions where the action is on a much finer temporal and spatial scale than is currently represented in anatomy ontologies. This is discussed in Section 11.5.

11.4.3 Developmental Stages

Developmental stages are states in the life history of the organism. They are defined by biologists as practical tools to choose comparable embryos for descriptive
and experimental studies and to communicate the results. Some are based on natural
punctuations, such as larval stages in Drosophila (flybase.org) and gastrulation in zebrafish (zfin.org), but most are arbitrary divisions of development. In this latter sense,
the concept of developmental stage rests on the remarkable fact that even unrelated
anatomical structures develop in a consistent temporal sequence in all individuals of
the same type (Figure 11.4a).

Developmental stages are defined for different 'species' of organism. In practice,
individuals from different inbred strains of laboratory animals may differ in the precise relative timing of anatomical development as well as in the morphology of some
features used to define stage. Similar differences may occur between outbred or natural populations and in this case there is also the possibility of variation within a population, particularly over a range of environmental conditions. With these caveats,
we simply refer to 'species' below. Here, we will use 'stage' to denote consecutive
states of development between defined start and end states. There are many ways to
define the state of an organism, but the widely used developmental staging systems
are based on anatomy.

Thus, a developmental stage is a series of anatomical states of the organism that
exist between two boundary states s_{n-1} and s_n separated in time by the stage interval
si_n. Boundary states are defined by the presence of a subset of anatomical structures
and their states. Successive stage intervals **meet**. The set of structures that exists at
any period during si_n is not identical to the set of structures that exists at any period
preceding or **preceded_by** si_n. We must include the states of anatomical structures
as well as the structures themselves in this definition because continously variable
qualities of anatomical structures are sometimes used to distinguish different stages
(for example, the length/width ratio of the limb bud in the chick [26]). The relationship between developmental stage and the intervals of existence of anatomical
structures is illustrated in Figure 11.4. Note that a developmental stage for a species
is characterised by a set of anatomical structures whose intervals of existence have
overlaps, overlapped_by, starts, started_by, during, contains, finishes, finished_by, or
equals relations with the corresponding stage interval (Figure 11.4a). Notice that in
this view of developmental stage each structure is considered independently and not
as a member of a developmental series so that the concept of developmental stage
is independent of *was* relations. The concept of stage does not require information
about the duration of intervals. It relies formally only on the order of events.

Staging systems have been devised to describe the development of the whole embryo for each model organism. Examples are: Amblystoma [27]; chick [26]; mouse
[50]; Xenopus [42]; medaka [32]; zebrafish [33]. In some cases, a particular structure has its own staging system; for example, the limb in mouse [53] and in newt

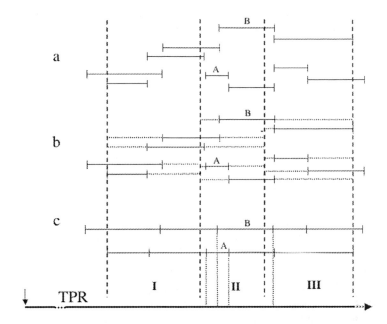

Fig. 11.4. An illustrative diagram of the relationship between developmental stages and intervals of existence of anatomical structures. Only a small part of the anatomy of the organism is shown. a) The order relationship between intervals of existence of anatomical structures and developmental stage boundaries. b) The same intervals in the context of anatomy ontologies that list structures present at each stage. Solid lines represent actual intervals of existence, dashed lines represent the extension of these intervals to stage boundaries to indicate the loss of temporal resolution inherent in this approach. c) The same anatomical structures in their respective developmental series. Stage interval boundaries are represented by vertical dashed lines. A,B, etc indicate anatomical structures the intervals of existence of which are shown as horizontal lines with interval boundaries marked by vertical bars. Intervals can be related using interval calculus: for example, i_A overlaps i_B; Stage II interval contains i_A. The relation between intervals and TPR, time post reference, as it applies under standard conditions, is illustrated by vertical dotted lines. This simplified view does not show possible temporal variation.

[54]. Staging systems are based on features visible by external examination, generally without compromising subsequent observation or experiment. Individual stages are defined by a few characteristics that may mark one or both stage interval boundaries, although the characteristics of mid-stage states are also sometimes used. These are features that can be decisively assayed and not necessarily those that mark ontologically significant changes. Some are easily measured quantitative criteria, as for example in the chick limb bud, some mark a brief event – for example closure of the lens vesicle for the mouse [50]. Several features are used to define each stage. Different developmental states of the same anatomical system are often used to define successive stages. In vertebrates, for example, somite number provides an al-

most ideal metric during early development; limb bud, hand-plate, visceral arches, eye, otic structures and integumental development are used during organogenesis and later development.

The remarkable consistency of development is useful, but not perfect. The relative rates at which different organs develop may vary from one individual to the next, even under optimal conditions. For example, in the chick embryo the branchial arches may be more, or less, advanced compared with the limbs [26]. Staging systems should be viewed in this context. Each system draws on expert observations made on hundreds of individual embryos and assimilates a significant amount of variation to provide a workable standard for use at the bench. This variation should be borne in mind when any computational analysis is made using information from 'staged' material. Indeed, a number of questions arise when one relates any developmental event to embryonic stage. Suppose, for example one is interested in comparing two successive states of development of the heart. Are the standard characters for staging whole embryos well enough defined to distinguish embryos with these different states of heart development? Do the staging characteristics behave in the same way in the strain being studied as in the strain in which the standards stages were defined? Are the stages fine enough? Is the temporal development of the heart sufficiently tightly linked to that of the staging criteria or is there variation that makes staging unreliable? Many standard staging systems are useful only for broad categorisation of embryos. Beyond this, most workers use informal, but refined, criteria to assess the stage of development of the characters they are interested in. For some purposes, for example in mathematical modelling, time is important as well as stage. Here the question arises, how well do even refined staging characteristics indicate the actual time of development; that is, how variable are developmental rates under the conditions being studied?

As well as providing a measure of the developmental state of the organism as a whole, developmental stages provide an approximate temporal relation between anatomical structures, including those in different developmental series. How useful is this approach as a representation of time in anatomy? Some ontologies have represented the intervals of existence of individual structures by listing structures present at each stage. Anatomy ontologies for zebrafish and mouse have used this approach, see Figure 11.4b. (ZFIN; http://zfin.org; the Edinburgh Mouse Atlas (EMAP) anatomy ontology for the mouse embryo: http://genex.hgu.mrc.ac.uk; [5, 10, 37], http://www.gudmap.org: the Adult Mouse Anatomy Dictionary, http://www. informatics.jax.org/; [28]; see also [1]). The EMAP ontology can be viewed as a series of separate directed acyclic graphs one for each of the standard Theiler stages. This approach simplifies the use of the ontology for annotation of gene expression patterns in staged embryos. Listing structures present at each stage is the principal way that temporal relations across different developmental series are represented in anatomy ontologies. However, this approach has two limitations. First, it is difficult to integrate this system with other staging systems that rely on the presence or absence of structures at each stage. Second, the relations between the real intervals of existence are lost, making it difficult to achieve a temporal resolution better than the stage interval (Figure 11.4b). This difficulty is less severe in the zebrafish where stages are finer than in the mouse. One solution is to recognise ever finer stages. Developmental biologists refer to early, mid and late parts of a stage, but this has not yet been formalised in anatomy ontologies. Subdividing stages in this way is certainly practicable and matches the level of resolution of many anatomy ontologies.

Will this be good enough in the future? It is difficult to be sure. Certainly, one can foresee advantages in a scheme that represents temporal relationships between developmental series more finely, to provide a framework that can support temporal comparisons between annotated data sets from different sources and accommodate pointers to studies in systems biology such as the two case studies at the end of this chapter. One might also suspect that such a representation would begin to reveal unexpected constraints in the temporal relations between apparently independent developmental trajectories, perhaps leading to insights into underlying mechanisms of control. Understanding the temporal coupling of developmental series from data on variability, perhaps in different environments, may also give us insights into the role of heterochrony in the evolution of anatomical structure.

How might such a representation be achieved? A better approach than listing structures at each stage would be to project stage boundaries onto an independent temporal representation of anatomy (Figure 11.4c). Formally, events in different series can be related temporally by order, duration and placement in time. Placing events with respect to one another, for example using only interval relations, is impractical due to the familiar limitation of a rapidly expanding arithmetic progression. The only practical approach is to place events independently on a continuous reference TPR time-line. Hours post fertilisation in zebrafish and medaka are examples. Placing events on such a timeline may still require some binary relations (event x **before** event y, or interval relations) to capture obligatory ordering of events in the face of variation in actual timing between individual specimens. Perhaps the ideal TPR framework would be generated by using data from live imaging, though this is difficult in many organisms, particularly in the mouse.

Beyond the creation of a reference time-line, there are two problems of actually acquiring data about the timing of development. First, different developmental series are often represented with very different temporal resolution. Structure-based ontologies aim at broad coverage using morphological distinctions that can be applied by those who annotate data, typically in situ gene expression data. There are more such morphological distinctions in some developmental series than in others. Second, there is almost always much more information about the order of events within than between series. This is an area where ontology builders might well put useful effort.

11.5 Representing Process in Anatomy

11.5.1 Process Ontology

Processes have traditionally not been considered part of anatomy ontologies. In fact, ontologies have taken a rather static, 'dissection-based' view. Even in purely structural terms, however, it is difficult to represent anatomy without some means to articulate continuous change of state. This holds true at all levels of scale. Morphogenesis, growth, cell migration, changes in cell shape and changes in the distribution of cells of a clone within an anatomical structure are all central to anatomy. There is no reason to exclude biochemical processes. The processes of interest here are simply those in which an anatomical structure is not only a participant, but is the subject of change.

There are numerous practical benefits to representing processes in anatomy ontologies. One example is to be able to map information about intermediate states of normal and variant development (mutants and evolutionarily related forms) to otherwise discrete, named anatomical structures. For example, by mapping a continuous, morphometric description of changes in the shape and size of the limb hand-plate to limb hand plate and hand in an anatomy ontology, the shape of the hand-plate at any time during the intervals of existence of these structures can be represented. A second benefit is to identify and order sub-processes within a larger-scale process. For example, GO terms referring to processes within a named anatomical structure may be further divided into sub-processes and these represented in temporal order and perhaps with input/output relations. A third example is to relate systems biology models to processes occuring within and between anatomical structures.

It is beyond the scope of this chapter to review the literature on the representation of process in biology other than give a very brief overview of those examples relevant to anatomy. The most important 'process ontology' is the Gene Ontology (GO: www.geneontology.org) – which identifies and classifies processes and is used to annotate gene products with terms denoting process, molecular function and subcellular location. The ontological aspect of the GO is concerned with the taxonomy of processes rather than with their temporal structure. Formal schemes have been advanced that represent relationships between processes, particularly at the molecular and cellular levels. For example, Peleg et al [44] have used workflow and Petri net representations to articulate the structural, dynamic and functional interactions between named processes. In particular, they showed how high-level biological processes can be broken down into their component molecular processes interacting in a time-ordered fashion. A different approach has been taken by Cook et al [14] in building a Foundational Model of Physiology. This model deals directly with states of objects rather than with named processes and these states have cause/effect relationships. This model is closely related to anatomy ontologies through association with the FMA. The models of Peleg et al and Cook et al have similarities to the general concepts of process discussed below. A rule-based, machine-learning approach to represent molecular interactions and reactions has been developed by Calzone and colleagues and implemented in the biochemical abstract machine BIOCHAM software environment [11]. This approach uses temporal logic to reason over sets of different processes that constitute a model system (a well-studied model of the cell cycle in this study) in order to detect missing rules for molecular interaction or to generate new ones.

There are also several general representations of the topology of molecular interactions within a system. The familiar molecular 'process' diagram is an example, with arrows and other symbols to indicate different kinds of process according to their qualitative outcomes (inhibition, stimulation, translocation, etc.) [34]. Powerful computational representations of gene regulatory networks and signal networks with graphical interfaces have also been developed (a good example is BioTapestry http://labs.systemsbiology.net/bolouri/software/BioTapestry/; [8]). These representations can be expressed mathematically in dynamical models that include time as an independent variable. The Systems Biology Markup Language (SBML) provides a standard language to represent dynamical systems at the molecular level ([31]; see also [48]).

None of these approaches, however, gives a general ontology of process that fully integrates time into anatomy ontologies. Here we can briefly consider anatomical processes from an ontological viewpoint and attempt to identify the main concepts.

A process is a change of state, from start state s_s to an end state s_e. In a single instance of a process, s_s, s_e, are unique. The time taken to pass from s_s to s_e in a particular process is the interval $\Delta t = t_e - t_s$. Where there is more than one route from s_s to s_e, different routes are clearly different processes and may take different times. Thus, at the class level, process P is change from S_s via S_n to S_e where S_n is one or more intermediate states: we can conveniently denote a process as $P(S_s, S_n, S_e)$. At all levels of scale relevant to anatomy, processes can be regarded as continuous though we may choose to represent them by a series of discrete states. By 'state' we mean the material component of the system and its arrangement (including anatomical structures) and the energy component of the system. For some purposes it is useful to consider the information content of a state. In defining a state it is necessary to simplify the situation and include only those aspects relevant to the purpose of the description. Neither time nor location are part of the definition of a state, but state can be mapped to times and locations in the organism. Processes may be related by *part_of* relations in order to define sub-processes; however, the *part_of* relation between processes is distinct from that which pertains between structures and the two cannot be mixed [47]. To articulate the links between processes, we must include descriptions of input and output. One tentative definition of input is 'material, energy and information that is not part of the start state of process p and becomes part of p at t and affects the direction, rate and/or extent of p'. A tentative definition for output is 'material, energy or information that is produced by p at t where t_s **precedes** t and that is not part of the end state of p'. Based on whatever definitions of input and output are preferred, instance level relations, p **has_input** x and p **has_output** y, can be defined and used in turn to define the important class level relations:

P has_input *X* = *for all p instance of P there is some x instance of X such that x is input into p.*

P has_output *Y* = *for all p instance of P there is some y instance of Y such that y is output from p.*

Process and structure relate to time in very different ways. Anatomical structures are continuants that exist as a whole at any moment during their interval of existence; processes are 'occurrents', they 'take time' and exist as a whole only over a period of time. This has been emphasised by Smith and colleagues ([47]; see also references therein). An instance of a process – for example, the deposition of a single collagen molecule in osteiod extracellular matrix, or a single cell cycle in a particular osteoblast precursor cell – has a duration, Δt. At the class level, the duration of a process, ΔT, is the class of Δt for all instances p of process P. An example is the cell cycle time of osteoblast precursor cells. It is important to be able to represent the interval S_s to S_e (Figure 11.5) because this allows us to construct internal temporal relations within and between processes, particularly between one process and its subprocesses. This, in turn, allows us to capture the timing of events and mechanisms at the instance level as well as the stoicheiometric aspects of the flow of material, energy or information across a system. Importantly, the timing and duration of input and output can also be represented in relation to the interval S_s to S_e.

For processes, as for anatomical structures, there is another useful interval, the interval over which there is some instance of the process is occurring. As was the case for structures, the interval relating to the universal class P is not very useful. For structures, we found it useful to distinguish the class referring to a type of structure in an individual organism. For processes, we need to refer not only to the individual organism, but also to locations within an organism, that is, to named anatomical structures or locations within them. Informally, iP, the interval of occurrence of P, is the interval starting at the same time as the start time of the earliest-occurring instance p of P and ending at the same time as the end time of the latest-occurring instance p of P at the location l in a particular organism with reference to a general period under consideration (see below). Notice that iP is continuous and that this definition allows that there is some time during iP when there is no p occurring. At the class level, the interval of occurrence of P at L, IP, denotes the interval during which the process P occurs at a particular location, L. An example is the period during which the division of osteoblast precursors occurs at a particular location in the ulna. (Notice that for simplicity we gloss over the distinction between left and right ulna.) Since these events may themselves be repeated at separate times during the life cycle, it is often useful to restrict the meaning of IP to the particular period under consideration for example, to a particular cycle of bone remodelling.

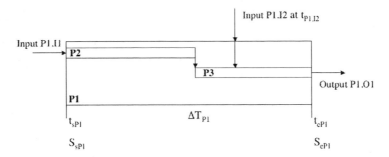

Fig. 11.5. A graphical representation of the general description of anatomical process discussed in the text. P1, P2 and P3 are processes. States and times are shown for P1: S_{sP1}, start state; t_{sP1}, start time; S_{sP1}, end state; t_{eP1}, end time; T_{P1} duration. The description of processes in terms of states and times is extensible to include any level of detail. Thus P1 can be sub-divided into sub-processes, P2 and P3. Here, P2 and P3 have a 'chaining' relationship as described in the text. Input and output are shown: for example, input to P1 is shown as input to the P3 sub-process. Inputs and outputs have times (in relation to the process or to the whole organism) and duration (not shown). This representation of process provides a 'black box' to which independent descriptions, for example, mathematical models or live imaging data can be 'mapped' or 'indexed' via states of the system. One challenge for such an extensible system is to integrate different levels of granularity and different time-scales.

Cyclical processes additionally entail the temporal concepts of frequency and change of frequency, for example, the cell cycle, the cyclic signalling mechanisms involved in somitogenesis [4, 18]. The interval of occurrence is important in mapping processes to structures in an anatomical ontology, for example, in mapping a process denoted by a GO term to a structure, and in assessing the possibility for interactions between processes on the basis of timing.

A process can be assigned a probability that the end state will occur given the initial state. Where we are considering many instances of a process repeated in parallel and where more than one end state is possible, the composition of the global end state will be a mixture according to the probabilities of the individual possible processes. For example, if we model the development of each cell in a tissue where each can either divide, remain undivided or die, the net result will be a mixture of states according to the probabilities of the processes of division and death in each cell given its initial state. There is a stochastic as well as a determinate element in developmental anatomy. The inclusion of a probabilistic component accommodates this stochastic element and, indeed, is necessary in order to represent expressivity and penetrance in anatomical development in mutants.

To gain the benefits of a process ontology in anatomy it is important to be able to relate processes to structures at particular times. Smith et al [47] proposed as a primitive relation p **has_participant** c **at** t and defined the reciprocal relation *partic-*

ipates_in in two alternative forms:

C sometimes_participates_in P = *for all c there is some t and some p such that c is an instance of C at t and p is an instance of P and p* **has_participant** *c at t.*

C always_participates_in P = *for all c, t, if c is an instance of C at t then there is some p such that p is an instance of P and p* **has_participant** *c at t.*

Smith et al also use the **has_participant** relation to define **occurring_at** and **has_agent**. For example (from [47]),

p **occurring_at** *time t* = *there exists a structure c such that process p* **has_participant** *c at t.*

Importantly, this definition relates a process not only to a time point, but necessarily to a structure existing at that time. This is an expression of the fact that process necessarily involves structures as part of the definition of a state.

The relation p **occurring_at** t is distinct from, and complementary to, the representation of the duration of a process or its interval of occurrence IP at location L. Here, interval relations or mapping to a common TPR may be used to achieve a common mapping with the intervals of existence of anatomical structures. In addition, the **has_participant** and **occurring_at** relations are complementary to any mapping of the states involved in the process (particularly the start and end states) to one or more anatomical structures. Thus, process descriptions complement the temporal relations discussed in Section 11.4. Whereas the relation D *was* C simply asserts a temporal relationship, P (C,D) denotes a process by which C becomes D. Processes can be related in this way to states within a structure, P(C,C') or to transitions between structures for example, P(C,D) or P(A and B, C) representing a fusion process. Any representation of a process, including those referred to at the beginning of Section 11.5.1, can be reciprocally linked to the relevant structures, temporal relations and times represented in anatomy ontologies.

11.5.2 Relating Serial and Parallel Processes with output_input and Temporal Relations

Processes in serial, parallel and network arrangements can be related by their temporal relations and by their output_input relations. Indeed the intervals of existence of those anatomical structures that are participants in processes clearly constrain the timing of processes. In some cases, output or input will be restricted to a small interval of time. In other cases, particularly in dynamical models, output_input relations between processes may occur, and take varying values, over the duration of the process.

The instance level relation **output_input** is a binary relation connecting elements in two processes or two elements in the same process. At the class level, we can define the *output_input* relation between processes P1 and P2 as follows:

*P1*output_input *P2* = *for all p1 instance of P1, p1* **has_output** *x AND there is some p2 instance of P2 such that p2* **has_input** *x.*

Series of processes in which the end of one is beginning of the next can be represented by a special 'chaining' relation similar to the output_input relation.

Process P1 ends_during process P2: P1 $S_e\_S_s$ P2 = for all p2 instance of P2 there is some p1 instance of P1 such that the end state of p1 is the start state of p2.

Clearly, the output_input and chaining relations might be combined with *part_of* relations (e.g. P1 *part_of output_input* P2 to indicate that only part of the output of P1 is input to P2). The *part_of* relation is underspecified here since material, energy and information are potentially involved.

One intuitive representation of *output_input* relations is as directed edges (arcs) in a 'process graph' in which the vertices represent processes. Each process within the graph may have several different inputs and outputs; the number corresponding to the degree of the vertex. Many graphs representing biological processes are directed cyclic graphs containing feedback loops such that it is possible to trace at least one cyclical path along the arcs. Output_input relations can be quantitative, for example, concentration of a signal ligand, or qualitative, for example inhibitory, stimulatory, permissive, or neutral with respect to the process receiving input. The output / input connections between processes in the models of Peleg et al [44] and Cook et al [14] are similar to the general output_input relation described here though without the temporal component.

11.5.3 The Systems Biology of Anatomy: Dynamic Stability and Change

Quantitative models have shown how complex dynamical systems can drive visible anatomical development. For example, models of segment polarity development in Drosophila [15, 38], have suggested how the topology of interactions between processes has been optimised in evolution to achieve modular design and robust emergent behaviour. Computational models of intercellular signalling have been used to explore the formation of anatomical structure. Systems of coupled differential equations that describe the dynamics of signalling can account for both spatial and temporal discontinuities in anatomy (tissue boundaries and cell fate switches respectively). These models will play an increasingly important part in developing our concepts of anatomy. As models become more sophisticated and more extensive, data annotated using standard anatomical ontologies will play an increasing role in setting up new models and fine-tuning existing ones. Ultimately, dynamical models themselves may become part of a descriptive framework for data annotation based on a consistent ontology.

Two case studies illustrate these and other points made in this chapter. In particular, they illustrate the importance of representing continuous processes in anatomy.

Saha and Schaffer [46] have modelled the formation of an anatomical boundary between MN and V3 neurons in the ventral part of the chick neural tube in response to Shh signals from the floorplate. In a directed graph of the kind shown in Figure 11.2, this transition is a vertex of degree 3: a new boundary is formed in undifferentiated neuronal precursor tissue in the neural tube, tissue on the dorsal side of the boundary becoming fated to form MN neurons and on the ventral side to form V3 neurons. In a notional trasverse section through the embryo, the process modelled by Saha and Schaffer starts as the neural tube is closing and spans the interval between Hamburger & Hamilton stages 10 and 26. In order to build their dynamical model, Saha and Schaffer expressed time in the TPR measure, hours after egg laying, using times taken from the literature. Using a system of partial differential equations to model the rate of change of the concentration of Shh signal ligand and interacting molecules in the neural epithelium as a function of distance from the floorplate, these authors were able to model the formation of the tissue boundary. This boundary was formed as a spatial discontinuity in the concentration of Gli1 protein which drives an intracellular cell-fate switch from V3 to MN phenotype. The model was assessed by comparison with results from the literature showing the time and position of the earliest markers of the prospective V3 and MN boundary (expression of Nkx2.2 and Pax6 respectively). Saha and Schaffer were also able to simulate the results of experimental manipulations of the Shh signalling pathway reported in the literature. This is a typical example of how computational modelling relies on information about actual and relative times in anatomical development. This example also illustrates how discrete anatomical structure can be formed from temporally and spatially continuous variables, as a result of the bistable behaviour of intracellular regulatory circuits. Such a complex system can only be understood by computational modelling which is able to simulate processes that are occuring on a wide range of timescales (ligand diffusion over a cell diameter over tenths of seconds, ligand binding and internalisation over minutes and gene expression and protein turnover over several hours). Two further points of interest arise from this example. First, Saha and Schaffer found that the anatomical boundary forms many hours before a steady-state ligand concentration is reached. The sensitivity of the control networks that respond to signal concentration means that boundary formation can be more rapid than was previously thought. Secondly, by studying the effects of molecules outside the core Shh signalling pathway, Saha and Schaffer were able to show that the position of the V3/MN boundary is sensitive to at least 3 additional mechanisms that interact with the core pathway. These are centered respectively on the activities of the genes Hip, and Dis and on heparan sulphate proteoglycans in the extracellular matrix. Like the core pathway, each of these modules can be regarded as a sub-process with its own interval of occurrence and kinetics within the intricate dynamics of the overall process. As Saha and Scaffer point out, these modules are candidates for modifiers in evolution. It may be through understanding the dynamics of such process on a developmental timescale that we can approach the proximal causes of evolutionary differences in anatomical form.

The dynamical behaviour of complex processes is also important for the stability of the adult anatomy. Martin and Buckland-Wright [40] have modelled the process of bone deposition during remodelling in human cancellous bone. Bone remodelling occurs in a highly organised 'basic multicellular unit (BMU)' (about 1mm long and 03mm wide) containing osteoclasts and osteoblasts and its own vascular and nerve supplies and connective tissue. The BMU tracks across the trabecular surface at about 10-25 μm/day, excavating a trench by the action of osteoclasts which is then filled by the activity of osteoblasts, first by collagen deposition to form osteoid then mineralisation to complete the formation of new bone. The lifespan of a BMU is 6-9 months; at any particular location the remodelling process takes about 200 days and the intervals between successive remodelling events at any one location has been estimated at 2-5 years; see Manogalas [39] for a review. Martin and Buckland-Wright modelled the role of osteoblasts at a fixed location traversed by the BMU. They took as a start state for the overall process 5 cells in the osteoblast lineage and the substate materials for osteoid formation and mineralisation; the end state comprised mineralised bone and differentiated osteocytes and lining cells. Martin and Buckland-Wright decomposed the overall process into three sub-processes linked by output/input: the proliferation of precursor cells, osteoid deposition, and mineralisation. The authors included in their model the reduction in osteoblast activity as a result of the known decrease in size of the osteoblasts and their apoptosis or differentiation to osteocytes. There is known to be a lag period (about 15 days) before new osteoid is mature enough for mineralisation and this was also modelled, not as a time interval, but as a requirement for the formation of a given thickness of new osteoid (16 μm, a figure obtained, like many of the other values in this model from published histomorphometric studies). Martin and Buckland-Wright validated their model against data from the literature on the time-course of osteoid apposition and mineralisation (depth of osteoid versus time in days measured from the origination of modelling). Notice that at all levels, this model and the one of Saha and Schaffer require a detailed knowledge of the intimate relationship between time and anatomical structure. Using the model to explore the sensitivity of the overall process to changes in the value of different variables, Martin and Buckland-Wright showed that the amount of bone formed at a microsite is more sensitive to factors that affect precursor cell proliferation and the number of mature osteoblasts than to factors that determine cell activity. This result clearly has significance for understanding the anatomy of healthy bone and distinguishing those factors that may be important in preventing or treating bone disease. In the present context, the example clearly illustrates the challenges that modelling will bring to anatomy ontologies. As in the example, wide-ranging timescales are involved, from hours (cell cycle and cell activity) through days (the buildup of osteoblast precurser cells), months (the lifetime of a mature osteoblast about 3 months) and years (interval between modelling events). There is also an enormous difference in spatial scale between the BMU and the whole skeleton: an estimated 1 million BMU are active at any moment in a healthy adult human. Equally, these examples illustrate how modelling relies on published data. As large data-sets become more common and as the number of models increases, the quality of temporal annotation of raw data will become increasingly important.

11.6 Summing up

In this Chapter, I have attempted to highlight the importance of time in anatomy and explore basic ideas about how time can be represented in anatomy ontologies. In particular, I have tried to show how an interval representation of time follows naturally from the concepts of anatomical structure and developmental stages in anatomy ontologies. This view of anatomy allows us to represent development as a directed graph and apply purely temporal relations between discrete anatomical structures. I have attempted to show some of the limitations of this discrete view and to explore ways to incorporate a continuous view of time into anatomy ontologies, both from the viewpoint of a continuous, time-post-reference time scale for locating events and from the viewpoint of an ontology to represent processes. There is a great deal of advanced work in other areas that can be applied to discrete and continuous representations of time in anatomy, particularly work on temporal reasoning in artificial intelligence. Readers are referred to the work of Fages, Calzone and colleagues [21] and of Bittner [7] as well as to older work on reasoning about action and temporal constraints [16, 12]. De Beule [6] has described a stimulating dynamic approach to building ontologies of time. However, we are still in a phase of basic ontology development as far as anatomy is concerned. The big challenges for the immediate future are likely to come, first from a demand for increasingly refined temporal analysis of large data sets, particularly gene expression data and second from the need to represent cell lineage data. These and other challenges will need more effort (and funding) simply to populate and maintain anatomy 'ontologies' as well as a more powerful representations of anatomy in time.

Acknowledgements

I would like to thank Albert Burger for helpful discussions.

References

1. S. Aitken. Formalising concepts of species, sex and developmental stage in anatomical ontologies. *Bioinformatics*, 21:2773–2779, 2005.
2. J.F. Allen and G. Ferguson. Actions and events in interval temporal logic. *J. Logic and Computation*, 4(5):513–579, 1994.

3. C.G. Arques, R. Doohan, J. Sharpe, and M. Torres. Cell tracing reveals a dorsoventral lineage restriction plane in the mouse limb bud mesenchyme. *Development*, 134:3713–3722, 2007.

4. A. Aulehla and B. Herrmann. Segmentation in vertebrates: clock and gradient finally joined. *Genes and Dev.*, 18:2060–2067, 2004.

5. J.B.L. Bard, M.H. Kaufman, C. Dubreuil, R.M. Brune, A. Burger, R.A. Baldock, and D.R. Davidson. An internet-accessible database of mouse developmental anatomy based on a systematic nomenclature. *Mechanisms of Development*, 74:111–120, 1998.

6. J. De Beule. Creating temporal categories for an ontology of time. In R.Verbrugge, N. Taatgen, and L. Schomaker, editors, *BNAIC-04*, pages 107–114, 2004.

7. T. Bittner. Approximate qualitative temporal reasoning. *Annals Math. Artificial. Intelligence*, 36:39–80, 2002.

8. H. Bolouri and E.H. Davidson. Modeling dna sequence-based cis-regulatory gene networks. *Develop. Biol.*, 246:2–13, 2002.

9. M. Brochhausen. The derives_from relation in biomedical ontologies. In *MIE 2006 Studies in Health Technology and Informatics*, volume 124, 2006.

10. A. Burger, D. Davidson, and R. Baldock. Formalization of mouse embryo anatomy. *Bioinformatics*, 20:259–267, 2004.

11. L. Calzone, N. Chabrier-Rivier, F. Fages, L. Gentils, and S. Soliman. Machine learning bio-molecular interactions from temporal logic properties. In G. Plotkin, editor, *Proceedings of Computational Methods in Systems Biology (CMSB)*, 2005.

12. L. Chittaro and A. Montanari. Temporal representation and reasoning in artificial intelligence: issues and approaches. *Baltzer Journals*, July 2 2002.

13. R.L. Chow and R.A.Lang. Early eye development in vertebrates. *Annu. Rev. Cell Dev. Biol.*, 17:255–296, 2001.

14. D.L. Cook, J.L.V. Mejino, and C. Rosse. Evolution of a foundational model of phisiology: symbolic representation for functional bioinformatics. In M. Fieschi et al, editor, *MEDINFO*, Amseterdam, 2004. IOS Press.

15. G. Von Dassow, E. Meir, E.M. Munro, and G.M. Odell. The segment polarity network is a robust developmental module. *Nature*, 406:188–192, 2000.

16. R. Dechter, I. Meiri, and J. Pearl. Temporal constraint networks. *Artificial Intelligence*, 49:61–95, 1991.

17. G.R. Dressler. The cellular basis of kidney development. *Annu. Rev. Cell Dev. Biol.*, 22:509–529, 2006.

18. J. Dubrulle and O. Pourquié. Coupling segmentation to axis formation. *Development*, 131:5783–5793, 2004.

19. P. Ducy, T.Schinke, and G. Karsenty. The osteoblast: a sophisticated fibroblast under central surveillance. *Science*, 289:1501–1504, 2000.

20. P.L. Williams et al. *Gray's Anatomy*. Churchill Livingstone, London, 38th edition, 1995.

21. F. Fages. From syntax to semantics in systems biology towards automated reasoning tools. In C. Priami et al, editor, *Trans. On Comput. Syst. Biol. IV*, number 3939 in LNBI, pages 68–70, 2006.

22. J.B. Gross, J. Hanken, E. Oglesby, and N. Marsh-Armstrong. Use of a ROSA26:GFP transgenic line for long-term xenopus fate-mapping studies. *J. Anat.*, 209:401–413, 2006.

23. P. Haas and D. Gilmour. .chemokine signaling mediates self-organizing tissue migration in the zebrafish lateral line. *Developmental Cell*, 10:673–680, 2006.

24. A-K. Hadjantonakis and V.E. Papiaoannou. Dynamic in vivo imaging and cell tracking using a fluorescent protein fusion in mice. *BMC Biotechnology*, 4(33), 2004.

25. M.A. Haendel, F. Neuhaus, D.S. Osumi-Sutherland, P.M. Mabee, J.L.V. Mejino Jr., C.J. Mungal, and B.Smith. Caro - the common anatomy reference ontology. In A. Burger, D. Davidson, and R. Baldock, editors, *Anatomy Ontologies for Bioinformatics: Principles and Practice*, New York, In press. Springer.

26. V. Hamburger and H.L. Hamilton. A series of normal stages in the development of the chick embryo. *J. Morph.*, 88:49–92, 1951.

27. R.G. Harrison. Harrison stages and description of the normal development of the spotted salamander, ambysoma punctatum (linn). In S. Willens, editor, *Organisation of the embryo*, pages 44–66. Yale University Press, 1969.

28. T.F. Hayamizu, M. Mangan, J.P. Corradi, J.A. Kadin, and M. Ringwald. The adult mouse anatomical dictionary: a tool for annotating and integrating data. *Genome Biol.*, 6(3):R29, 2005.

29. Y. Hirose, Z.M. Varga, H. Kondoh, and M. Furutani-Seiki. Single cell lineage and regionalisation of cell populations during medaka neurulation. *Development*, 131:2553–2563, 2004.

30. J.R. Hobbs and J. Pustejovsky. Annotating and reasoning about time and events. In *Proceedings of AAAI Spring Symposium on Logical Formalizations of Commonsense Reasoning*, March 2003.

31. M. Hucka, A. Finney, and H.M. Sauro et al. The systems biology markup language (SBML): a medium for representation and exchange of biochemical network models. *Bioinformatics*, 19:524–531, 2003. See also http://www.sbml.org/ for more recent updates.

32. T. Iwamatsu. Stages of normal development in the medaka orysias latipes. *Zool. Sci.*, 11:825–839, 1994.

33. C.B. Kimmel, W.W. Ballard, S.R. Kimmel, B. Ullmann, and T.F. Schilling. Stages of embryonic development of the zebrafish. *Dev. Dyn.*, 203:253–310, 1995.

34. H. Kitano, A. Funahashi, Y.Matsuoka, and K. Oda. Using process diagrams for the graphical representation of biological networks. *Nature Biotechnology*, 23:961–966, 2005.

35. W. Kritz and L. Bankir. A standard nomenclature for structures of the kidney. *Kidney Int.*, 33:1–7, 1988.

36. R.Y.N. Lee and P.W Sternberg. Building a cell and anatomy ontology of caenorhabditis elegans. *Comparartive and Functional Genomics*, 4:121–126, 2003.

37. M.H. Little, J. Brennan, K. Georgas, J.A. Davies, D.R. Davidson, R.A. Baldock, A. Beverdam, J.F. Bertram, and B. Capel. A high-resolution anatomical ontology of the developing murine genitourinary tract. *Gene Expression Patterns*, 7:680–699, 2007.

38. W. Ma, L. Lai, Q. Ouyang, and C.Tang. Robustness and modular design of the drosophila segment polarity network. *Molecular Systems Biology*, 2:70, 2006.

39. S.C. Manolagas. Birth and death of bone cells: basic regulatory mechanisms and implications for the pathogenesis and treatment of osteoporosis. *Endocrine Rev.*, 21:115–137, 2000.

40. M.J. Martin and J.C Buckland-Wright. A novel mathematical model identifies potential factors regulating bone apposition. *Calcif. Tissue Int.*, 77:250–260, 2005.

41. J.L.V. Mejino, A.V. Agoncillo, K.L. Rickard, and C. Rosse. Representing complexity in part-whole relationships within the foundational model of anatomy. In *Proc AMIA Symp*, pages 450–454, 2003.

42. P.D. Nieuwkoop and J. Faber. *Normal Table of Xenopus laevis*. 3rd edition, 1994.

43. Y.J. Passamaneck, A.DiGregorio, V.E. Papaioannou, and A-K. Hadjantonakis. Live imaging of fluorescent proteins in chordate embryos: From ascidians to mice. *Microscopy Research and Technique*, 69:160–167, 2006.

44. M. Peleg, I. Yeh, and R.B. Altman. Modelling biological processes using workflow and petri net models. *Bioinformatics*, 18:825–837, 2002.
45. C. Rosse and J.V.L. Mejino. A reference ontology for biomedical informatics: the foundational model of anatomy. *J Biomed Inform.*, 36:478–500, 2003.
46. K. Saha and D.V. Schaffer. Signal dynamics in sonic hedgehog tissue patterning. *Development*, 133:889–900, 2006.
47. B. Smith, W. Ceusters, B. Klagges, J. Köhler, A. Kumar, J. Lomax, C. Mungall, F. Neuhaus, A. Rector, and C. Rosse. Relations in biomedical ontologies. *Genome Biology*, 6(5):r46, 2005.
48. L. Strömbäck and P. Lambrix. Prepresentations of molecular pathways: an evaluation of SBML, PSI MI and BioPAX. *Bioinformatics*, 21:4401–4407, 2005.
49. S.L. Teitelbaum and F.P Ross. Genetic regulation of osteoclast development and function. *Nature Reviews Genetics*, 4:638–649, 2003.
50. K. Theiler. *The House Mouse: Atlas of Embryonic Development.* Springer-Verlag, New York, 1989.
51. P. Tomancak, A. Beaton, R. Weiszmann, E. Kwan, S.Q. Shu, and S.E. Lewis. Systematic determination of patterns of gene expression during drosophila embryogenesis. *Genome Biology*, 3(12), 2002.
52. M. Vilain, H. Kautz, and P.van Beek. Constraint propagation algorithms for temporal reasoning: a revised report. In *Readings about qualitative reasoniung about physical systems.*, 1989. cs.rochester.edu.
53. N. Wanek, K. Muneoka, G. Holler-dinsmore, R. Burton, and S.V. Bryant. A staging system for mouse limb development. *J. Exp. Zool.*, 249:41–49, 1989.
54. C.J. Wong and R.A. Liversage. Limb developmental stages of the newt notophthalmus viridescens. *Int. J. Dev. Biol.*, 49:375–389, 2005.

The Edinburgh Mouse Atlas

Richard Baldock and Duncan Davidson

Summary. The Edinburgh Mouse Atlas Project (EMAP) Anatomy Ontology is a hierarchically organised list of histologically distinguishable tissues visible at each Theiler stage of development. The ontology is held in the EMAP Anatomy Database freely available from http://genex.hgu.mrc.ac.uk/. The ontology was developed to be both a standard reference for describing normal and mutant tissue anatomy, and a mechanism to allow textual descriptions of gene expression patterns submitted to the Edinburgh Mouse Atlas of Gene Expression (EMAGE) database. It has also been adopted by the Mouse Genome Informatics (MGI) Group for use in their GXD gene expression database. The ontology uses 'part-of' relationships and is based primarily on anatomical structure rather than function. Presentation of the ontology as a hierarchy (tree) for each developmental stage displays the structural relationships between the anatomical entities within each stage as well as during the process of development. This part-of hierarchy defines how the ontology is used in the annotation of gene expression patterns, specifically how logic relating to annotated regions is propagated up and down the tree.

The anatomy ontology is an integral part of the EMAP Mouse Atlas. The Atlas also includes three-dimensional models of mouse embryos, one or more for each developmental stage. Selected anatomical terms are represented by domains in the corresponding model. These domains link anatomical concepts with space in the embryo and thus give a structural definition to the corresponding terms in the ontology. Current developments include the provision of additional parent anatomical terms that can be envisioned as standing above the basic tree to provide alternative groupings of the underlying tissues. The ontology is also being expanded to include tissue derivation relationships. The anatomy hierarchy for each stage represents instances of stage-independent concepts. Future versions of the ontology will provide this stage-independent view of the entire mouse anatomy.

12.1 Introduction

In 1992 Baldock *et al* [2] outlined a proposal to build 3D reconstructions of the mouse embryo to provide a spatial framework for mapping spatially organised data such as *in situ* gene-expression patterns. This culminated in a collaborative gene-expression database project to develop a database [16] for capturing spatially and text annotated patterns of expression. The text annotation used a controlled vocabulary of anatomical terms organised as a hierarchy to reflect the transition from collective large scale tissue groupings through to small discrete structures. This hierarchy was implemented using the relationship "part-of". This collaboration between the MRC Human Genetics Unit, Edinburgh University and the Jackson Laboratories, USA, is managed under the Edinburgh Mouse Atlas Project (EMAP) at the MRC Human Genetics Unit in Edinburgh and the Gene-Expression Database (GXD) component of the Mouse Genome Informatics (MGI) database at the Jackson Laboratory, Bar Harbor, USA. Additionally, at the MRC, we have used the spatio-temporal framework [9] provided by the atlas to develop a database of spatially mapped gene-expression patterns [3, 7]. This is the Edinburgh Atlas of Gene-Expression (EMAGE) and provides a graphical query and analysis capability.

The original development of the anatomy of embryonic tissues was as a controlled vocabulary. Since then the relationships included have been formalised [4, 5] and we now refer to this as the EMAP anatomy *ontology*. Future development will extend the simple stage-based treatment of temporal sequence to the interval-algebra based approach proposed by Davidson (see chapter 11 which will admit more reasoning capability and a more accurate linking of temporal events). Futher to this we plan to align the relationships within the ontology to match the proposed standards [17] for biomedical ontologies so that maximal interoperability with other anatomy ontologies can be achieved. In this chapter we describe how the EMAP ontology has been developed and deployed in the context of the GXD and EMAGE.

12.2 Mouse Embryo Development

The mouse embryo develops from the fertilised egg to the new-born pub in about 18-20 days. The single cell of the fertilised egg divides and sub-divides to form a ball of cells which then begin to organise and differentiate to form the embryo and extra-embryonic material such as the amnion, umbilical tube and part of the placenta. This process is an orchestration of cellular events such as growth (including cell division), movement and cell death (apoptosis). Cells respond to their environment by detecting levels and gradients in morphogens and by direct cell-cell signalling. The biochemical processes are complex and include gene-activation and control. Especially in the early embryo it is clear that the spatial organisation and location of cells is critical to their behaviour and response and the original purpose of EMAP was to provide a common spatio-temporal framework to enable the collation, query and analysis of gene-expression *in situ* patterns.

In this context an anatomy of development was required to:

1. provide a "simple" mechanism to annotate gene-expression patterns, particularly for published data and

2. provide a bridge between a purely iconic representation of embryo development, namely reconstructed 3D models of embryo histology and the textual descriptions e.g. in published papers.

For this purpose a simple controlled vocabulary in the form of a "part-of" hierarchy or tree was developed with development represented as a series of anatomy trees a increasing complexity and detail.

During development genes are active (expressed) at different times and places and if the data is to be compared clearly it is necessary to define the developmental state. In most cases this is simply recorded as the elapsed time since presumed conception. This has a number of problems. The first is that the moment of egg fertilisation can not be easily established and even with controlled mating in which the male has access to the female for only a short period the error can be significant in terms of the rate of development - hours can be critical. Secondly not all embryos will develop at precisely the same rate with natural and strain variation providing a degree of heterochrony. Because of these issues "staging" systems have been developed which define the developmental state in terms of specific development events. This allows a more accurate comparison of embryo data. A number of staging systems have been developed for the mouse embryo. The most commonly used because it spans the entire development from fertilised egg through to new-born mouse-pup is that defined by Karl Theiler [19]. This provides a low-temporal resolution, i.e. to about half a day, but reliable series of developmental events which can be used to stage embryos. More refined staging systems are available for particular periods of development, e.g. cell-count, somite-count and the Downes and Davies [10] staging for early embryo development but only Theiler provide a set of stages for the entire embryonic period. We therefore have adopted Theiler as the primary staging framework and define the anatomy ontology in the context of these stages. Figure 12.1 shows a series of diagrams representing the development of the mouse embryo with each figure selected to be typical of each successive Theiler stage [12] . Figure 12.2 shows the approximate relationship between Theiler stages and other staging systems including estimated elapsed time since conception.

In terms of the ontology therefore, development is represented as a series of stages. Each stage is bounded by developmental events, e.g. eye closure, so that in principle an embryo can be examined and reliably assigned to a given stage. The features selected are typically those visible from the outside and the link between these stages and more detailed internal anatomy is not defined. Kaufman does give more detail and this has been used to develop the stage anatomy trees, attempting to identify the stage range for every tissue component. Needless to say this will always be approximate. Furthermore each stage is actually a time and development interval therefore tissues will differentiate within the bounds of a single stage. This means that a given stage tree will actually include tissues that are not present at the same time in a given individual. The level of detail to represent time within stage has not been included except in the rather ad hoc way of annotating some tissues as "early

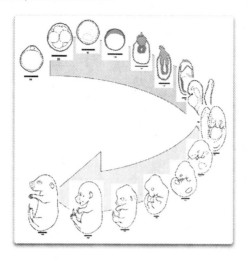

Fig. 12.1. Developmental stages of the mouse embryo as defined by Theiler - or at least a picture something like this.

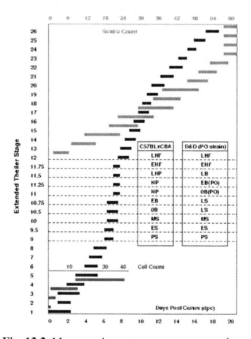

Fig. 12.2. Mouse embryo stage system comparison.

stage" or "late stage". A new representation such as the interval calculus proposed by Davidson 11 is needed. Tables 12.1 and 12.2 provide a synopsis of the Theiler staging system and its approximate relationship with other staging/timing systems.

Theiler Stage	DPC (range)	Somite Number	Cell Count	(C57BLxCBA)F1 Mice	PO Mice
1	0-0.9 (0-2.5)		1	One-Cell egg	
2	1 (1-2.5)		2-4	Dividing Egg	
3	2 (1-3.5)		4-16	Morula	
4	3 (2-4)		16-40	**Blastocyst**, Inner cell mass apparent	
5	4 (3-5.5)			**Blastocyst** (zona free)	
6	4.5 (4-5.5)			**Attachment of Blastocyst**, primary endoderm covers blastocoelic surface of inner cell mass	
7	5 (4.5-6)			**Implantation and formation of egg cylinder** Ectoplacental cone appears, enlarged epiblast, primary endoderm lines mural trophectoderm	
8	6 (5-6.5)			**Differentiation of egg cylinder.** Implantation sites 2x3mm. Ectoplacental cone region invaded by maternal blood, Reichert's membrane and proamniotic cavity form	
9	6.5 (6.25-7.25)			**Pre-streak (PS)**, advanced endometrial reaction, ectoplacental cone invaded by blood, extraembryonic ectoderm, embryonic axis visible,	PS
				Early streak (ES), gastrulation starts, first evidence of mesoderm	ES
10	7 (6.5-7.75)			**Mid streak (MS)**, amniotic fold starts to form	MS
				Late streak, no bud (LSOB), exocoelom	LS
				Late streak, early bud (LSEB), allantoic bud first appears, node, amnion closing	
11	7.5 (7.25-8)			**Neural plate (NP)**, head process developing, amnion complete	OB
				Late neural plate (LNP), elongated allantoic bud	EB/LB
				Early head fold (EHF)	EHF
				Late head fold (LHF), foregut invagination	LHF
12	8 (7.5-8.75)	1-4 / 5-7		**1-4 somites**, allantois extends, 1st branchial arch, heart starts to form, foregut pocket visible, preotic sulcus (at 2-3 somite stage)	
				5-7 somites, allantois contacts chorion at the end of TS12	
				Absent 2nd arch, ¿7 somites	
13	8.5 (8-9.25)	8-12		**Turning of the embryo**, 1st branchial arch has maxillary and mandibular components, 2nd arch present	
				Absent 3rd branchial arch	

Table 12.1. Synopsis of Theiler staging characteristics including alignment with other staging systems for early stages up to the turning of the embryo. Reproduced from the Mouse Atlas web-site with permission from the Medical Research Council.

1. Days post conception, with the morning after the vaginal plug is found being designated 0.5 dpc (or E0.5). For detailed discussion see Kaufman [12] pp. 515-525.
2. The figure given refers to the number of the most caudal somite. No account is taken of somites partitioning into dermomyotomes and sclerotomes, nor of their subsequent differentiation.
3. Adapted from Theiler [19] and Kaufman (1994); detailed staging for Theiler stages 9-12 courtesy of K. Lawson [personal communication].
4. From Downes, K.M. and Davies, T. (1993). Staging of gastrulating mouse embryos by morphological landmarks in the dissecting microscope. Development, 118, 1255 - 1266.

Theiler Stage	DPC (range)	Somite Number	Cell Count	(C57BLxCBA)F1 Mice	PO Mice
14	9 (8.5-9.75)	13-20		**Formation & closure of ant. neuropore**, otic pit indented but not closed, 3rd branchial arch visible / **Absent** forelimb bud	
15	9.5 (9-10.25)	21-29		**Formation of post. neuropore**, forelimb bud, forebrain vesicle subdivides / **Absent** hindlimb bud, Rathke's pouch	
16	10 (9.5-10.75)	30-34		**Posterior neuropore closes**, Formation of hindlimb & tail buds, lens plate, Rathke's pouch; the indented nasal processes start to form / **Absent** thin & long tail	
17	10.5 (10-11.25)	35-39		**Deep lens indentation**, adv. devel. of brain tube, tail elongates and thins, umbilical hernia starts to form / **Absent** nasal pits	
18	11 (10.5-11.25)	40-44		**Closure of lens vesicle**, nasal pits, cervical somites no longer visible / **Absent** auditory hillocks, anterior footplate	
19	11.5 (11-12.25)	45-47		**Lens vesicle completely separated from the surface epithelium**. anterior, but no posterior, footplate. Auditory hillocks first visible / **Absent** retinal pigmentation and sign of fingers	
20	12 (11.5-13)	48-51		**Earliest sign of fingers** (splayed-out), posterior footplate apparent, retina pigmentation apparent, tongue well-defined, brain vesicles clear / **Absent** 5 rows of whiskers, indented anterior footplate	
21	13 (12.5-14)	52-55		**anterior footplate indented**, elbow and wrist identifiable, 5 rows of whiskers, umbilical hernia now clearly apparent / **Absent** hair follicles, fingers separate distally	
22	14 (13.5-15)	56-60		**Fingers separate distally**, only indentations between digits of the posterior footplate, long bones of limbs present, hair follicles in pectoral, pelvic and trunk regions / **Absent** open eyelids, hair follicles in cephalic region	
23	15			**Fingers & Toes separate**, hair follicles also in cephalic region but not at periphery of vibrissae, eyelids open / **Absent** nail primordia, fingers 2-5 parallel	
24	16			**Reposition of umbilical hernia**, eyelids closing, fingers 2-5 are parallel, nail primordia visible on toes / **Absent** wrinkled skin, fingers & toes joined together	
25	17			**Skin is wrinkled**, eyelids are closed,umbilical hernia is gone / **Absent** ear extending over auditory meatus, long whiskers	
26	18			**Long whiskers**, eyes barely visible through closed eyelids, ear covers auditory meatus	
27				**Newborn Mouse**	

Table 12.2. Synopsis of Theiler staging characteristics including alignment with other staging systems from limb development to birth. Reproduced from the Mouse Atlas web-site with permission from the Medical Research Council.

1. Days post conception, with the morning after the vaginal plug is found being designated 0.5 dpc (or E0.5). For detailed discussion see Kaufman [12] pp. 515-525.
2. The figure given refers to the number of the most caudal somite. No account is taken of somites partitioning into dermomyotomes and sclerotomes, nor of their subsequent differentiation.
3. Adapted from Theiler [19] and Kaufman (1994); detailed staging for Theiler stages 9-12 courtesy of K. Lawson [personal communication].
4. From Downes, K.M. and Davies, T. (1993). Staging of gastrulating mouse embryos by morphological landmarks in the dissecting microscope. Development, 118, 1255 - 1266.

12.3 The Anatomy Ontology

For the Mouse Atlas anatomy we adopted the Theiler staging system as the primary organisation of development. Within each stage the anatomical terms were first established by collating all terms from [12] and then adding as required. These terms where then organised into a tree structure using part-of relationships to define a part-of hierarchy or *partonomy*. Clearly top-level terms near the *root* of the tree[1] provide terms for larger scale structures which can be expanded tor terms representing smaller structures i.e. from lower to higher resolution. The basic desiderata were:

1. On a structural basis, include all anatomical terms to a level of detail that can be discerned with light microscopy and with standard histological preparations.
2. Define terms to cover the entire embryo.
3. For each node in the tree define a complete set of parts that are complete and non-overlapping.
4. All relationships in the tree are "part-of".
5. The organisation should be simple and intuitive.

This provides a simple ontology which can be used to infer part-hood relations and overlap between terms in the tree. The part-of relationship can be any of the structural sub-types discussed in more formal approaches (see e.g. [11]) including arbitrary collections of non-connected tissues. Basically a term can be included if it is useful and used for describing anatomy. This means that we have introduced some terms for convenience to represent particular groupings of terms to enable easier search of the ontology and as a convenience for annotation. These terms form part of the strict tree hierarchy. An example of this type of convenient term is "organ system" which provides a useful classification but is not strictly needed.

When the anatomy was first developed capturing a usable hierarchy for each stage for the purpose of annotation was the primary goal. In the first instance that was a simple part-of tree for each stage with room for synonyms and some sub-stage qualification. The set of anatomical terms plus the structural part-of links for each stage were developed first as simple indented lists then converted to an object-oriented database to manage the data-set and enable incremental editing and curation. The design of the database identified each stage dependent term as a "timed-component" and a hidden "abstract mouse" which was the global set of terms and part-hood relations. Only the timed components were made accessible for use in annotation and at this point only the timed-component unique IDs were preserved from version to version. At this point the anatomy was adopted as the basic ontology for annotating data in the Jackson GXD database of published gene-expression patterns.

This basic set of hierarchies provided a basis for annotation but needed extension in a number of ways ways. Some of these are complete, others are work in progress.

---

[1] Curiously it is common practice to refer to being closer to the *root* of a tree as *higher* in the hierarchy contrary to the normal way a tree grows.

1. The first is the simple idea that each tissue or structure component should have a unique "printname". The full name of a tissue is the set of terms from the root of the tree to the individual term and is unique. The name of the "leaf" term however may not be unique e.g. "mesenchyme" may appear in a number of places. Using the full name uniquely defines the tissue but is rather cumbersome and a more concise name would be useful in may situations for example lists of tissues or embedded in text. Print names have therefore been implemented for terms which would otherwise be ambiguous.

2. The second was to include some derivation information to enable queries to be made across stages. Derivation in this case is defined as a link between anatomical terms that captures at least part of the known developmental relationships. Much of this is of course inferred from histological observation and may represent only part of the cell-lineage story. Given the more tentative nature of the derivation assertions in the database it was implemented with a mandatory attribution to a published work. It also carried the date and authorship of any changes to the lineage links. In the context of the set of stage-trees any terms that did not change from stage to stage, i.e. had the identical full-path in the abstract mouse were automatically linked. Any lineage link between terms with different names needed to be put in manually.

3. An extension that rapidly became necessary was to provide alternative composite structures or *groups*. With the experience of using the ontology for annotation it quickly became apparent that to annotate certain experiments alternative groupings where required. This requirement arise from a few different scenarios for using the ontology in the context of annotating a gene-expression pattern:

   a) A gene-expression pattern is annotated with the anatomical term at the most appropriate resolution and it is asserted that the gene expression is detected *somewhere* within the structure but it is not implied that is is expressed in *all* of the tissue [6]. This is especially true of experiments where the detailed spatial location may be lost by the experimental procedures, for example for a microarray study. In this case the annotation does not propagate down the hierarchy i.e. to more detailed tissues.

   b) When annotating a series of terms, for example all the muscles, the editor may need a group term to make the process more efficient and accurate, i.e. if there is a muscle group term that can be used as a short-cut to electing each of the muscles in turn. In this situation the editor requires that the annotation is propagated to each of the component parts.

   In these cases the group terms are introduced that break the original exclusiveness of the terms and converts the stage-tree into a directed acyclic graph (DAG). In order that the reasoning logic for querying the database remains consistent (see formalisation discussion below) the part-of parent of a new group term is determined as the nearest common-parent of its component children.

4. It is now clear that the future development of the anatomy ontology will depend on the input of many experts for detailed aspect of development and embryonic anatomy. For this to be possible the centralized approach to curation and extension is unlikely to be successful. It will fail because the required expertise will

not be available to any one group and practically it will not be funded. A possible approach is based on community effort and to support this model we have transformed the ontology to a more supportable format based on openly available relational database technology coupled with a standard input/output format developed by the OBO consortium. This is work in progress.

With some of these changes in place the anatomy ontology has been in routine use within the EMAGE and MGI/GXD databases. The same ontology has been adopted and extended for use within a number of large-scale collaborative projects such as GUDMAP, EURExpress and EuReGene. In addition the ontology has been used as the basis for a corresponding ontology of human embryo development and used in the context of the EADHB prototype database. In particular extensions to the genito-urinary system part of the ontology have been implemented [13] by the GUDMAP [2] project. This was led by Professor Mellisa Little and involved a group of developmental biologist expert in genito-urinary development. Further extensions in the developing brain are now being implemented by Professor Luis Puelles, University of Murcia, Spain.

12.3.1 Simplified Views

The ontology captures a view of the structural anatomy of the developing embryo to a certain level of detail. This can be extended to a greater depth as usage demand but it has also proved necessary to develop a simplified set of anatomical terms for certain application areas. A good example is the use of anatomical terms for annotating microarray data. Here the depth and resolution of the ontology acted as a barrier to usage and a simplified list that could be applied within the context of microarray experiments for both mouse and human derived tissue was required. In this context a set of terms has been developed under the auspices of the Standard Ontologies for Functional Genomics Group (SOFG). This set is termed the SOFG Anatomy Entry List (SAEL), is available for use within the MIAME microarray data standard and available on-line (www.sofg.org). In this context it was discovered that there was a requirement to be able to refer to a part independently of a reference to a *stage* which may not be known. Therefore the hidden abstract mouse described earlier was made available and the unique IDs used internally for these terms where made accessible as persistent, tracked IDs for purposes of annotation and interoperability. Each of the terms of the simplified view provided by the SAEL has been associated or mapped with a specific abstract mouse term in the EMAP anatomy. This provides a simple mechanism to pass from the simplified view to the more complete anatomy provided by the full ontology.

12.3.2 Definitions

In the view of more formal definitions of an ontology each term must be properly and unambiguously defined in order for the ontology to be meaningful. This raises

---

[2] GenitoUrinary Development Molecular Anatomy Project - http://www.gudmap.org/

a real problem in practice. Anatomical terms in current use may have slightly different meanings and interpretations for different experts. A very good example of this is the fact there are a number of detailed structural descriptions of the developing and adult rat brain such as [1], [18] and [14]. In these atlases it is possible to find the same terms used in different ways and different terms used for the same structures. Furthermore recent work [15]shows that there is still significant debate over the proper interpretation of e.g. mesencephalon (midbrain). Nevertheless the anatomical terms serve as a useful mechanism for communication, annotation and navigation in the context of the mouse embryo. How then can an ontology be developed that will be fit for purpose for annotating biological data, capturing structural and derivation knowledge and yet is not so closely bound by definitions that may make it unusable by many practitioners?

One answer to this question is to not try and provide detailed and therefore contentious definitions, this however could lead to an ontology unable to serve for unambiguous communication. A second possibility, which we propose, is that the definitions of structure should be based on the embryo itself. This is the same approach taken by the rat brain atlases where the anatomical components or brain regions and structures are identified by direct marking of histological sections images appropriately stained (e.g. silver staining). In this case the definition is provided by the atlas and is completely unambiguous (to within the resolution of the images). This is the approach we have taken. In parallel with the development of the ontology we have built full 3D models of embryo histology from serial sections. By "painting" these models with regions or anatomical domains a definition is established. As in any ontology the meaning of terms is with respect to an external reference. In this case the external reference is a particular instance of a mouse embryo which can be used to confirm the part-of ontological structure but also represents a much richer set of possible relations arising from the topology and geometry of the embryo. This can in some sense be considered an extension of the ontological structure based on logical relationships to topology and geometry. This idea is developed further in chapter 10.

In this ontology we have not provided precise textual definitions of each term and in effect appeal to the common understanding of these terms amongst developmental biologists. The database structure are extended to include such definitions which could be included (or imported from other ontologies) in the future.

12.3.3 Formalisation

The EMAP anatomy ontology has served as the basis for annotation (its original purpose) in a number of databases. We have also established a formal description of the ontology in terms of predicate logic [4]. In this view we use the following terminology. A **full name** of an anatomical structure is given as an n-tuple: $(t_0, t_1, \ldots, t_n)$. The **path name** of the structure is $(t_0, t_1, \ldots, t_{n-1})$. The **component name** or **part name** is t_n. For example, given the structure name (using a file directory style notation):

$/embryo/branchial\ arch/3rd\ arch/branchial\ pouch/endoderm/dorsal$

its full name is:

$(embryo, branchial\ arch, 3rd\ arch, branchial\ pouch, endoderm, dorsal),$

its path name is:

$(embryo, branchial\ arch, 3rd\ arch, branchial\ pouch, endoderm)$

and its path name is:

$dorsal.$

It should be noted that the names defined here are not simple in the sense they only provide a unique string to identify a structure, the name actually encodes the structural hierarchy i.e. it include the part-of relationships identified by the order of the terms.

Each tissue also has a unique **identifier**, a Mouse Atlas accession number. We use predicate $tissue(X, FN)$ to state that the tissue with identifier X has FN as its full name. Predicate $hasPart(X, Y)$ represents the fact that tissue Y is part of tissue X; X and Y are unique tissue identifiers.

Let predicates $pName(FN, PN)$ and $cName(FN, CN)$ represent the fact that PN and CN are the *path name* and *component name* of the *full name FN*, respectively. The following constraints must hold for all primary anatomy trees, i.e. without the addition of groups:

1. A full name uniquely denotes a tissue, i.e. there are no two tissues with the same full name.
   $tissue(X, FNx) \land tissue(Y, FNy) \land FNx = FNy \rightarrow X = Y$
2. The full name of a node is the path name of all its immediate sub-part nodes:
   $hasPart(X, Y) \land tissue(X, FNx) \land tissue(Y, FNy) \land pName$
   $(FNy, PNy) \rightarrow FNx = PNy.$

We say tissue X is a *super-part* of tissue Z if there exists a $hasPart$ path from X to Z, e.g. $hasPart(X, Y)$ and $hasPart(Y, Z)$. X is a *sub-part* of tissue Y, if Y is a super-part of X. Formally we define $superPart(X, Z)$ and $subPart(X, Z)$ recursively as follows:

$hasPart(X, Z) \lor (hasPart(X, Y) \land superPart(Y, Z)) \rightarrow superPart(X, Z)$
$superPart(Z, X) \rightarrow subPart(X, Z)$

As previously discussed, there is a need to complement the primary anatomy hierarchies with *groups*. In general, primary and group nodes can be treated equally. Hence, the same predicates $tissue$ and $hasPart$ are used to represent groups in the ontology. However, for some of the reasoning we need to be able to distinguish them.

The predicate $primary(X)$ is true, if X is a primary node. Predicate $group(X)$ is true, if X is a group node. All nodes are either primary or group, but not both:

$primary(X) \land group(X) \rightarrow \bot$

There are a number of constraints that groups must adhere to; too many to list them all, so we will only give one example, the definition of a *minimal group*.

Assume tissue a has parts b and c, and someone wishes to create a new group tissue g which consists of b, c and d (d is not part of a). An obvious way to achieve this would be to add $hasPart(g, b)$, $hasPart(g, c)$ and $hasPart(g, d)$. However, we would like to keep the graph *minimal*, i.e. place $hasPart$ links at the highest appropriate level. In our example, instead of adding $hasPart(g, b)$ and $hasPart(g, c)$ we should add $hasPart(g, a)$. We generalise this idea into the *minimal group* constraint.

Before giving a formal definition for this constraint, the concept of *shared parts* is introduced. Predicate $sharedParts(X, Y)$ states that X and Y have at least one common part:

$$hasPart(X, Z) \ \wedge \ hasPart(Y, Z) \rightarrow sharedParts(X, Y)$$

Definition 1. *Group G is* **minimal***, if for every tissue T it shares some part with, at least one of the parts of T is not also a part of G:*

$$\forall T \cdot sharedParts(T, G) \cdot \ \exists X \cdot hasPart(T, X) \ \wedge \ \neg hasPart(G, X) \ \rightarrow minGroup(G)$$

Being able to formulate constraints such as the one above, is one of the benefits of formalising an ontology. We have also formalised aspects of the abstract mouse and tissue derivation, but omit the details here.

12.3.4 Ontology Management

Many biology ontologies exist as simple files which can be read into a suitable browser for editing and curation. For the EMAP ontology we have used an object-oriented database to manage the terms and maintain the unique IDs. The database is accessible using the standard remote object interface CORBA and the viewers are provided as Java applications. In addition there are now a number of web-browser based viewers directly in html or using Javascript. A C++ object-oriented DBMS was selected so that it was easy to build in directly the required images (models and mapped-data) and image-processing software.

Up to this point the ontology has been curated using bespoke java interfaces interacting directly with the EMAP database which impose the required logic and check for inconsistencies. The next version of the EMAP database will be based on a relational model and the editing will use a standard ontology editor such as OBO-Edit or Protégé.

The ontology is available in a number of flat-file formats. We have a bespoke dump-format based loosely on the Bibtex style with named records with

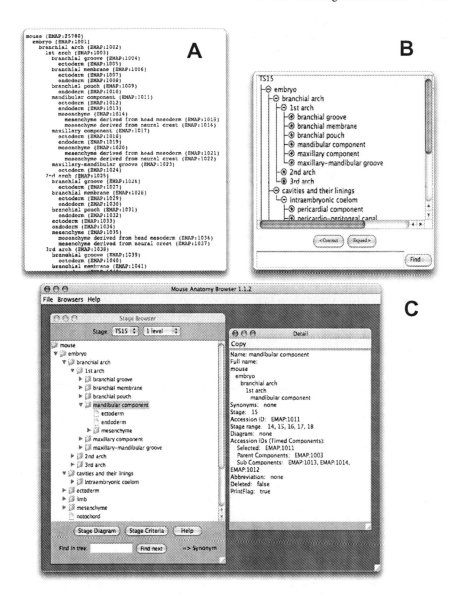

Fig. 12.3. Anatomy browsers currently available with EMAP. A: simple indented text list with EMAP IDs. The list is non-iinteractive and search is via the "find" function of the browser. B: Java applet viewer, this provides interactive expand and collapse and a search function and has been embedded into a query interface for EMAGE enabling direct search of EMAGE and other gene-expression databases. C: Jave application providing full access and query capability over the ontology. Search includes synonyms and the details page includes the IDs of all parent, child and derivation links as well as the stage range for the tissue structure.

(name,value) pairs to record details. We can also dump to plain indented text or to the basic GO format. We are currently developing I/O to the new OBO standard which will become our standard interchange format. In addition to this file format options we also provide direct access to the DB via CORBA and web-services (WSDL).

12.4 Viewers and Browsers

Within EMAP we provide three basic browsers for the anatomy ontology:

Text List: Simple indented text list which can be viewed with a standard web-browser. The list has no interactive elements and search is via the test search options available with the browser. The lists are presented per Theiler stage. Location: `http://genex.hgu.mrc.ac.uk/Databases/Anatomy/new/`

Applet: For most web-browsers the applet allows active browsing of the anatomy tree including node expansion and contraction and item selection. There is search of the anatomy terms and on selection the user can query a number of gene-expression databases including EMAGE and MGI/GXD. This applet also displays a number of fixed sections from the corresponding atlas models and if available the anatomy domains. Location: `http://genex.hgu.mrc.ac.uk/Atlas/SBFrames.html?14`

Java application: The most sophisticated access is via a download-able java application which connects directly to the database. This provides full interactive access to the anatomy and will display all details for a given node including the unique IDs, synonyms and lineage (if defined). In addition the search will allow partial/wildcard match and synonyms. The functionality of this browser is also available in the EMAGE java interface for query and display of the gene-expression submissions.

Figure 12.3 shows a screen-shot of each of these browsers in their current form. The more powerful browse capabilities provided by the applet and java application are at the cost of less portability. The applet will not work properly on all combinations of web-browser and operating system and the java application needs certain software installed (WebStart) and configuration of the users local firewall (certain "ports" need to be open).

In addition to these basic browsers other projects have delivered alternative mechanisms for the same data, for example the GUDMAP project has developed a presentation based using the browser-based technology JavaScript which solves many of the portability and firewall issues. The anatomy display has also provided an exemplar for studying how more complex data should be presented to the user. For visual analysis of complex data [8] have developed usability evaluations that show that the use of the third dimension with semantic focus provides significant benefit to the user. Figure 12.4 shows an example of these alternative methods.

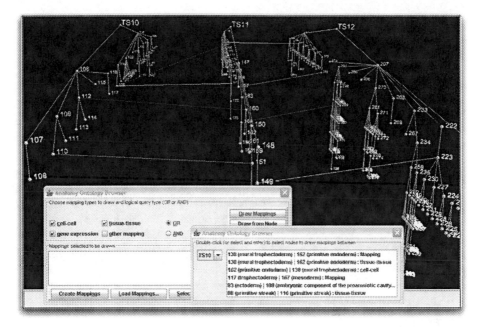

Fig. 12.4. Displaying multiple anatomy ontologies using the 3rd dimension. Here a viewer has been developed to show how using 3D visualisation can provide a view much easier to understand and follow. The application is from XSPAN (www.xspan.org) and provides a view of the association between anatomy ontologies of differant species. Links can defined and selected with additional textual detail displayed by double-clicking on any link or choosing the appropriate item from the View menu. Figure reproduced with permisssion from the PhD thesis of Aba-Sah Dadzie.

12.5 Discussion

In this chapter we have presented the EMAP anatomy ontology, how it has been used in the context of the Edinburgh Mouse Atlas and how it has been used for annotating gene-expression data. The ontology as a useful tool is now well established but to remain useful it must be able to develop to met the needs of new research. This will require extension of the ontology in a number of ways:

1. Developing the detail and depth of the ontology for example the level of detail in the brain is very sparse.
2. Gradual editing and correcting the hierarchies to meet requirements, introducing new groups is a good example.
3. Extending the relationships available within the hierarchy to include at least the class relationship "is-a".
4. Complete the definition of tissue derivation.
5. Extend the notion of time to include time intervals with the required number of time-interval relations (see chapter 11).

6. Formalise the links between the various anatomical ontologies in order to foster interoperability.

Curating and maintaining an ontology in public use can be a very demanding process. Thus far the EMAP group in Edinburgh has managed the underlying database and Jonathan Bard has acted as the ontology editor. There has been significant effort over the years developing the database itself, the user and editor interfaces and of course the definition of the ontology content. The ontology is now in wide use and managing the process in this way is unlikely to succeed. We believe the way forward is to enable more direct community input so that modifications and changes can be implemented more quickly but with appropriate safeguards so that existing use and databases (e.g. annotations) are protected, i.e. so that use of "old" terms remains valid. In addition we need to using more standard and community supported editing tools, e.g. OBO-Edit. Finally it is very important for the ontology to respond to the demands of biological research for example with extensions to support the annotation of mutant phenotype and the integration of tissue type and disease data.

References

1. Alvarez-Bolado, G. and Swanson, L. W. (1995). *Structure of the Embryonic Rat Brain.* Elsvier Science Publishers B. V., Amsterdam.
2. Baldock, R., Bard, J., Kaufman, M., and Davidson, D. (1992). A real mouse for your computer. *BioEssays*, 14:501–502.
3. Bard, J. B. L., Baldock, R. A., Kaufman, M., and Davidson, D. (1997). Graphical gene-expression database for mouse development. *European Journal of Morphology*, 35(1):32–34.
4. Burger, A., Davidson, D., and Baldock, R. (2004a). Formalization of mouse embryo anatomy. *Bioinformatics*, 20:259–267.
5. Burger, A., Davidson, D., Yang, Y., and Baldock, R. (2004b). Integrating multiple partonomic hierarchies in anatomy ontologies. *BMC Bioinformatics*, 5:184.
6. Christiansen, J. Emage: database structure and rules for the annotation and querying language.
7. Christiansen, J. H., Yang, Y., Venkataraman, S., Richardson, L., Stevenson, P., Burton, N., Baldock, R. A., and Davidson, D. R. (2006). Emage: a spatial database of gene expression patterns during mouse embryo development. *Nucl. Acids Res.*, 34:D637.
8. Dadzie, A.-S. and Burger, A. (2005). Providing visualisation support for the analysis of anatomy ontology data. *BMC Bioinformatics*, 6(1):74.
9. Davidson, D. and Baldock, R. (2001). Bioinformatics beyond sequence: Mapping gene function in the embryo. *Nature Reviews Genetics*, 2:409–418.
10. Downes, K. and Davies, T. (1993). Staging of gastrulating mouse embryos by morphological landmarks in the dissecting microscope. *Development*, 118:1255–1266.
11. Gerstl, P. and Pribbenow, S. (1995). Midwinters, end games, and body parts: a classification of part-whole relations. *Int. J. Hum.-Comput. Stud.*, 43(5-6):865–889.
12. Kaufman, M. (1994). *The Atlas of Mouse Development.* Academic Press., London.
13. Little, M., Brennan, J., Georgas, K., Davies, J., Davidson, D., Baldock, R., Beverdam, A., Bertram, J., Capel, B., Chiu, H., Clements, D., Cullen-McEwen, L., Fleming, J., Gilbert, T., Houghton, D., Kaufman, M., Kleymenova, E., Koopman, P., Lewis, A., McMahon, A.,

Mendelsohn, C., Mitchell, E., Rumballe, B., Sweeney, D., Valerius, M., Yamada, G., Yang, Y., and Yu, J. (2007). A high-resolution anatomical ontology of the developing murine genitourinary tract. *Gene Expr Patterns*.

14. Paxinos, G. and Franklin, K. B. J. (2001). *The Mouse Brain in Stereotaxic Coordinates*. Academic Press, San Diego, 2nd edition.

15. Puelles, L. and Rubenstein, J. L. R. (2003). Forebrain gene expression domains and the evolving prosomeric model. *Trends Neurosci*, 26(9):469–476.

16. Ringwald, M., Baldock, R., Bard, J., Kaufman, M., Eppig, J., Richardson, J., Nadeau, J., and Davidson, D. (1994). A database for mouse development. *Science*, 265:2033–2034.

17. Smith, B., Ceusters, W., andJacob Köhler, B. K., Kumar, A., Lomax, J., Mungall, C., Neuhaus, F., Rector, A. L., and Rosse, C. (2005). Relations in biomedical ontologies. *Genome Biology*, 6(R46).

18. Swanson, L. (1998). *Brain Maps: Structure of the Rat Brain 2nd Edition*. Elsevier Science Publishers B.V., Amsterdam.

19. Theiler, K. (1989). *The House Mouse*. Springer-Verlag, New York.

13

The Smart Atlas: Spatial and Semantic Strategies for Multiscale Integration of Brain Data

Maryann E. Martone, Ilya Zaslavsky, Amarnath Gupta, Asif Memon, Joshua Tran, Willy Wong, Lisa Fong, Stephen D. Larson, and Mark H. Ellisman

Summary. This chapter focuses the application of brain cartography to the problem of multi-scale integration of brain data in the context of the Biomedical Informatics Research Network (BIRN) project. The BIRN project focuses on creating a grid infrastructure for integrating data on brain morphology and function obtained by different researchers to support comprehensive understanding of the mechanisms and developing treatment for schizophrenia, Alzheimer's, Parkinson's and other dementias. One of the project goals is to create an online environment where brain data produced by different groups across multiple techniques can be integrated, accessed and queried. In this chapter, we describe the use of geographical information system technology to create a spatial database of the brain to which diverse data, primarily but not restricted to imaging data, is registered and queried. We discuss the role of terminological ontologies in the Smart Atlas for multiscale queries and for overcoming some of the limitations of purely spatial integration.

Neuroscience from its inception has been an interdisciplinary science, drawing upon a multitude of technologies and approaches to tackle the complexity of the nervous system. The interdisciplinary nature of neuroscience is certainly one of its strengths, but it also presents a serious weakness when trying to build informatics systems to make sense of all of the data. The unique difficulties in dealing with neuroscience data were one of the main motivations behind the conception of the Human Brain Project [26], an attempt to create tools and databases for integrating neuroscience data. With its complex and voluminous data types, and lack of a simple unifying framework, this task has proven difficult. Nevertheless, progress has been made, if not in creating fully functional systems, at least in understanding at a deeper level the issues involved [21].

Despite the plethora of in vivo and in vitro preparations that characterize the neuroscience experimental arsenal, most preparations are at some level referenced back to nervous system anatomy, whether it be gross or cellular. It is thus not surprising, then, that most early neuroinformatics efforts looked to neuroanatomy to provide the framework for uniting neuroscience data. The brain is perhaps the most structurally and molecularly complex tissue in the body, and the means to describe this structure has been under study for centuries. Brain atlases have provided the principal means for identifying brain regions and for localizing signals. These atlases typically consist of a collection of 2D plates or images on which major brain structures have been identified. Brain atlases ideally come with a coordinate system to provide a standard reference across members of a species, that may either be based on internal, e.g., Talaraich coordinates, or external features, e.g., stereotaxic coordinates.

With attempts to map human and animal brains nearly as old as the mapping of the earth, brain cartography is certainly not a new field. What has matured, however, is the technologies available for creating these atlases and the types of systems that are equipped for managing, viewing and interrogating cartographic data. As new imaging technologies and digital representations have become available, many atlas projects moved away from a sampling of 2D histological sections to full 3D methods for revealing brain anatomy. Such methods include magnetic resonance imaging (MRI: [25]) and block face serial sectioning [9] [18]. The trade-off for this third dimension is usually seen in the granularity of delineations, with most of the purely 3D atlases failing to achieve the level of annotation of the more sparsely sampled atlases. The lack of granularity stems both from the relatively poor resolution and more limited set of contrast protocols for MRI, and from the amount of time required for detailed annotations of higher resolution data. Successful atlas efforts have generally required dedicated manpower for several years to produce high quality delineations for even sparsely sampled 2D data.

The transition to digital atlases has provided new opportunities for expanding the scope of the brain atlas beyond a static reference system. Various computational tools have been developed to allow users to add data to an atlas through template or coordinate-based registration, essentially turning the atlas into a database for brain data and a computational tool for comparing both structural and functional features [7] [16]. Much of this data is in the form of 2D and 3D images, showing the topography of cell distributions, gene expression patterns or physiological signals. To date, the majority of electronic brain atlases are developed as standalone databases for one or two types of brain data, e.g., MRI or Nissl stains or gene mapping techniques, with desktop or web application interfaces [18] [16]. Relatively little emphasis has been given to how data from different groups taken across scales can be integrated to support development of a comprehensive picture of brain morphology and function. One can speculate on the reasons for this, but some of the limitations can be traced directly to the variety of representational forms and techniques used in the development of brain images, the variety of resolutions and scales of imaging and the limitations of converting 3D brain volumes into 2D representations.

As part of the Biomedical Informatics Research Network (BIRN; http://nbirn.net) project, we have been working to created infrastructure for building multiscale views of mouse models of neurological disease. This work involves the characterization of genetically modified mouse across multiple scales, using diverse techniques such as light and electron microscopy, MRI-based imaging and microarray analysis. The multiscale mouse project was inaugurated to serve as one of the biological drivers for the construction of collaborative infrastructure for biomedical science built upon high performance networking, shared data and computational resources, and the means by which information contained in diverse and distributed resources can be linked together [11].

Like other neuroinformatics projects, the mouse BIRN looks to neuroanatomy to provide the framework for linking information taken at different scales with often incompatible techniques. The goal of this project is to provide the means to query and retrieve different types of data based on its location in the brain and ultimately to provide the means to make the relationships among these diverse data machine processable. In the following, we describe our efforts in creating and employing cartographic approaches to integration of diverse brain data, and describe the necessary interplay between spatial and terminological referencing that will be necessary to achieve this goal.

13.1 Brain Cartography: Atlas to Brain GIS

The Smart Atlas project was begun several years ago to bring the infrastructure and approaches developed for geographical information to the realm of neuroinformatics [20]. The Smart Atlas ("Spatial Mark Up and Rendering Tool") was one of the first projects to employ Geographical Information Systems (GIS) for query and storage of multiscale brain data referenced to the coordinate system of the atlas. GIS are computerized systems specialized for storage, query and visualization of spatially distributed information and are now standard for many geographical applications. At the time this project was begun, electronic atlases were still in their infancy, and quite a lot of effort was directed towards turning what was essentially a digitized paper atlas into a spatially aware GIS tool.

To create the Smart Atlas, we utilized a commercial brain atlas of the mouse brain Paxinos and Franklin, 2000 [22] as its interface. The Paxinos and Franklin atlas was chosen both because it is a standard reference for mouse brain anatomy, and because the 2000 edition included a set of computerized vector drawings of each plate. Thus, it was one of the first of the published brain atlases to provide a suitable substrate for GIS. However, the drawings provided with the Paxinos and Franklin atlas were created for humans and not machines. Significant challenges had to be overcome to turn these drawings into a topologically correct database that would support spatial queries (See [31], for details).

To import the plates into the Smart Atlas required first aligning the plates to the coordinate system, reconciling the two sets of plates that were provided for the coronal and sagittal planes of section. Second, the lines drawn to delineate brain regions in the atlas needed to be grouped into a set of closed polygons enclosing a defined

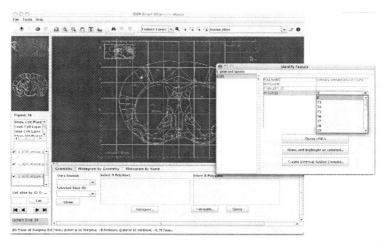

Fig. 13.1. Smart Atlas interface: a GIS system created from a commercial brain atlas. The current slice is shown in the main window. Users can identify brain structures and search for them across slices. Navigation is provided for both coronal and sagittal slices (upper left panel). The Smart Atlas is written in Java and launched via a Java Webstart application.

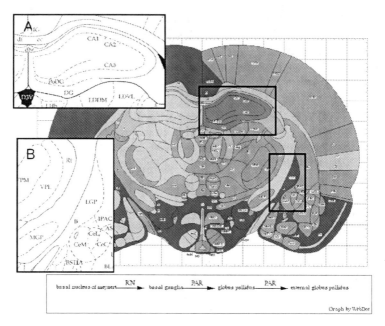

Fig. 13.2. Results of the first pass in the automatic polygon delineation procedure for a single section from the Paxinos and Franklin (2000) brain atlas. The boxed regions are shown enlarged to the right (note that orientation is flipped right to left). The different shades reflect the number of anatomic feature labels within a closed polygon. The corresponding terms and relations from UMLS are shown in the bottom panel that were used to define the relationships between multiple labels found in one polygon (B and LGP in the lower right enlargement).

brain region. This latter task was made more challenging by the fact that many regions defined in the atlas have incomplete or missing boundaries (Fig. 13.2). Many regions have multiple labels assigned, displaced labels or no label at all (Fig. 13.2) . The original version of the Smart Atlas [20] required a human neuroanatomist (MM) to painstakingly select each line segment in the atlas and assign it to a polygon with a label. Obviously, this approach was extremely time consuming and rather boring.

To provide an automated solution for defining polygons from the vector drawings, at the same time reconciling polygon labeling conflicts, Zaslavsky et al (2004) [31] followed planar enforcement algorithms used in GIS for constructing topologically correct polygon coverages, and enhanced them by utilizing the anatomical information contained within the Unified Medical Language System (UMLS), a large metathesaurus of biomedical terms [14]. The UMLS contains as one of its source vocabularies the Neuronames [4], an anatomical nomenclature resource for primate brain anatomy. In the simplest cases, the algorithm extracted the set of lines that surrounded a single label on the brain atlas and mapped that polygon to both the brain abbreviation and the UMLS concept identifier, if one was available. If two or more labels were contained within a single polygon, the algorithm searched for the concepts in the UMLS and looked at the relationships among them (Fig 13.2A and B). For example, a polygon was defined that enclosed labels for CA1, CA2, CA3, DG and PoDG (Fig. 13.2A). All of these terms could be mapped to a common parent in UMLS, the hippocampus, and so the polygon was assigned the name hippocampus.

The Paxinos and Franklin atlas employs a convention whereby fiber tracts (white matter) are labeled on one side of the brain in small letters and nuclei (gray matter) are labeled on the opposite side in capital letters. This convention led to a significant number of polygons not having a label on one side of the brain or the other. The algorithm identified polygons in the same relative location and assigned the appropriate labels to both sides. In the case where a structure was labeled with both a white matter label and a gray matter label, e.g., the medial forebrain bundle (mfb) and the lateral hypothalamus (LH), we assigned the name of the gray matter structure as the primary label and the white matter structure as a secondary, which was assumed to be passing through the gray matter region.

The automated approach was successful in defining approximately 80% of the polygons and associating them with correct or appropriate labels. In the remaining 20%, either no consistent region label could be assigned, or the correct polygon could not be defined. In the former case, many of these errors arose because the Paxinos and Franklin (2000) atlas does not close many polygons for areas in which there are no clearly defined boundaries. In the example shown in Fig. 13.2B, for example, the ventral extent of the LGP extends into the basal forebrain and cerebral peduncle. Some of these large polygons extend through many brain regions with disparate labels that could not be related using the UMLS. In the case where polygons were not defined properly, generally a close inspection of the vector drawings revealed that one or more lines were not closed given the tolerances specified for the polygon coverage. As of the writing of this chapter, elimination of these errors involves manual intervention.

13.2 Spatial Query of Diverse Data through the Smart Atlas

Once the polygons and labels were defined for the atlas plates, each was stored in a spatial database (Oracle 9i spatial). By utilizing a spatial database, we can issue queries not only on text strings, e.g., find all slices containing the mfb, but also perform spatial operations on both atlas features and the brain image data that are referenced to them. For example, users may utilize standard GIS spatial filters such as "contained in" or "overlaps with" to query the interaction of a geometrical probe with underlying brain structures. In the example shown in Fig. 13.3, utilizing the sphere probe, a user draws a circle on one atlas plane and the Smart Atlas calculates which structures overlap with a 3D sphere across the atlas planes. The planes containing overlapping structures are highlighted in green in the navigation window. This function is useful to query the spread of tracer injections or functional signals arising from a given brain region.

Current data in the Smart Atlas ranges from individual slices generated from MRI to high resolution 3D reconstructions of subcellular structures such as dendrites and neuropil using electron tomography to microarray data from defined brain locations (Fig. 13.3). Each type of data is displayed in a separate layer within the Smart Atlas interface. The current Smart Atlas has four data layers representing the different types of data currently stored: 1) image layer; 2) cell layer; 3) electron microscopy layer; 4) microarray layer. The image layer is somewhat of a misnomer, as data types 1-3 all involve images of some sort. However, it is used to refer to images of sufficient scope that they can be spatially aligned with multiple atlas features (Fig. 13.3). These images are then shown in the context of the atlas, with the atlas delineations superimposed over the image. Large images are stored on an image server so that they can be efficiently accessed over the internet. In contrast, the cell and EM layers are at much higher resolution than the coarse anatomy shown in the atlas plates. The feature layers for these data types provide a symbol representing the location at which the data were acquired (Fig. 13.3). The actual datasets are stored in the Cell Centered Database (CCDB; http://ccdb.ucsd.edu/), an on-line resource for high resolution light and electron microscopy data. Clicking on the symbol takes the user to the dataset in the CCDB.

The current interface of the Smart Atlas allows a user to browse through brain slices and retrieve data that are registered to a particular slice or region within a slice. Users may select data by category: image, filled cell or electron microscopic dataset and display all data of a particular type found at a location. For image data, the Smart Atlas will return a list of images that have been registered to a particular location. Selecting the image from the dialog box will place the image in the context of the atlas.

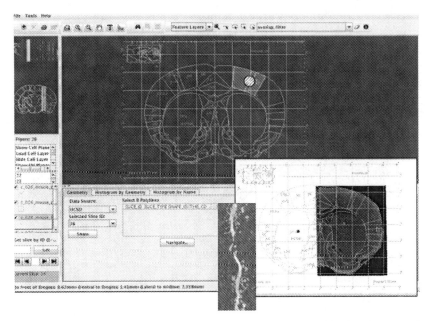

Fig. 13.3. Spatial query through the Smart Atlas. In the upper panel, the user defines a geometric probe (hatched sphere) and selects the overlap filter. The Smart Atlas calculates which structures overlap with the probe on that plane (shaded polygons) and highlights the overlapped structures on additional planes (highlighted in yellow in navigation window in the upper left). Users may also request to see what data is registered to this location (lower right panel). In this case, data registered to anywhere in the slice was requested. The location of electron tomographic dataset is indicated by a dot on the left side of the atlas plane and a brain section stained for alpha synuclein is returned on the right. Clicking on the dot retrieves the tomographic dataset (small inset) from the Cell Centered Database (see text).

The Smart Atlas has been developing a set of functions to allow users to query the content of images based on the location and type of signal contained in a particular image. For example, many of the images registered to the Smart Atlas reveal the distribution of proteins or other molecular targets through application of special staining techniques. The Smart Atlas has implemented a signal query function that allows a user to define an arbitrary polygon on an atlas plate and return a set of images that contain a signal in that area (Fig. 13.4). Users may set conditions on the search such as the amount of labeling as a percentage of area, the intensity of signal (high, medium or low), the type of signal e.g., protein or the specific molecular target, e.g. alpha synuclein. The search function relies on a unique data type developed using the Oracle 9i spatial data cartridge [30] that permits efficient calculation of the percentage of labeling in a given location. By storing an abstraction of the signal contained in an image, the Smart Atlas overcomes the difficulties of querying images that use very different contrast mechanisms and look up tables (e.g., Fig. 13.4).

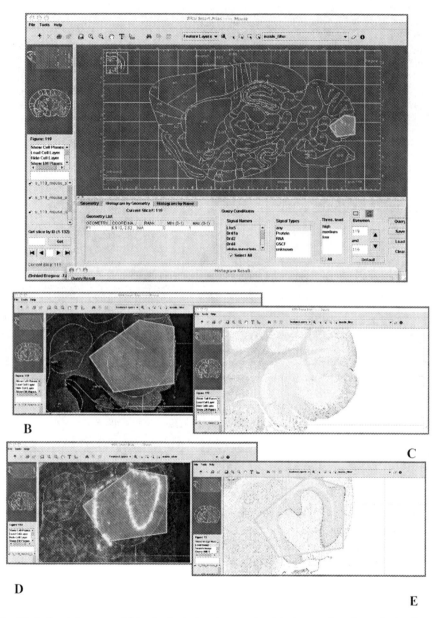

Fig. 13.4. Spatial query of histological signals contained in image data through the Smart Atlas. A) User begins by drawing any geometrical probe at a particular location, in this case cerebellum. All images containing signal in that area are returned. For this query, 4 different images showing the expression pattern of the transcription factor Lhx5 are shown B) immunofluroescent localization of Lhx5 protein; C) Image from Gensat database registered to the Smart Atlas showing cells expressing the Lhx5 promoter; D) Image from Gensat database showing radioactive in situ hybridization to Lhx5 mRNA; E) Image from Allen Brain Atlas showing non-radioactive in situ hybridization to Lhx5 mRNA.

The power of spatial tools such as the Smart Atlas is the ability to provide a means of information retrieval independent of the level of annotation or the precise anatomical terminology employed. Because of the cellular and molecular heterogeneity of the brain compared to other organs, efforts to provide manual annotation of labeling patterns for images that span significant amounts of brain tissue or contain multiple scales of labeling are bound to be inadequate. The same is true of large scale gene mapping efforts such as the Gensat project [10] [13] and Allen Brain Atlas [16]; with thousands of genes and images involved, the level of annotation by trained individuals is still likely to be sparse compared to the richness of the data. When the data are spatially registered, they inherit the anatomical delineations of the atlas, on the one hand, allowing query for labeling within anatomical structures to the level of granularity of the atlas. On the other hand, spatial registration allows users to transcend the anatomical nomenclature to retrieve and compare signals based solely on the properties of the signal [7]. This function is useful, because while there may be disagreement among neuroscientists about the identity of a brain area giving rise to a signal, its location in terms of spatial coordinates is at least quantifiable.

Of course, the ability to integrate and compare data according to its spatial location depends on the accuracy with which we can register images to the coordinate system of the atlas. Many groups are developing algorithms for spatial normalization [1], but currently, we recognize that spatial normalization is far from perfect. Indeed, because of individual variation, distortions introduced during processing of histological sections and oblique planes of section compared to the atlas, we will likely never achieve perfect registration.

Currently, the Smart Atlas interface provides the users with some utilities that can compensate for disparities in registration. For example, users may set a range of slices over which to run a query to compensate for spatial uncertainty along the anterior-posterior or medial-lateral axes of the atlas. Users may also adjust the size of the probe to account for likely difficulties in correctly aligning image features with the atlas. For example, when looking for signals associated with a small structure like the anterior commissure, the probe size can be set larger than the target structure to ensure that neighboring areas are sampled. For large structures such as the caudate nucleus, the probe size can be set smaller and sample the middle of the structure so that only images with label in the caudate nucleus are returned. The purpose of spatial registration is to bring images in close enough alignment that a spatial query has a reasonable chance of returning a result. However, we admit that with the current level of registration, a human needs to intervene and interpret the signals once they are returned. The challenge for the future will be to provide metrics of uncertainty that are more amenable to algorithmic analysis.

13.3 Neuroanatomical Nomenclature: Location, Location, Location

The Smart Atlas was constructed as a tool for determining spatial relationships among diverse signals generated by different methods of probing the brain. These signals may be gene expression patterns from in situ hybridization, microarray or immunocytochemistry, histological staining patterns, activation patterns, lesions, pathological features, representations of cellular or tissue archiecture, and essentially run the gamut of the types of data that are produced in neuroscience. For this purpose, the important components of the atlas are its spatial features: the coordinate system along with the set of landmarks, i.e., the anatomical delineations, that can be used to register data to the coordinate system. The terminology employed as labels for these delineations provides one set of annotations that can be used to interpret signals arising from a particular area. However, terminology is also important as scientists do not communicate with each other in terms of coordinates, rather, they use terms like "hippocampus", "cerebral cortex". Names link signals to greater scientific literature.

The Smart Atlas utilizes the terminology that comes from the Paxinos and Franklin (2000) brain atlas [22]. Much as with the atlas delineations themselves, the terminology applied to this atlas was not designed for ease of machine access. The terms as they appear in the atlas are not tied to an explicit hierarchy. For example, the terms "cerebellum" and "cerebral cortex" do not appear in the atlas as structures. These higher order, yet common, terms must be constructed from a list of their parts in the atlas. As part of the BIRN project, we have been struggling with the question of how best to tie these parts together.

Because of its maturity as a discipline, there is a rich legacy of anatomical nomenclature, much of it based purely on descriptions of macroscopic features. Most of the structures of the brain were named before any function could be ascribed to them. The nomenclature of regional brain anatomy is generally quite useful for describing location in the brain. A statement by a neuroscientist that "we cut sections at the level of anterior hippocampus" generally evokes a fairly good idea of location in the brain, although one certainly can't define with great precision the exact boundaries of the "anterior hippocampus" from this description. But our interest in brain anatomy has never been purely structural; we assume that structural differences between brain regions also reflects functional differences as well and have gone to great lengths to subdivide the brain based on variation in cell type, number, arrangement, biochemistry, etc. Thus, the same anatomical terms that originally described gross anatomical features have come to designate functional systems. A good example is the basal ganglia, a term originally applied to all non-cortical nuclei deep in the brain [28], but which has come to mean a set of interconnected brain nuclei involved in motor planning and sequencing. Because of the conflation of structure and function, the anatomical nomenclature can be very confusing, as anatomists wrestle over the boundaries of nuclei and their functional significance. If you ask a neuroscientist to point to the amygdala in a brain slice, most will point to the gray matter mass just anterior to the hippocampal formation in the temporal lobe with no difficulty. If you ask neuroscientists to describe the functional system of the amygdala and how it should be divided, there will be much less consensus and in some cases downright hostility [29]. Thus, there is a dual nature to the anatomical nomenclature that must be recognized when designing neuroinformatics systems based on anatomy [12].

The Smart Atlas takes an agnostic view of neuroanatomy, that is, the exact labels applied to regions within the atlas are not important, so long as they correspond to an identifiable feature that can be used to localize a signal. We have adopted this agnostic view because the BIRN project is dealing with so many different techniques that have inherently different contrast mechanisms and resolution. The important thing for comparing microarray data obtained from gross dissection of the amygdala to immunocytochemical data obtained from brain imaging of the amygdala is not whether the amygdala is part of the limbic system, whether the parcellations should follow neuroanatomist X or Y or whether the stria medullaris should be considered as part of the amygdalar system. The important point of nomenclature for multimodal integration is that the microarray data can be reliably compared to the immunocyto-chemical data. Therefore, we must ensure that the two definitions of amygdala are pointing to the same region. In addition, because these two techniques have fundamentally different resolutions, we must ensure that any definition of the amygdala can accommodate different levels of granularity. An MRI image or gross dissection of the brain may not be able to resolve subdivisions or tracts passing through a structure, whereas microscopic images may be able to differentiate such structures clearly. Signals reported for the "amygdala" for low resolution techniques may thus contain multiple systems not directly related to the amygdala while microscopic data may be more precise in its localization.

As part of the BIRN project, we are creating BIRNLex, a lexicon of biological and experimental and biological entities utilized by BIRN (http://nbirn.net/birnlex/). In BIRNLex neuroanatomy, we are seeking to provide a set of standard definitions for high level brain regions such as cerebral cortex, hippocampus, etc., that can be used to reliably locate data in the brain. This lexicon is built upon the Neuronames hierarchy [4], which takes a strictly structural point of view regarding part of relationships. Structures are characterized as predominantly white or gray and they are localized within the part of brain where they reside in the adult. This volumetric hierarchy is critical for the type of multimodal and multiscale data contained in BIRN, because it provides a set of possible structures from which a signal in brain can arise. For example, because the deep cerebellar nuclei are volumetrically contained within cerebellar white matter, if an MRI experiment localizes a signal to the cerebellar white matter, then we know that the deep cerebellar nuclei may be involved in producing this signal, even if we can't resolve these structures on the MRI.

The combination of a spatial tool like the Smart Atlas with a well structured volumetric hierarchy of brain parts should, in theory, allow practitioners of different techniques to be fairly precise about the location of their data. Consistent spatial and terminological definitions for brain regions should make it easier to interpret signals arising from a given region, regardless of the resolution of the technique.

13.4 Space Limitations: Interpretation of Multimodal and Multiscale Data

The *a priori* assumption of spatial normalization of data is that co-localization of signals implies some sort of relationship between the signals, e.g., two genes that are found in the same metabolic pathway, or some relationship between the signals and a feature that was not visualized, e.g., two proteins that are found in dendritic spines. Thus, integration across different types of data occurs to the extent that we can perform feature matching and specify the spatial relationships among data. This assumption has proven to be very powerful for those analyzing signals that come from a single modality and scale (e.g., Fox and Lancaster, 2005 [7]. However, close examination of the types of heterogeneous data in the BIRN project illustrate some significant limitations in the spatial approach when trying to integrate data that comes from very different techniques and scales of resolution. This problem is particularly acute for the brain, because of the nature of the nerve cell with its extensive and wide ranging projections.

Consider the example shown in Fig. 13.4. In this figure, a spatial query to the Smart Atlas for signals relating to gene expression in the cerebellum returned four images. All four images relate to the same gene, the transcription factor Lhx5. However, all four images come from different techniques: immunocytochemistry (protein; CCDB), promoter driven expression of GFP (Gensat; [10] [13]), radioactive in situ hybridization (mRNA; Gensat) and non-radioactive in situ hybridization (mRNA; Allen Brain Atlas [16]). A user who understands the cellular anatomy of the cerebellum will note that two signals (protein and promoter-GFP) are localized to the cerebellar molecular layer while two signals (in situ hybridization) are localized to the Purkinje cell layer. However, the protein signal is found in the dendrites of Purkinje cells, as higher magnification examination reveals (not shown) while the promoter-GFP signal is found in cerebellar stellate cells (not shown). Despite the different pattern of labeling, a neuroanatomist would conclude that the protein and mRNA signals are consistent, because mRNA tends to be localized in the cell soma in neurons while protein can be localized throughout the cell.

This example illustrates two truths well known to neuroanatomists: there is no single magic technique for localizing the true distribution of a gene; each technique has its pitfalls and limitations. "Ground truth", if there is such a thing in neuroanatomy, is arrived at by consensus among multiple techniques, not the application of a single technique. Second, this example illustrates that additional knowledge is required to interpret the significance of labeling patterns. In this example the knowledge required is both technical and biological; that mRNA is localized by in situ hybridization and is primarily located in the cell soma in neurons and protein is localized by imunocytochemistry and may be found anywhere in the nerve cell. Because processes can extend long distances from the cell soma, mRNA and protein may have very disparate localizations.

Limits with purely spatial integration are also encountered when trying to integrate data across anatomical scales. Figure 13.5 shows a sampling of images from the rodent cerebellum taken at different spatial scales. While all of these images can be registered to the same location in the Smart Atlas, the relationship among these images is not clear through spatial localization alone. Although we are sampling the same tissue in each case, each image represents only a part of cerebellum and the relationships among the parts depicted in each scene are not obvious without additional knowledge. The disconnect arises in part because of the large number of distinct cell parts packed into a single region, but also because of the different techniques used to reveal aspects of cellular anatomy.

Despite major advances in specimen preparation techniques and imaging methods, we still do not have a single technique that is capable of resolving all cellular and subcellular constituents present in a tissue across a significant expanse of tissue. With emerging techniques in electron microscopy, e.g. cryoelectron tomography, we can resolve macromolecular structures within a portion of a single cell [17] and subcellular structures across a single cell [19]. With super-resolution techniques in light microscopy, we can detect single molecules across a population of cells or tissue, but we can only detect a small number of molecules at a single time because of the need to introduce fluorescent tags into the molecules. Thus, to build multiscale views of the nervous system, or any tissue for that matter, requires that we be able to combine techniques across scales. Although methods for correlated microscopy, that is, imaging the same specimen using light and electron microscopy, are available [8], these studies are laborious and in many cases the specimen preparation techniques required by different techniques are incompatible.

In summary, while we can assign images obtained by different methods to the same locations in space, considerable human expertise is needed to draw relationships among these data and the exact relationship of an image to the structures contained in a location may be difficult to discern by spatial co-localization alone.

13.5 The SAO: A Cell-centered Ontology for Multiscale Anatomy

As the preceding discussion illustrates, interpretation of image data and integration of diverse data requires additional semantic layers through which image content must be filtered. Some of the semantic knowledge is biological, e.g., connectivity information between brain regions; the relationship between mRNA and protein; some is methodological, e.g., the nature of signals from in situ hybridization vs immunocytochemistry. From its inception, the Smart Atlas has looked to ontologies to provide

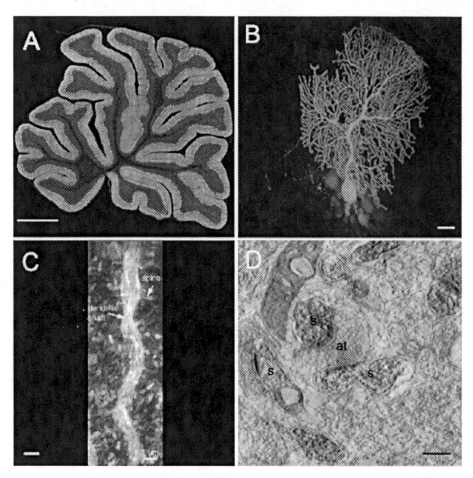

Fig. 13.5. Multiscale views of the cerebellum. A) Large scale mosaic of cerebellar cortex imaged with triple label multiphoton microscopy [23]; B) Single Purkinje neuron from the cerebellar cortex injected with Lucifer Yellow and imaged using confocal microscopy; C) Tomographic reconstruction of a single dendrite from an intracellularly injected Purkinje neuron prepared for electron microscopy; D) Single computed slice of a tomographic volume through the cerebellar molecular layer neuropil. Spine (s) are labeled with F-actin; at = axon terminal. All images are available through the Cell Centered Database (http://ccdb.ucsd.edu/). Scale bars: A = 1 mm; B = 15 μm; C = 1 μm; D = 0.25 μm.

some of this required knowledge. Several mature terminological resources for neuroanatomy, e.g., BAMS[3] and Neuronames[4], have appeared in recent years. Both of these resources are dedicated to the very real and significant problem of reconciling different nomenclatures developed over the centuries for neuroanatomy. The BAMS divides the brain into gray and white matter, and defines terms as a "part of" hierarchy[3]. The lowest level of granularity of the BAMS is the cell, although the bulk of terminology covers gross anatomy. NeuroNames also identifies brain structures as predominantly white or predominantly gray and defines a volumetric hierarchy but does not cover cellular structures. Other ontology/terminology resources for neuroanatomy include the neural portion of the Foundational Model of Anatomy [24].

All of the above mentioned resources take a top down view of anatomy, dividing the brain up according to fundamental developmental divisions and defining the major white and gray brain regions contained within them. This same approach is inherent in the Foundational Model of Anatomy, where a high level parcellation of the body is given and the scales of resolution are traversed down to the cell and some cell parts. The brain, however, differs significantly from other parts of the body in that its cells are not confined to a given location. Instead, as discussed above, neurons possess many long processes that traverse the anatomical boundaries that we set from gross anatomy. As far as we know, this situation does not obtain on such a large scale in any other organ. If we were to take a tiny micro-scalpel capable of cutting cleanly between cells, we could drill into heart, liver or lung and clearly dissect out a defined regional part of the tissue, e.g., the left ventricle, in which the cells were wholly self-contained. In the brain, if we dissected out one of the classic regional parts, e.g., the caudate nucleus, we would be severing parts of neurons that are projecting to the caudate, those that are projecting from it and those that are passing through. When ontologies for gross anatomy claim that a neuron is "part of" a brain region, it usually means that the cell soma is localized to that region.

The top down approach creates several problems when trying to reconcile different views of the nervous system. First, a signal coming from a brain region may not derive from a cell that has been classified as part of that brain region, e.g., a signal coming from the caudate nucleus may arise from the axons of neurons in the cortex projecting to or passing through the caudate nucleus. Second, many anatomically identified brain regions, e.g., the amygdala, contain many different brain systems intermixed [29], some of which extend into other regional parts of the brain [5].

Without a model of the cell in a neuroanatomical ontology, the process of traversing scales between the gross anatomical delineations given by atlases like the Smart Atlas, and the detailed subcellular information contained in most imaging studies of the nervous system, becomes more challenging. Signals don't arise from brain regions; they arise from cells or cellular processes and the extent of these signals will reflect the extent of these processes. Macroscopic anatomy was defined largely based on relative amounts of cells and unmyelinated processes (gray matter) vs myelinated axons (white matter). Most histological stains used to parcellate the brain reveal only parts of the cell, either the cell soma or the myelinated processes. Thus gross anatomical delineations are generally not drawn based on the extent of cellular processes, but based on differences in cell soma size, packing density or density of myelinated axons.

In order to create an ontology that will encompass the various anatomical scales under study in Mouse BIRN, we have been building an ontology for neuroanatomy that starts with a model of the cell, rather than a model of gross brain anatomy. The Ontology for Subcellular Anatomy of the Nervous System (SAO) describes the parts of neurons and glia and how these parts come together to define supracellular structures such as synapses and neuropil [6]. Molecular specializations of each compartment and cell type are identified.

The ontology is constructed using the Basic Formal Ontology (BFO) as a foundation [27]. Cells are divided into regional parts and component parts, similar to the way the Foundational Model of Anatomy divides anatomical structures. Fig. 13.6 shows the parcellation of a neuron into its regional parts: cell body, dendrite, axon and spine. Each of the parts of a cell can be further divided into regional parts and component parts, e.g., a spine can be divided into head and neck. Each part of the cell is connected to its parent part through the relationships "continuous with". Thus a dendrite is continuous with the cell somata; the dendritic spine is continuous with the dendritic shaft.

We developed the SAO with the goal of providing a bridge between more coarse anatomical scales, represented by the delineations in the Smart Atlas, and macromolecular distributions revealed by high resolution mapping techniques such as immunocytochemistry. Because the SAO is built on a model of the cell, both molecular constituents and anatomical location are assigned to the subparts of cells, rather than to the cell itself (Fig. 13.6). The SAO utilizes the "located in" relationship to situate cellular parts into higher order brain regions. These higher order brain regions can be drawn from standard nomenclatures like BAMs; however, they can also be a set of coordinates from the Smart Atlas. The relationship "located in" rather than "part of" was deliberately chosen to reflect the view of the Smart Atlas that anatomical regions represent landmarks by which neurons and their processes can be located, rather than a set of functional brain regions.

The SAO was constructed using OWL (Web Ontology Language), a first order description logic that supports reasoning. We have constructed the SAO in a way that allows inferencing to be performed across scales so that molecules and higher order connectivity may be inferred from local interactions. Ideally, the observation that F-actin is located in the head of a dendritic spine from a Purkinje cell dendrite (Fig. 13.5) found in the molecular layer of the cerebellar cortex should allow for the following statements to be inferred:

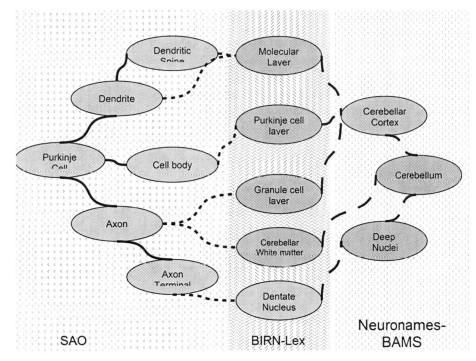

Fig. 13.6. Representation of a Purkinje cell in the SAO and the location of the different parts of the SAO with respect to histological (middle panel) and gross (right panel) divisions of the cerebellum. Histological divisions are currently delineated in BIRNLex while gross anatomical terms come from Neuronames or BAMS.

- *Purkinje cells express f-actin*
- *The cerebellar molecular layer expresses f-actin*
- *The cerebellar cortex expresses f-actin*
- *The cerebellum expresses f-actin*

Similarly, from the observation that the Purkinje cell spine receives a synaptic contact from a bouton from a parallel fiber, we can infer statements like:

- *There exists an axon to which the bouton belongs*
- *There exists a cell to which the axon belongs*
- *If the axon is a parallel fiber, then the cell type is a cerebellar granule cell*

And from these statements, we should be able to traverse scales to conclude that:

- *Granule cell axons project to Purkinje cell dendrites*
- *Granule cells project to Purkinje cells*
- *The cerebellar granule cell layer projects to the cerebellar molecular layer*

We have just begun to utilize the SAO in combination with rules to demonstrate exactly this type of inferencing [15]. Thus, through the integration of multiscale ontologies, with the SAO as the foundation, we can begin to bridge across the multiple scales at which we record observations. Perhaps more importantly, as the SAO is populated with reference to brain regions defined by the Smart Atlas, we will begin to build up a sense of where parts of neurons are located within the brain, e.g., cerebellar molecular layer *has part* Purkinje cell dendrite, Purkinje cell spine, granule cell axon. The SAO will allow the subcellular anatomy to inform interpretations of signals arising in spatially registered data. The SAO is currently being deployed in some of our annotation tools for electron tomographic data so that this type of population can occur [6].

13.6 Conclusions

The Smart Atlas project is combining both spatial and terminological approaches for integrating and interpreting diverse signals within the brain. The anatomical delineations provided in the atlas provide a "top-down" view of neuroanatomy, by parcellating the brain into regions that can be recognized in MRI or Nissl stained preparations. Underlying these regions is a subcellular ontology which creates a model of the nerve cell with its extended processes, in order to build a "bottom up" view by mapping the different pieces of individual neurons across multiple brain regions. The SAO provides a finer grained substrate in which to localize molecular constituents, yet one which supports reasoning across different resolutions. Our next challenge is to develop or deploy ontologies that can interpret the different techniques employed and so reconcile views obtained with different techniques. For this, we are looking to the Ontology for Biomedical Investigations (http://obi.sourceforge.net/).

While integration across the different types of data registered to the Smart Atlas still requires an informed human to interpret returned results, we are hoping that with the development of the SAO and additional formal ontologies for experimental techniques, tools like the Smart Atlas at the least will provide naïve users with the necessary knowledge to interpret data themselves. In the best case, the Smart Atlas will be able to perform these human tasks automatically through tools like the BIRN mediator, a tool for semantic integration of distributed data that can utilize knowledge contained in ontologies to answer queries across distributed data [2].

The current Smart Atlas is built upon a commercial brain atlas which limits its availability only to BIRN participants. However, the functionality and strategies for heterogeneous data integration developed through the Smart Atlas is being engineered into a freely available tool, MBAT, that uses a 3D MRI-based atlas as the interface and is available for download at the BIRN website (http://nbirn.net/downloads/mbat/index.shtm).

Acknowledgments

Supported by NIH grants NIDA DA016602 (CCDB), NCRR RR04050 and RR08605. The Bioinformatics Research Network is supported by NIH grants RR08605-08S1 (BIRN-CC) and RR021760 (Mouse BIRN).

References

1. Ardekani BA, Guckemus S, Bachman A, Hoptman MJ, Wojtaszek M, Nierenberg J. (2005) Quantitative comparison of algorithms for inter-subject registration of 3D volumetric brain MRI scans. J Neurosci Methods. 142: 67-76.
2. Astakhov, V., Gupta, A., Santini, A., Grethe, J. S.: Data Integration in the Biomedical Informatics Research Network (BIRN). DILS 2005: 317-320, 2005
3. Bota M, Dong HW, Swanson LW. Brain architecture management system. Neuroinformatics. 2005;3(1):15-48.
4. Bowden DM, Dubach MF (2003) NeuroNames 2002. Neuroinformatics. 1(1):43-59.
5. de Olmos JS, Heimer L. The concepts of the ventral striatopallidal system and extended amygdala. Ann N Y Acad Sci. 1999 Jun 29;877:1-32.
6. Fong, L. L, Larson, S. D., Gupta, A., Condit, C., Bug, W., Chen, L., West, R., Lamont, S., Terada, M. and Martone, M. E. An ontology-driven knowledge environment for subcellular neuroanatomy. OWL: Experiences and Directions workshop, submitted.
7. Fox PT, Laird AR, Lancaster JL.(2005) Coordinate-based voxel-wise meta-analysis: dividends of spatial normalization. Report of a virtual workshop. Hum Brain Mapp. 25: 1-5.
8. Giepmans, B. N., Adams, S. R., Ellisman, M. H. and Tsien, R. Y. (2006) The fluorescent toolbox for assessing protein location and function. Science, 312: 217-224.
9. Gustafson C, Bug WJ, Nissanov J. (2007) NeuroTerrain–a client-server system for browsing 3D biomedical image data sets. BMC Bioinformatics. 8:40.
10. Gong S, Zheng C, Doughty ML, Losos K, Didkovsky N, Schambra UB, Nowak NJ, Joyner A, Leblanc G, Hatten ME, Heintz N. (2003) A gene expression atlas of the central nervous system based on bacterial artificial chromosomes. Nature. 425: 917-25.
11. Grethe, J. S., Baru, C., Gupta, A., James, M., Ludaescher, B., Martone, M. E., Papadopoulos, P. M., Peltier, S. T., Rajasekar, A., Santini, S., Zaslavsky, I. N. and Ellisman, M. H. (2005) Biomedical informatics research network: building a national collaboratory to hasten the derivation of new understanding and treatment of disease. Stud Health Technol Inform, 112: 100-109.
12. Hayamizu TF, Mangan M, Corradi JP, Kadin JA, Ringwald M. (2005) The Adult Mouse Anatomical Dictionary: a tool for annotating and integrating data. Genome Biol. 6: R29.
13. Heintz N. (2004) Gene expression nervous system atlas (GENSAT). Nat Neurosci. 7:483. PMID: 15114362 [PubMed - indexed for MEDLINE]
14. Humphreys BL, Lindberg DA, Schoolman HM, Barnett GO. The Unified Medical Language System: an informatics research collaboration. J Am Med Inform Assoc. 1998 Jan-Feb;5(1):1-11.
15. Larson SD, Martone ME, Rule-Based Reasoning With A Multi-Scale Neuroanatomical Ontology. OWL: Experiences and Directions workshop, submitted
16. Lein ES, Hawrylycz MJ, Ao N, Ayres M et al. (2007) Genome-wide atlas of gene expression in the adult mouse brain. Nature 445: 168-176.
17. Lucic, V., Forster, F. and Baumeister, W. (2005) Structural studies by electron tomography: from cells to molecules. Annu Rev Biochem, 74: 833-865

18. MacKenzie-Graham A, Lee EF, Dinov ID, Bota M, Shattuck DW, Ruffins S, Yuan H, Konstantinidis F, Pitiot A, Ding Y, Hu G, Jacobs RE, Toga AW. (2004) A multimodal, multidimensional atlas of the C57BL/6J mouse brain. J Anat. 204: 93-102.

19. Marsh, B. J., Volkmann, N., McIntosh, J. R. and Howell, K. E. (2004) Direct continuities between cisternae at different levels of the Golgi complex in glucose-stimulated mouse islet beta cells. Proc Natl Acad Sci U S A, 101: 5565-5570.

20. Martone M. E., Gupta A., Ludscher B., Zaslavsky I. and Ellisman M. H. (2002b) Federation of brain data using knowledge guided mediation, in Neuroscience databases: a practical guide, R. Kotter, Knopf, New York, pp. 275-92.

21. Martone ME, Gupta A, Ellisman MH. (2004) E-neuroscience: challenges and triumphs in integrating distributed data from molecules to brains. Nat Neurosci. 7: 467-72.

22. Paxinos, G. and Franklin, K. (2000) The Mouse Brain in Stereotaxic Coordinates. San Diego, Academic Press.

23. Price DL, Chow SK, Maclean NA, Hakozaki H, Peltier S, Martone ME, Ellisman MH. (2006) High-resolution large-scale mosaic imaging using multiphoton microscopy to characterize transgenic mouse models of human neurological disorders. Neuroinformatics. 2006 Winter;4(1):65-80.

24. Rosse C, Mejino JL Jr. A reference ontology for biomedical informatics: the Foundational Model of Anatomy. J Biomed Inform. 2003 Dec;36(6):478-500.

25. Sharief AA, Johnson GA. (2006) Enhanced T2 contrast for MR histology of the mouse brain. Magn Reson Med. 56: 717-25.

26. Shepherd GM, Mirsky JS, Healy MD, Singer MS, Skoufos E, Hines MS, Nadkarni PM, Miller PL (1998) The Human Brain Project: neuroinformatics tools for integrating, searching and modeling multidisciplinary neuroscience data. Trends Neurosci. 21: 460-8.

27. Smith B and Rosse C. The role of foundational relations in the alignment of biomedical ontologies. Medinfo, 2004 11(Pt 1):444-8.

28. Swanson, L. W. (2002) Brain architecture: understanding the basic plan, New York: Oxford University Press.

29. Swanson LW, Petrovich GD. (1998) What is the amygdala? Trends Neurosci. 21: 323-31.

30. Wang, Y., Santini, S., and Gupta, A. (2005) Efficiently Querying Spatial Histograms. SPIE International Symposium on Electronic Imaging 2005. San Jose, California, USA. January 16-20, 2005.

31. Zaslavsky I, He H, Tran J, Martone ME, Gupta A (2004) "Integrating Brain Data Spatially: Spatial Data Infrastructure and Atlas Environment for Online Federation and Analysis of Brain Images " Biological Data Management Workshop (BIDM 2004), pp. 389-393.

Part IV

Anatomy Ontologies — Modelling Principles

14

Modelling Principles and Methodologies – Relations in Anatomical Ontologies

Fabian Neuhaus and Barry Smith

Summary. It is now increasingly accepted that many existing biological and medical ontologies can be improved by adopting tools and methods that bring a greater degree of logical and ontological rigor. In this chapter we will focus on the merits of a logically sound approach to ontologies from a methodological point of view. As we shall see, one crucial feature of a logically sound approach is that we have clear and functional definitions of the relational expressions such as '*is_a*' and '*part_of*'. While this chapter is mainly concerned with the general issues of methodology, chapter 15, on 'Spatial Representation and Reasoning', will apply the methodology to the specific case of spatial relations. Although both chapters are self-contained, we recommend that they be seen as forming a unity.

14.1 The Semantic Content of Type Terms

The reason why logical rigor is crucial for the development and use of biomedical ontologies becomes clear if we consider their purpose and mode of operation. The term 'ontology' is used very ambiguously, but in the life sciences 'ontology' means roughly: 'controlled vocabulary in computer interpretable form' and in this chapter we will restrict ourselves to this reading of the term. For a more detailed account of what ontologies are, see [2]. Since ontologies must be not just computer readable but also computer interpretable, an ontology is more than a list of terms stored in a computer parsable format; it comprises also the semantic content associated with these terms – or at least it is supposed to do so.

Before we take a closer look at how this works, we need to make some terminological distinctions. First, we shall use the term 'type' in what follows to refer to those entities in reality which terms in ontologies designate. Second, it is important that we distinguish between terms (like 'female pelvis') and '*part_of*' on the one hand, and the semantic content of such terms on the other. In biomedical ontologies there are two kinds of terms: those denoting types (to be more specific, biomedical types) and those denoting relations. Thus in an anatomy ontology the term 'female pelvis' denotes the type *Female Pelvis* and the term '*part_of*' denotes the parthood relation. (Throughout this chapter, we use italics and initial capitals when using type terms to denote types, and quotation marks when we need to talk about the terms themselves. Further we will assume that all terms denote types of the human anatomy, if not explicitly stated otherwise.) Terms are linguistic entities that are created by humans; they satisfy linguistic conventions created by humans; and they can be used to create sentences that express statements about the world. It is terms that are the bearers of semantical content, which means they have denotations – which for present purposes are types and relations.

Note that the types and relations that we are talking about are not the familiar entities that we know from set theory. In [9] we propose a distinction between types (for example the type *Ear*, which is what particular ears share in common) and the sets which are the extensions of such types (the collection of all particular ears). This distinction should be borne in mind to avoid certain sorts of confusion. The types of the biomedical domain (also sometimes called 'universals' or 'kinds') are the patterns in reality that scientists study and describe in their theories.

Sets and types behave similarly in one important respect: just as sets have *members*, so types have *instances*. However, there are important differences between the two. First: for any arbitrarily chosen group of individuals there is a corresponding set, but there need not be a corresponding type. For example, there is the set whose members are exactly: Barbara Bush, Bill Clinton's left foot, and a given red blood cell of my dog; but there is no corresponding type in reality of which exactly these entities are instances. Types are contrasted with such arbitrary collections by the fact that they can serve as objects of scientific investigation and play a role in scientific generalizations (some of which are then captured in ontologies). A second important distinction turns on the fact that the membership relation is timeless, whereas the instantiation of types is time-dependent. For example, an animal that instantiates the type *Adult Frog* now used to instantiate the type *Tadpole* at some time in the past. Hence types like *Adult Frog* and *Tadpole* gain and lose instances over time, where sets cannot gain or lose members. (If you 'add' a member to a set, then the result will be a different set.) Individuals, similarly, can gain and lose parts. However individuals, like this blood cell or that heart are distinguished from the types *Blood Cell* and *Heart* by the following criterion: At any time of its existence an individual necessarily occupies a unique spatial location; a type, in contrast, can be (through its instances) fully present at multiple locations. For example you are necessarily present at exactly one location, whereas the type *Humanity* is currently located at about six billion different locations.

With this background it is easy to formulate the problem that ontologies address. As mentioned above, an ontology is not just a list of syntactical strings like 'pelvis', 'urinary bladder', 'body' (its type terms); its goal is to comprehend also the semantical content of these terms – that is, the types which they denote – in a machine readable form. This is more problematic than one might think. If humans do not understand the meaning of a term, we can use dictionaries. For example, assume that you don't speak German and you are wondering what the term 'Handwurzelknochen' means. If you look it up you will come to the following conclusion:

1 *The German term 'Handwurzelknochen' denotes the type Carpal Bone.*

Note that in (1) the German expression is in quotation marks, whereas the corresponding English term on the right hand site is not. This is because in (1) we are using the English term in order to explain the semantic content of the corresponding German term. This strategy works very well – at least for those who understand the expression 'carpal bone'. But imagine that a young child were to ask you what 'Handwurzelknochen' means. Because the child does not know what a carpal bone is, (1) would not be very helpful. In the best case the child would memorize (1) and would afterwards be able to say 'carpal bone' whenever somebody asks what 'Handwurzelknochen' means. But obviously she would not know the semantic content of either term. For this reason (1) would not be an appropriate answer in the given case. It would be better to explain the term to the child for example with the help of pictures in an anatomy textbook. But while we can explain the semantical content of terms to people with the help of examples, paraphrases, pictures, and translations into other natural languages, these strategies won't work for computers. A computer has no better understanding of the term 'carpal bone' than of 'Handwurzelknochen', so statement (1) will provide the computer with no assistance at all in grasping the semantical content of the latter. Of course we could create a digital dictionary which links 'Handwurzelknochen' to the term 'carpal bone'. Such a dictionary might be useful, because it allows human users to find the appropriate translation; but in this case the computer would be in a similar situation as the child who knows that the two terms have the same meaning, but does not understand either of them. And while we can use pictures to educate the child, this strategy, too, will not work for computers. For them we need an alternative approach – ontologies.

In a first, rough formulation, the main idea of ontologies is the following. The semantical content of expressions that denote types is captured in ontologies through the assertion of relations between the types. This can be easily seen if we conceive an ontology as a graph, whose nodes are labeled with type terms like 'myelin' or 'lipoprotein' and whose edges are labeled with relation terms like '*is_a*', '*part_of*' and '*located_in*'. A graph with many labels is a syntactic entity, but it is important to notice that in an ontology each edge of the graph is equivalent to a statement about the corresponding entities in the biomedical domain. For example, if a node labeled 'myelin' and another node labeled 'lipoprotein' are connected by an edge labeled '*is_a*', then the ontology expresses the statement '*Myelin is_a Lipoprotein*'. Hence the ontology contains claims about the relations that hold between the types that are the denotations of the type terms of the ontology – and it is here that the ontology gains semantic traction.

It is important to notice that in a well-constructed ontology the type terms will be connected by many links to other terms, thus creating a semantical network each link of which represents a statement about some ontological relation between the types in reality represented by its nodes. The idea is that the semantic content of a term is not determined by any one specific link, but rather by its connection with many other terms that charges it with semantic content. This holistic approach to semantics is not new; it is a central feature of de Saussure's structuralism [5] or of Trier's word field theory [11]. Both de Saussure and Trier identified the meaning of a term with its position in a semantic network of terms. Note however that the holistic thesis in this radical form, although accepted by many contemporary computer science ontologists, is not plausible. For consider the following statements:

2 *Shkart is_a Trkarp.*

3 *Brajhn is_a Trkarp.*

4 *Trkarp part_of Xriprg.*

If the semantical content of a type term in an ontology were completely determined by its relation to other terms, we would have a good understanding what 'trkarp' means by looking at (2-4). However, no semantical content is fixed by (2-4), and even adding further similar statements would not bring about a change in this respect. This does not mean that additional statements would not make a difference. On the contrary: each of them puts additional restrictions on the use of the corresponding terms, and thus helps to narrow down their semantic content. For example, (2-4) allow for the possibility that 'trkarp' denotes *Pelvis*, 'shkart' denotes *Female Pelvis*, 'brajhn' denotes *Male Pelvis*, and 'xriprg' denotes *Body*. However, if we add the additional statement (5), then this possibility is eliminated.

5 *Shkart is_a Brajhn.*

Imagine we were to add hundreds of additional statements to our list by using the terms 'shkart', 'trkarp', 'brajhn', 'xriprg' together with a few dozen other similar fantasy type terms. Each connection in the resultant semantic network would restrict the possible interpretations of 'trkarp' and the other terms and thus provide us with extra semantical content. However, even with hundreds of additional statements it would still not be possible to determine which type is denoted by 'trkarp', for there will still be many possible interpretations left. The meaning of the terms will thus not be completely determined, and so the holistic thesis in its radical version is false.

It is important to realize the falsity of radical holism, because this has an important consequence: since an ontology can't completely fix the semantics of a type term, the semantic content of a term is in this sense not an all-or nothing matter but a matter of degree: the more a term is connected to other terms, the more its semantical content is determined. A sparse ontology that consists of only loosely connected terms provides these terms with very little in the way of semantic content. In particular, type terms in an ontology that are not distinguished by their connections within the network of terms are semantically indistinguishable with respect to that ontology. For example, assume that we have an ontology that consists of (2-4) and (6).

6 *Shkart and Brajhn are disjoint.*

Statement (6) guarantees that 'shkart' and 'brajhn' do not denote the same type. However, even with (6) the terms 'shkart' and 'brajhn' would still not be distinguishable: all we know about them is that they denote subtypes of *Trkarp* and that their denotations are disjoint. This limits the value of the given ontology for applications. For example, assume that two scientists use the ontology to annotate their data and that one of them believes that 'shkart' denotes *Male Pelvis* while the other believes that it denotes *Female Pelvis*. Since the ontology does not contain any information about the difference between male and female pelvises, an automatic reasoner would never be able to detect that the scientists are using the term 'shkart' in a crucially different way. This example shows why sparse ontologies are inferior to rich ontologies; the latter convey a greater amount of semantic content.

14.2 The Semantic Content of Relation Terms

Let us recap the results so far. An ontology is more than a list of type terms, it is designed to encapsulate also the 'meaning' of these terms in a computer parsable form. Biomedical ontologies consist of statements that involve type terms and a relation term; since ontologies are often visualized as graphs, it is helpful to think of their type terms as labels attached to nodes and of the relation terms as labels attached to edges. Since there is no way to tell a computer directly which type is determined by a given type term, ontologies seek to do this indirectly. The basic idea is that the semantic content of a type term is captured by its position in the network of type terms of which it is a constituent. Since each statement expresses a relation between the denotations of its type terms, each statement limits the possible interpretations of its type terms. For example, (7) expresses the thesis that the denotations of 'trkarp' and 'xriprg' are related by the parthood relation, thus limiting the possible interpretations of 'trkarp' and 'xriprg'.

7 *Trkarp part_of Xriprg.*

The approach will not be sufficient to single out some specific type as denotation of each given type term, but if the term is connected to a multitude of other terms, then the possibilities will be correspondingly restricted. Note that, according to this approach, the semantical content is determined by restricting the possible interpretations of the type terms via the relations between their respective denotations – in our example the parthood relation between 'trkarp' and 'xriprg'. However, while humans know that the relation term '*part_of*' is supposed to express the parthood relation, computers do not. For a computer '*part_of*' is just a string like any other, and for the computer (7) is itself a string which is not intrinsically different from a string such as (8):

8 *Trkarp cxzc Xriprg.*

How, then, do we bridge the gap between relation terms and the relations themselves? When we describe the links between the types we use relation terms like '*is_a*' and '*part_of*'; but how do we capture the denotations of the latter in a machine interpretable way? This question is important for two reasons. First, it is relations which form the principal vehicle for interoperability of ontologies. Thus if the same relations can be used in all members of a given set of ontologies, then to this degree these ontologies form an interoperable family – an idea which forms one central pillar of the OBO Foundry initiative (http://www.obofoundry.org). The use of common relations when creating a system of ontologies is equivalent to the use of a common gauge when creating an international railway system.

Second, the relation terms used in ontologies typically denote rather abstract relations. If we use expressions like '*part_of*', '*located_in*', and '*develops_from*' as unanalyzed primitives, these expressions are semantically underspecified. As shown in [1] and [10], the result is that they are used in an ambiguous way. For example, in the FMA we find:

9 *Female Pelvis part_of Body.*

10 *Urinary Bladder part_of Female Pelvis.*

11 *Urinary Bladder part_of Body.*

Statement (9) is used to assert that every female pelvis is part of a human body, but it does not imply that every body has a female pelvis as part. In contrast, (10) is used to assert that every female pelvis has a urinary bladder as a part, but not that every urinary bladder is part of a female pelvis. The parthood relation between the types denoted in (11) is the strongest of the three: Every urinary bladder is a part of a body and every body has a urinary bladder as part.

Another example is the use of 'contains' in GALEN, where we find:

12 *Pelvic Cavity contains Ovarian Artery.*

13 *Male Pelvic Cavity contains Urinary Bladder.*

14 *Tooth Socket contains Tooth.*

The different statements express different states of affairs, because the relation term 'contains' is used ambiguously. For every ovarian artery there is a pelvic cavity such that the pelvic cavity **contains** the ovarian artery. However, not every pelvic cavity **contains** an ovarian artery. This is expressed by (12). In contrast (13) states that every male pelvic cavity **contains** a urinary bladder, but it does not say that every urinary bladder is contained in a male pelvic cavity. In (12) and (13) 'contains' denotes distinct relations holding, respectively, between a type of immaterial entity (a cavity) and types of material objects (arteries, urinary bladders). In both cases the material objects are completely located in the cavities. In contrast, 'contains' in (14) relates two types of material objects (tooth sockets and teeth). Further the teeth are only partially contained in the tooth sockets. Hence 'contains' in (14) expresses a relation that is different from the relations expressed by the same term in (12) and (13).

The fact that '*part_of*' and '*contains*' in statements (9-14) are used ambiguously would be less problematic if the statements were to appear in a text that is intended to be read by humans with some knowledge of anatomy. A human can use background knowledge to disambiguate the statements in appropriate ways. However, a computer is not able to handle ambiguity in the way a human can, so that it is crucial for an ontology that relation terms are used in a clear-cut way; otherwise automatic reasoning is bound to lead to false conclusions (see [1] for examples). In addition, since the relations are used in ontologies to determine the semantical content of the terms in the ontology, a lack of clarity with respect to the relations will contaminate the whole ontology. For this reason, too, therefore it is essential for the use of an ontology that the semantics of the relation terms be made explicit in a non-ambiguous way.

The first step in solving this problem is to distinguish between relations that hold between types and those that hold between the instances of those types. Ontologies are about types: Statement (9) asserts that a parthood relation holds between the type *Female Pelvis* and the type *Body*, statement (12) that a containment relation holds between the type *Pelvic Cavity* and the type *Ovarian Artery*, etc.

Since the type terms in an ontology denote types and the relation terms like '*part of*', '*is a*', '*develops from*' denote relations between types, instances might seem to be not important for an ontologist. However, an anatomist is not able to study the types directly. We have epistemic access to types only via their instances. Hence the only way to evaluate a statement concerning types – for example '*Pelvic Cavity contains Ovarian Artery*' – is to look at instances of *Ovarian Artery* and their locations; there is no way to look at the type *Ovarian Artery* directly. Similarly, the only way to evaluate a statement like '*Appendix part_of Body*' is to look at instances of the type *Appendix* and to check whether they are part of some instances of the type *Body*. Note that the parthood relation between the instances differs from the various parthood relations on the type-level that we have considered above. One major difference is that the parthood relation between anatomical structures (i.e. between the different types of anatomical structures) holds in a time-dependent way. For example, it might be the case that Bill's appendix is part of his body at 6 am, but that it is not part of his body at 8pm on the same day. The relation expressed in '*Appendix part_of Body*' is however a timeless relation between types. Arguably, another difference is that the fact that Bill's appendix is part of his body at a given time entails that the location of his appendix and the location of his body overlap at this time (where it is not clear what it would mean for the type *Appendix* to have a location that overlaps with the location of the type *Body*). To capture the differences we will henceforth distinguish relations between types, for which we use *italic font*, from relations of other kinds, picked out by using **bold**.

We begin by studying the **part_of at** t relation between instances (where t stands in for times). Only by studying this and similar relations on the instance level can we gain insight into the parthood relations on the type-level [1, 7]. Although the latter are of first importance for ontologies, they are actually secondary to the former in an epistemic sense. We can use the tight connections between **part of at** a given time and the parthood relations on the type-level to disambiguate the use of '*part of*' in the problematic cases mentioned above. In cases (9-11) the relation term '*part of*' can be read as denoting three different relations; hence we have to distinguish at least three different parthood relations that hold between types. In the following we will use the term '*part of*' only to denote one of these relations; for the others we will use the terms '*is part*' and '*integral part of*'.

Let C and C_1 be types of anatomical entities, let x, y, z be anatomical entities (instances), and t a time. Further, let 'Cyt' be the abbreviation for 'y is an instance of C at time t' and 'C_1zt' the abbreviation for 'z is an instance of C_1 at t'. We can now define:[1]

d 1 C *part of* C_1 $=_{def}$ *for all* y, t, *if* Cyt *then there is some* z *such that* C_1zt *and* y **part of** z **at** t.

d 2 C *is part* C_1 $=_{def}$ *for all* z, t, *if* C_1zt *then there is some* y *such that* Cyt *and* y **part of** z **at** t.

d 3 C *integral part of* C_1 $=_{def}$ C *part of* C_1 *and* C *is part* C_1.

These definitions provide an example of how we can define relations between types in terms of the relations between the corresponding instances. One major advantage of these definitions is that they provide us with a better understanding of the type-level statements that form an ontology. With the help of the definitions (d 1 - d 3) it is easy to see that (15) is true, but (16) false, in virtue of the fact that there are (human) bodies that have no female pelvis (because they have a male pelvis).

15 *Female Pelvis part of Body.*

16 *Female Pelvis is part Body.*

In addition, the definitions (d 1 - d 3) allow us to check the logical properties of the type level relations and their logical connection. Without the definitions it might not be obvious whether 'A *part of* B' implies 'B *is part* A' and vice versa, or in other words whether *part of* is the inverse of *is part*. With the help of (d 1 - d 3) it is easy to see that this is not the case.

Let us consider another example. Does (15) entail (17)?

---

[1] The terms '*part of*' and '*integral part of*' are defined as in [7], *is part* is the inverse of *has part* as defined in [7], which means that C *is part* C_1 is logically equivalent to C_1 *has part* C. The relations *part of*, *is part*, and *integral part of* are equivalent to P_1, P_2, and P_{12} as defined in [1] and in chapter 15 of this book.

17 *Body is_part Female Pelvis.*

According to the definition (d 2) the statement (17) means: For any instance of *Female Pelvis* at any time, there is some instance of *Body* such that that instance of *Body* is part of that instance of *Female Pelvis* at that time. Since human bodies are never parts of pelvises, this is obviously false – hence we have shown that *part_of* is not the inverse of *is_part* [1, 7].

Let's consider two other examples. Since men have urinary bladders, some urinary bladders are not part of a female pelvis. Hence (18) is false. In contrast, (19) is true, because female pelvises have urinary bladders as parts:

18 *Urinary Bladder part_of Female Pelvis.*

19 *Urinary Bladder is_part Female Pelvis.*

The definitions (d 1 - d 3) provide us with a clear understanding of the relations which allows us to use the corresponding assertions to draw logical inferences. To give a very primitive example, from (d 1 - d 3) it follows immediately that (20) entails (21) and (22). Such logical connections facilitate automatic reasoning (see chapter 15 of this book).

20 *Urinary Bladder integral_part_of Body.*

21 *Urinary Bladder part_of Body.*

22 *Urinary Bladder is_part Body.*

In the beginning of this section we addressed two problems: (a) Humans use relation terms like '*part_of*' ambiguously, which undermines the quality of ontologies and leads automatic reasoners astray. And (b) the relations are used in ontologies to determine the semantic content of the type terms, hence we need to capture the denotation of relation terms like '*part_of*' in a machine interpretable form. These problems were addressed by defining relations on the type-level (in our example *integral_part_of*, *part_of*, and *is_part*) with the help of a relation between individuals (**part_of at**). The definitions (d 1 - d 3) allow us to resolve the ambiguities in existing uses of the term '*part_of*' and it is easy to translate the definitions above into a formal language, hence the approach that was embraced in the last section was a step in the right direction.

However, it did not solve the problems completely. The definitions (d 1 - d 3) involve the parthood relation between individuals. Hence the denotation of the terms '*part_of*', '*is_part*', and '*integral_part_of*' depends on the denotation of the term '**part_of at**' that we used in these definitions. Thus in order to get a clear understanding of these terms we need to determine the denotation of '**part_of at**'. This can be done via an axiomatization of this relation, i.e. by providing a set of axioms which amount to a so-called 'contextual definition' of '**part_of at**'. A contextual definition is not really a definition in the strict sense, but the axioms serve to capture our intuitions about the logical properties of the relation that is axiomatized and thus they restrict the possible interpretations of the term '**part_of at**'.

It would have been possible to axiomatize the various parthood relations on the type-level directly instead of defining them with the help of the parthood relation on the instance-level. There are however two reasons why it is better not to do this, but to use the **part_of at** relation as we have done. One reason is that we could use the single parthood relation **part_of at** on the level of individuals to define the three relations on the type-level. Thus we needed only one primitive notion instead of three. Further, since we have access to types only via their instances, our intuitions about the logical properties of the relations on the instance-level are much more developed. Moreover, much of our digital data about anatomical and other entities in the biomedical domain comes in the form of the instance data contained, for example, in clinical records.

Actually, since 'part' is not a technical term but an expression we use in daily life (we talk about engine parts, or the parts of former Yugoslavia, or about cellulose as part of wood) one might suspect that we have very strong intuitions about the parthood relation and thus that it would be easy to develop a theory of wholes and their parts. Indeed it is true that people have strong opinions on mereological questions; unfortunately the intuitions governing our daily talk about wholes and their parts are quite heterogeneous (if not plainly inconsistent). For this reason mereology is a controversial field in philosophy. Hence it is important to give an explicit account of **part_of at**, otherwise type-level terms like '*integral_part_of*', '*part_of*', and '*is_part*' will themselves be used ambiguously. This is not the place to present a full axiomatization (see [6]), but some examples of axioms that many people would embrace are:

Ax. 1 *At any time t, every x* **part_of** *x at t.*

Ax. 2 *For any x, y, t: if x is* **part_of** *y at t and y* ***part_of*** *x* ***at*** *t, then x = y.*

Ax. 3 *For any x, y, z, t: if x is* ***part_of*** *y* ***at*** *t and y is* ***part_of*** *z* ***at*** *t, then x is* ***part_of*** *z* ***at*** *t.*

These axioms express time-relativized versions of the reflexivity, antisymmetry, and transitivity of parthood, respectively.

The approach that we have considered in this section allows us to restrict the semantical content of relation terms. This is achieved in two steps. We define the type-level relation with the help of a relation between individuals and we then give an axiomatization of the latter relation. This approach has been presented by means of by appealing to just a few examples and is still rather sketchy. In chapter 15 of this book it will be covered systematically and in greater depth.

14.3 Canonicity

So far we did not discuss one important objection to the above approach.[2] Let's assume that we encounter statement (23) in a textbook on human anatomy.

23 *Appendix is_part Body.*

Statement (23) is true: the human body has an appendix. However, according to the definition (d 2) statement (23) is equivalent to (24):

24 *For every x and time t, if x is a Body at t, then there is an instance of Appendix y at t such that y is* **part_of** x **at** t.

Statement (24) is plainly false: there are plenty of people who live happily without an appendix. Each of them provides a counterexample to the claim in (24) that every body has an appendix. Since (23) is true, but (24) is false, the statements (23) and (24) cannot be equivalent. Does that mean that our analysis of statements like (23) is incorrect? Is definition (d 2) inappropriate?

In order to understand the root of the problem we need to distinguish between canonical anatomy and instantiated anatomy [4, 8]. Instantiated anatomy concerns the anatomical entities represented for example in data about actual cases generated in clinical practice. Canonical anatomy is the result of generalizations deduced from qualitative observations that are implicitly sanctioned by their accepted usage by anatomists. While instantiated anatomy and canonical anatomy are both founded in empirical observations, only instantiated anatomy contains empirical statements about human bodies and their anatomical parts. In contrast, the relation between canonical anatomy and human bodies is in some respects similar to the relation between a technical drawing and the artifacts that are built with the help of the drawing. As anybody who has assembled a piece of Swedish furniture knows, many existing artifacts do not exactly match their technical drawings. That does not make the technical drawing 'false'; a technical drawing is not an empirical description of the composition of the existing artifacts; rather it tells us how the artifacts should be composed. Analogously, a canonical anatomy gives an account of the 'prototypical' composition of the male or female human body. For example, (23) does not assert that all human bodies have an appendix, but rather that a human body is supposed to have an appendix. Thus (23) cannot be refuted by the fact that some people lack an appendix. This example shows that a canonical anatomy consists of statements that describe how the anatomical entities of a given organism are supposed to be composed (for example in light of the structure of the underlying genes); and it is this that distinguishes a canonical anatomy from an instantiated anatomy.

The distinction between instantiated and canonical anatomy is important since it allows us to analyze the source of the mismatch between (23) and (24). Statement

---

[2] We thank Cristian Cocos, Alan Rector, and Cornelius Rosse for their critical remarks and suggestions.

(24) would be an appropriate analysis of (23) if (23) would be an assertion about instantiated anatomy – and in this case (23) would be false, since it is an empirical fact that not all human bodies have an appendix. However, we have assumed above that (23) is a statement within a textbook on canonical anatomy. One way to make the force of statements of this kind explicit is to use an adverb as in (25):

25 *Canonically, Appendix is_part Body.*

Syntactically, the expression 'canonically' in (25) works like 'necessarily', 'possibly', 'it is permissible that' and other expressions that are – from a logical perspective – logical operators. However, while the semantics of the latter is well understood, the semantics of 'canonically' is not. Thus in this form (25) is a logical black box and for this reason useless for logical reasoning. This is why we will present a logical analysis of statements like (25) in the remainder of this section.

In [7] we have (implicitly) embraced the assumption that in the context of canonical anatomy the domain of discourse is restricted to canonical entities. In this case (25) would have the same meaning as (23), except for an implicit understanding that we consider only canonical entities – which can be made explicit by restricting the range of the variables in (24). Hence – according to this approach – (25) is equivalent to (26):

26 *For every x and time t, if x is a Body at t, then there is an instance of Appendix y at t such that y is **part_of** x **at** t; where the variables x and y range exclusively over canonical entities.*

The term 'canonical entity' can be defined as follows:

d 4 *An anatomical entity x is canonical with respect to a given anatomy A if and only if x is structured in the way it is supposed to be structured according to anatomy A.*

Since a human body without an appendix is not canonical, it follows that such bodies fall outside the domain of quantification, and thus the problematic cases are excluded.

Unfortunately, this way of understanding 'canonically' leads to new difficulties. For example, (27) would be equivalent to (28).

27 *Canonically, Appendix part_of Body.*

28 *For every x and time t, if x is an instance of Appendix at t, then there is an instance of Body y at t such that x **part_of** y **at** t, where the variables x and y range exclusively over canonical entities.*

Statement (28) expresses that every canonical appendix is part of a canonical body – which is obviously wrong: there are people who have a perfectly normal appendix, but are lacking teeth. Therefore the idea of restricting the domain of quantification to canonical entities does not work; we need to find an alternative way to analyze (25).

In order to come up with the needed analysis, we have to remember that canonical anatomy gives an account of how a male or female human body is supposed to be composed. Thus a statement that is part of a canonical anatomy expresses a requirement that a human body has to meet in order to conform to the given canonical anatomy. We can express this in the following way: Let '$I_A xt$' be the abbreviation for 'x is a human body that is in conformity with anatomy A at t'. (As mentioned above we assume that we deal with human anatomy; otherwise one has to modify the definition of '$I_A xt$' in the obvious way.)

d 5 *Canonically, C is_part C_1 =$_{def}$ for all x, t, necessarily, if $I_A xt$, then* FOR ALL z, IF $C_1 zt$ THERE IS SOME y SUCH THAT Cyt AND y **part_of** z **at** t; *where y and z are anatomical entities that are* **part of** x **at** t.

Definition (d 5) can be paraphrased as follows: if a statement of the form 'C *is_part* C_1' is part of a canonical human anatomy, then the following holds for any human body x at any given time: necessarily, if x is in conformity with the given anatomy (at this time), then, for any anatomical part of x that is an instance of C_1 (at this time), there is an anatomical part of x that is an instance of C (at this time) and the instance of C is part of the instance of C_1 (at this time).

Let's consider an example. Definition (d 5) entails that (29) is equivalent to (30):

29 *Canonically, Carpal Bone is_part Hand.*

30 *Necessarily, if x is a human body that is in conformity with A at time t, then for all y, if y is an instance of Hand at t, there is (at least) one entity z that is an instance of Carpal Bone at t and is* **part of** *y at t; where y and z are anatomical entities that are* **part of** x **at** t.

Analogously, we can define *part_of* for canonical anatomies:

d 6 *Canonically, C part_of C_1 =$_{def}$ for all x, t, necessarily, if $I_A xt$, then* FOR ALL y, t, IF Cyt, THERE IS SOME z SUCH THAT $C_1 zt$ AND y **part_of** z **at** t; *where y and z are anatomical entities that are* **part of** x **at** t.

Definition (d 6) expresses the following: if a statement of the form 'C *part_of* C_1' is part of a canonical human anatomy, then the following holds for any human body x at any given time: necessarily; if x is in conformity with A (at this time), then, for any anatomical part of x that is an instance of C (at this time), there is an anatomical part of x that is an instance of C_1 (at this time) and the instance of C is part of the instance of C_1 (at this time).

The definitions (d 6) and (d 5) are closely linked to (d 1) and (d 2), respectively: the parts of the definitions that are emphasized by using small caps are the right hand sides of the definitions (d 1) and (d 2). We chose this way of presenting the definitions because it shows that the original account of the last section is preserved, it is just that it is now embedded in a context that does justice to the fact that statements like (23) are part of a canonical anatomy.

Let's consider another relation, where time plays a more important role than in the examples above. The human body is supposed have deciduous teeth and the human body is supposed to have androgenic hair – but obviously not at the same time. We can express this fact with the help of a relation *excludes* in (31), where *excludes* is defined in definition (d 7).

31 *Canonically, Deciduous Tooth excludes Androgenic Hair.*

d 7 *Canonically, C excludes C_1 $=_{def}$ for all x, t, necessarily, if $I_A xt$, then* THERE ARE NO y, zt, SUCH THAT Cyt AND $C_1 zt$; *where y and z are anatomical entities that are* **part of** x **at** t.

The relation *excludes* serves here as a simple example that illustrates how time can play an important role for the definitions of type-level relations; this holds in particular for relations that concern the development of anatomical entities.

Since we have focused on parthood relations in this section so far, let's consider an example that involves the *contains* relation between types, e.g. (32). Further, let's assume that we have an account of the corresponding **contains** relation on the instance-level (see chapter 15). We can now define *contains* with the help of **contains** as in definition (d 8).

32 *Canonically, Male Pelvic Cavity contains Urinary Bladder.*

d 8 *Canonically, C contains C_1 $=_{def}$ for all x, t, necessarily, if $I_A xt$, then* FOR ALL y, IF Cyt THEN THERE IS SOME z SUCH THAT $C_1 zt$ AND y **contains** z **at** t; *where y and z are anatomical entities that are* **part of** x **at** t.

Hence (32) is equivalent to (33), which is itself a complicated way of expressing (34):

33 *Necessarily, if $I_A xt$, then for all y, if y is an instance of Pelvic Cavity at t then there is some z such that z is an instance of Urinary Bladder at t and y **contains** z at t; where y and z are anatomical entities that are* **part of** x **at** t.

34 *Necessarily, if x is a human body that is in conformity with A at t, and x has a pelvic cavity, then there is a urinary bladder that is contained in the pelvic cavity.*

We will now generalize our approach and define 'canonically'. In this section we have analyzed statements of the form (35), where *rel* stands in for '*is_part*', '*part_of*', '*excludes*', and '*contains*'.

35 *Canonically, C rel C_1.*

As we have mentioned above, the definitions (d 5) and (d 6) (where *rel* is *is_part* and *part_of*, respectively) are closely linked to the definitions (d 1) and (d 2), which define statements of the form '*C is_part C_1*' and '*C part_of C_1*'. Analogously, the definition (d 8) is closely linked to (d 9). (Again, the relevant parts of the definition (d 8) are in small caps.)

d 9 *C contains C_1 =$_{def}$ if Cyt, then there is some z such that $C_1 zt$ and y* **contains** *z* **at** *t*.

It seems that for any definition that defines statements of the form (35), there is a corresponding definition of the statements that does not begin with 'canonically'. We will use this connection in order to define 'canonically':

d 10 *Let rel be any binary type-level relationship, and C, C_1 any types, and assume we have a definition of the following form:*

$$C \; rel \; C_1 =_{def} \phi(y, z)$$

where $\phi(y, z)$ represents a formula that involves only relationships between individuals and that ensures that y and z are anatomical entities that are instances of C and C_1, respectively. In this case we can define:

$$Canonically, \; C \; rel \; C_1 =_{def} for \; all \; x, t, necessarily,$$

$$if \; I_A xt, then \; (\phi(y, z) \; and \; y \; \textbf{part of} \; x \; \textbf{at} \; t \; and \; z \; \textbf{part of} \; x \; \textbf{at} \; t)$$

Definition schema (d 10) provides us with a systematic link between the relations within a canonical anatomy and the use of the corresponding relations within an instantiated anatomy. The definition schema (d 10) works not only for relations such as those that we have considered in this section; it can be applied to many type-level relations and in particular to the type-level spatial relations that will be considered in chapter 15.[3]

14.4 Conclusions

One purpose of an ontology is to encapsulate the meanings of its terms in a computer parsable form. We analyzed how anatomical anatomies fulfill this purpose. An anatomical ontology consists of statements composed of two kind of terms denoting types and relations, respectively. Typically such statements involve two type terms, so that they are of the form '*A rel B*'.

We showed that there is no way to tell a computer directly, for any given type term, which type is denoted by that term. Thus ontologies must find ways to convey

---

[3] Note that the connection between the relations is not always as straightforward as in the examples considered above. For example, definition (d 11) does not capture the semantic content of 'excludes'.

d 11 *C excludes C_1 =$_{def}$ there are no y, z, t, such that Cyt and $C_1 zt$.*

A more appropriate definition of 'excludes' is:

d 12 *C excludes C_1 =$_{def}$ there are no u, y, z, t, such that y* **part of** *u at t, z* **part of** *u at t, Cyt, and $C_1 zt$.*

such information indirectly: broadly, it is the totality of the relations between the denotations of the type terms that determines the semantical content of the type terms taken individually. This works as follows. Each statement '*A rel B*' asserts that the denotation of '*A*' and the denotation of '*B*' are linked by the relation *rel*. Thus any interpretation of '*A*' and '*B*' according to which their denotations do not meet this requirement is ruled out. The possible interpretations of the terms '*A*' and '*B*' are in this sense limited by the statement '*A rel B*'. While this approach is not sufficient to single out any specific type as denotation of a given term, if the term is connected to a multitude of other terms, then the possibilities will be correspondingly restricted. Fortunately, in the domain of anatomy we are already in possession of high-quality representations of such multiple relations.

Since the semantical content of type terms is determined by the relations that are expressed by '*is_a*', '*part_of*', '*contains*' and other relation terms, it is crucial to make explicit which relations these terms denote. This analysis is important not only because of our aim to capture the semantical content of the terms of an ontology in a machine-readable form, but also because people tend to use relation terms ambiguously, in a way which reduces the quality of ontologies. We showed that many relations between types can be defined with the help of relations that hold between instances of these types, and an approach based on this recognition has the advantage that we typically have a better understanding of the relations between instances than of the relations between the corresponding types. Further, the approach has the virtue of economy, since it is often possible to define different relations on the type-level with the help of one relation on the instance-level.

On the given approach the meaning of a statement '*A rel B*' in an ontology is an empirical assertion about the instances of types A and B. Thus '*Embryo develops_from Zygote*' is true if and only if: for any instance of *Embryo* x there is an instance of *Zygote* y such that x **developed_from** y. Here **developed_from** is an instance level relation that holds between individuals. '*Embryo develops_from Zygote*' is thus an empirical assertion that can be falsified (by discovering that at least one embryo did not develop from a zygote).

In the case of canonical anatomical ontologies such as the FMA, in contrast, the situation is more complicated, since canonical anatomical ontologies do not consist of empirical assertions in this sense, but rather of statements that express how the corresponding entities are supposed to relate to each other (in virtue of the workings of the underlying structural genes). For this reason we analyzed statements of the form '*canonically, A rel B*' in such a way as to show how the semantic content of such statements is systematically linked to statements without the prefix '*canonically*'. Very roughly, a statement like '*canonically, A rel B*' expresses that, necessarily, any human body x that is in conformity with the given anatomy meets the requirement '*A rel B*', where '*A rel B*' can spelled out as in the non-canonical case and the domain of discourse is restricted to the anatomical entities that are part of x. The fundamental picture then remains the same: the semantic content of the type terms is provided by the network of relations between them. A profound understanding of these relations is thus a prerequisite for a non-ambiguous use of type terms of the sort which can support automatic reasoning. The next chapter will present a deeper and more systematic analysis of those specific sorts of spatial relations that are relevant for anatomical ontologies.

Acknowledgements

This work was funded in part by the National Institutes of Health through the NIH Roadmap for Medical Research, Grant 1 U 54 IIG004028. Information on the National Centers for Biomedical Computing can be found at http://nihroadmap. nih.gov/bioinformatics.

References

1. M. Donnelly, T. Bittner, and C. Rosse. A formal theory for spatial representation and reasoning in bio-medical ontologies. *Artificial Intelligence in Medicine*, 36(1):1–27, 2006.
2. N Guarino. Formal Ontology and Information Systems. In N Guarino (ed.), *Proceedings of the 1st International Conference on Formal Ontologies in Information Systems, FOIS'98*, IOS Press, Trento, 3-15, 1998.
3. I. Johansson. On the Transitivity of the Parthood Relations. In H. Hochberg and K. Mulligan (eds.), *Relations and Predicates*, 161-181, 2004.
4. C Rosse, JL Mejino , BR Modayur , R Jakobovits, KP Hinshaw, JF Brinkley. Motivation and organizational principles for anatomical knowledge representation: the digital anatomist symbolic knowledge base. *Journal of the American Medical Informatics Association*, 5(1):17-40, 1998.
5. F de Saussure. *Course In General Linguistics*. McGraw-Hill, New York, 1965.
6. P. Simons. *Parts, A Study in Ontology*. Clarendon Press, Oxford, 1987.
7. B. Smith, W. Ceusters, B. Klagges, J. Köhler, A. Kumar, J. Lomax, C. Mungall, F. Neuhaus, A. Rector, and C. Rosse. Relations in biomedical ontologies. *Gnome Biology*, 6(5):46, 2005.
8. B Smith, A Kumar, W Ceusters, C Rosse. On carcinomas and other pathological entities. *Comparative and Functional Genomics*, vol 6, 7-8, 379-387, 2005.
9. B Smith, W Kusnierczyk, D Schober, W Ceusters. Towards a Reference Terminology for Ontology Research and Development in the Biomedical Domain. Proceedings of KR-MED 2006, O. Bodenreider, ed., Proceedings of KR-MED, http://ceur-ws.org/Vol-222, 57-66, 2006.
10. B. Smith and C. Rosse. The Role of Foundational Relations in the Alignment of Biomedical Ontologies. In M. Fieschi, et al. (eds.), *Medinfo 2004*, IOS Press, Amsterdam, 444-448, 2004.
11. J Trier. *Aufsätze und Vorträge zur Wortfeldtheorie*. Berlin: DeGruyter, 1973.

15

Modeling Principles and Methodologies - Spatial Representation and Reasoning

Thomas Bittner, Maureen Donnelly, Louis J. Goldberg, and Fabian Neuhaus

Summary. Spatial relations include mereological relations such as parthood and overlap, topological relations such as connectedness and one-pieceness, as well as location relations. The location and the arrangement of an anatomical structure within the human body can be further specified by means of relations that express spatial orderings in a qualitative way, e.g. superior, anterior, lateral, etc. In this chapter we give an overview of the various kinds of spatial relations and their properties. We particularly focus on properties of spatial relations that can be exploited for automated reasoning. We also discuss the distinction between so-called individual-level and type-level spatial relations.

15.1 Introduction

The representation of spatial relations between body parts is a central component of anatomical ontologies. The spatial concepts most often used in anatomical ontologies are not the quantitative, point-based concepts of classical geometry, but rather qualitative relations among extended objects such as body parts. The purpose of this chapter is to review the formal foundations of the kind of qualitative spatial representation and reasoning techniques that are needed for anatomical ontologies. The content of this chapter is a compilation of material that was presented originally in a series of papers on spatial relations in biomedical ontologies: [5, 11, 12, 13, 14].

The representation of spatial relations in anatomical ontologies concerns types of individuals. By an individual (also called a particular or an instance), we mean an entity which, at each moment of its existence, occupies a unique spatial location. Individuals can be either material (my liver, your brain) or immaterial (the cavity of my stomach), where material individuals are here understood as those individuals with a positive mass and immaterial individuals are those individuals with no mass. Individuals are distinguished from types (also called universals or kinds) which may have, at each moment, multiple individual instances. (See also chapter 14.) Examples of types are *Liver* (the type whose instances are individual livers), *White Blood Cell* (the type whose instances are individual white blood cells), and *Human Temporomandibular Joint* (the type whose instances are individual human temporomandibular joints).

(Throughout this chapter, we use italics and initial capitals for type names.)

Anatomical ontologies are to a large degree about the specification of the semantics of spatial relations between types. However these relations between types are epiphenomena of spatial relations between individuals. For this reason a full analysis of the spatial relations between types presupposes an analysis of the relations between individuals. In the first part of this chapter we give an overview of the spatial relations that are significant for the description of anatomical structures. To illustrate the various spatial relations we will use a human temporomandibular joint (TMJ) and the spatial relations between its parts as a running example. In the second part of this chapter we describe a general methodology of how to define spatial relations between types in terms of spatial relations that hold between their instances.

We present the formal ontology of spatial relations in a sorted first-order predicate logic with identity. (See [8] for an introduction.) We use the letters $x, x_1, y, z, \ldots$ as variables ranging over individuals and capital letters A, B, C as variables ranging over types (or universals). The logical connectors $\neg, =, \wedge, \vee, \rightarrow, \leftrightarrow$ have their usual meanings (not, identical-to, and, or, if ... then, and if and only if (iff), respectively). We use the symbol $\equiv$ for definitions. We write (x) to symbolise universal quantification (for all $x \ldots$) and $(\exists x)$ to symbolise existential quantification (there is at least one $x \ldots$). All quantification is restricted to a single sort. Restrictions on quantification will be understood by conventions on variable usage. Leading universal quantifiers are omitted.

15.2 Mereology

A mereology is a formal theory of parthood, where parthood (symbolized as P) is the relation that holds between two individuals, x and y, whenever x is part of y. For example, my heart is part of my body, my finger is part of my hand, etc. Since parthood relations apply directly to concrete individuals and require neither quantitative data nor mathematical abstractions (points, lines, etc.), a mereology is a natural basis for qualitative spatial reasoning in anatomical ontologies. Other relations, such as overlap (having a common part) and discreteness (having no common part), are defined in terms of parthood [26, 30].

Consider the human temporomandibular joint (TMJ). In this running example [5] we focus on its gross-level anatomical parts, i.e., maximally connected anatomical parts of non-negligible size (thus cells and molecules are parts of anatomical structures but are not gross-level anatomical parts). At this gross anatomical level of granularity we distinguish two kinds of anatomical parts: material parts and cavities. The material gross-level anatomical parts of the TMJ are schematically depicted in Figures 15.1(a) and (c) [18, 5]. The figures show, in a sagittal section through the middle of the condyle, a TMJ in closed jaw position – Figure 15.1(a) – and in open jaw position – Figure 15.1 (c): temporal bone (1), head of condyle (2), articular disc

(3), posterior attachment (4), lateral pterygoid muscle (5). Immaterial anatomical parts (cavities) are the superior and inferior synovial cavities, which are depicted as white spaces above and below the articular disc and the posterior attachment.

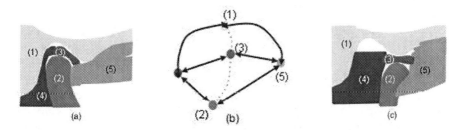

Fig. 15.1. Drawings of the major parts of a TMJ in jaw closed position (a) and open (c) positions. Structure (b) is a graph in which solid edges represent the relations of external connectedness between major material parts of the TMJ. The dotted edges represent the ternary relation of betweenness (the articular disc is between the temporal bone and the head of the condyle).

In the mereologies of [6, 10, 26], parthood is treated as a primitive relation. This means that, instead of being defined, axioms fixing the logical properties of the parthood relation are built into the theory. The parthood relation must then be interpreted in applications in a way that conforms to these axioms. Axioms that are included in nearly every mereology are:

(P1) Pxx (every object is part of itself)

(P2) $Pxy \land Pyx \rightarrow x = y$

  (if x is part of y and y is part of x, then x and y are identical)

(P3) $Pxy \land Pyz \rightarrow Pxz$

  (if x is part of y and y is part of z, then x is part of z)

(P1) tells us that P is reflexive, (P2) tells us that P is antisymmetric, and (P3) tells us that P is transitive. Thus, P is a partial ordering (a reflexive, antisymmetric, and transitive binary relation). Axioms (P1)-(P3) are not very strong. They cannot distinguish the parthood relation from other partial orderings such as the less-than-or-equal-to relation on the real numbers or the is-a-factor-of relation on the positive integers. For this reason, most mereologies include additional axioms which further restrict the parthood relation [3, 26].

The following relations among individuals are defined in terms of P: Individual x is a proper part of individual y, if x is any part of y other than y itself (D_{PP}); individuals x and y overlap, if there is some object, z, that is part of both x and y (D_O); individuals x and y are discrete if and only if x and y do not overlap (D_{DS}). Symbolically:

(D_{PP}) $PPxy \equiv Pxy \land x \neq y$ (x is a proper part of y)
(D_O) $Oxy \equiv (\exists z)(Pzx \land Pzy)$ (x and y overlap)
(D_{DS}) $DSxy \equiv \neg Oxy$ (x and y are discrete)

Consider Figures 15.1(a) and (c). All the depicted major anatomical parts (1-5) of the TMJ are proper parts of the TMJ and all of them are discrete from another.

From the definitions and the axioms the following can be proved as theorems $(PT1\text{-}7)$: If x is a proper part of y then y is not a proper part of x $(PT1)$, i.e., proper parthood is asymmetric; x is not a proper part of itself $(PT2)$, i.e., proper parthood is irreflexive; if x is a proper part of y and y is a proper part of z then x is a proper part of z $(PT3)$, i.e., proper parthood is transitive; if x is a part of y then x overlaps y $(PT4)$; overlap is symmetric, i.e., if x overlaps y then y overlaps x $(PT5)$; if x is a part of y and x overlaps z then y overlaps z $(PT6)$; and if x is a part of y discrete then x and z are discrete $(PT7)$.

$(PT1)$ $PPxy \rightarrow \neg PPyx$
$(PT2)$ $\neg PPxx$
$(PT3)$ $PPxy \land PPyz \rightarrow PPxz$
$(PT4)$ $Pxy \rightarrow O\,xy$

$(PT5)$ $O\,xy \rightarrow O\,yx$
$(PT6)$ $Pxy \land O\,xz \rightarrow O\,yz$
$(PT7)$ $Pxy \land DSyz \rightarrow DSxz$

We list these theorems as examples of how to make explicit the consequences of definitions and other assumptions using the deductive power of formal logic.

15.3 How Axioms and Theorems Support Reasoning

The properties of the relations that are made explicit by means of our axioms, definitions, and theorems can be used to support automated reasoning. As pointed out for example by [16], transitivity reasoning is critical in the biomedical domain. Transitivity reasoning has the form

if $a\,R\,b$ and $b\,R\,c$ then derive $a\,R\,c$,

where $a\,R\,b$ abbreviates that between a and b the relation R holds. Thus if we know that $a\,R\,b$ and $b\,R\,c$ hold then we can derive that $a\,R\,c$ holds. An important advantage of a logic-based anatomical ontology is that it makes explicit which relations have the property of being transitive and thus support transitivity reasoning. Thus PT3 tells us that we can validly derive 'my hand is part of my body' from 'my hand is part of my arm' and 'my arm is part of my body'.

Notice that many relations, like overlap, are not transitive. Consider Figure 15.2(a): the left region overlaps the region in the middle and the region in the middle overlaps the region on the right but the left region does not overlap the right region. Thus it is important to specify which relations do have the property of transitivity and which relations do not.

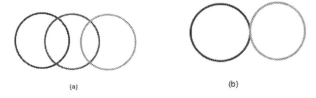

{a}   {b}

Fig. 15.2. (a) The non-transitivity of overlap. (b) Externally connected regions.

Transitivity reasoning, since it employs the transitivity property of a single relation, cannot be applied to a pair of premises of the form *a R b* and *b S c*, where *R* and
S are different relations, i.e., we cannot derive anything from the premises *a part-of b*
and *b is-discrete-from c* by means of transitivity reasoning, since *a part-of b* and
b is-discrete-from c are distinct relations. In this case a more general form of reasoning based on the composition of binary relations is required. Relation composition
has the form:

$$\text{if } a\ R\ b \text{ and } b\ S\ c \text{ then derive } a\ T\ c,$$

where *R*, *S*, and *T* are symbols referring to possibly distinct binary relations. The
importance of this kind of reasoning is that it provides means to combine information about different kinds of relations. Theorem (PT7) tells us that we can combine
information about parthood and discreteness in the sense that from (i) the head of
my condyle is a (proper) part of my condyle and (ii) my condyle is discrete from my
temporal bone, it follows that the head of my condyle is discrete from my temporal
bone. Thus theorem (PT7) supports a specific kind of composition of relations of
the form: if *a R b* and *b S c* then derive *a S c*. The importance of the composition of
relations in bio-ontologies was pointed out, for example, by [25, 29].

Consider theorems PT1-7. These theorems establish very basic properties of
proper parthood and overlap as well as interrelationships between parthood and overlap. Their importance is three-fold. Firstly, they show that the consequences of our
axioms and definitions conform with our commonsense intuitions which surely must
hold in any anatomical ontology: nothing can be a proper part of itself, if one object
is part of another then both objects overlap, if *x* overlaps *y* then *y* overlaps *x*, etc.
Secondly, TP4 and TP5 reflect the kind of reasoning an automated system trivially
should support. No medical information system should have to represent the fact
that *x* overlaps *y*, if it represents already that *x* is part of *y*. Thirdly, PT1 and PT2 are
examples of constraints a medical information system should check whenever new
anatomical facts are added.

Notice that there is an important difference between the use of first order logic as
a tool of formal ontology to specify the semantics of terms and the use of first order
logic as a tool for automatic reasoning. We use first order logic in the first sense
and not in the second. We use first order logic to prove (by hand) that relations have

certain properties. Once we have proven in our formal ontology that a relation has a property that can be used for automated reasoning, then we use other, more efficient tools (e.g., description logics) to actually implement this reasoning.

15.4 Location Relations

To be useful for anatomical ontologies, mereology needs to be further extended to include also location relations among individuals. We can already say something about the relative location of two objects using mereological relations: if x is part of y, then x is located in y in the sense that x's location is included in y's location. Also, if x and y overlap, then x and y partially coincide in the sense that x's location and y's location overlap. The location relations enable us to, in addition, describe the relative location of objects that may coincide wholly or partially without being part of one another or overlapping. A parasite in the interior of a person's intestine is located in the lumen of his intestines, but it is not part of the lumen of his intestines. Consider Figure 15.1(a): the articular disc is located in the synovial cavity. As another example, my esophagus partially coincides with my mediastinal space, but does not overlap (i.e. share parts with) my mediastinal space.

Human bodies have not only material parts (livers, hearts, etc.) but also immaterial parts such as passageways and spaces (the lumen of an esophagus, the cavities of the ventricles of a heart, an abdominal cavity) through which substances pass and in which anatomical structures are located. Since the material entities which are temporarily or permanently located in these spaces and passageways never share parts with them, mereological relations are not useful for describing the positions of material individuals relative to spaces and passageways. For these reasons, anatomical reasoning requires location relations distinct from mereological relations [11, 17, 20, 22, 23].

In [6], [10] and [14], all location relations are introduced in terms of a region function, r, that maps each individual to the unique spatial region at which it is exactly located at the given moment. Spatial regions are here assumed to be the parts of an independent background space in which all individuals are located. Because we are abstracting from temporal change and, in particular, from movement, we treat r as a time-independent primitive function. Axioms for the region function include:

(L1) $Pxy \rightarrow Pr(x)r(y)$ (if x is part of y, then x's region is part of y's region)
(L2) $r(r(x)) = r(x)$ (the region of x's spatial region is x's spatial region)

The location relations are defined using the region function and mereological relations: x is located in y if x's spatial region is part of y's spatial region (D_{LocIn}); x and y partially coincide if x's spatial region and y's spatial region overlap (D_{PCoin}). Symbolically:

$$D_{LocIn} \quad LocIn(x,y) \equiv Pr(x)r(y) \qquad D_{PCoin} \quad PCoin(x,y) \equiv Or(x)r(y)$$

For example: my brain is located in (but not part of) my cranial cavity. A parasite may be located in (but not part of) a patients intestinal lumen. My esophagus partially coincides with my mediastinal space. Notice that here the stronger relation *LocIn* does not hold. My esophagus region is not part of the region of my mediastinal space since part of my esophagus lies outside of my mediastinal space. As another example, a bolus of food that is just beginning to enter my stomach cavity partially coincides with (but is not located in) my stomach cavity.

From the axioms and definitions of we can derive the following theorems concerning the location relations:

$(LT1)$ $LocIn(x, x)$ (every individual is located in itself)

$(LT2)$ $LocIn(x, y) \land LocIn(y, z) \rightarrow LocIn(x, z)$
           if x is located in y and y is located in z, then x is located in z)

$(LT3)$ $Pxy \rightarrow LocIn(x, y)$ (if x is part of y, then x is located in y)

$(LT4)$ $PPxy \rightarrow LocIn(x, y)$(if x is a proper part of y, then x is located in y)

$(LT5)$ $LocIn(x, y) \land PPyz \rightarrow LocIn(x, z)$
           (if x is located in y and y is a proper part of z, then x is located in z)

$(LT6)$ $PPxy \land LocIn(y, z) \rightarrow LocIn(x, z)$
           (if x is a proper part of y and y is located in z, then x is located in z)

$(LT7)$ $PCoin(x, x)$ (partial coincidence is reflexive)

$(LT8)$ $PCoin(x, y) \rightarrow PCoin(y, x)$ (partial coincidence is symmetric)

$(LT9)$ $Oxy \rightarrow PCoin(x, y)$
           (if x and y overlap, then x and y partially coincide)

$(LT10)$ $LocIn(x, y) \rightarrow PCoin(x, y)$
           (if x is located in y, then x partially coincides with y)

Some theorems establish very basic and important properties of the location relations and their interrelationships with other relations like parthood, overlap, etc. Other theorems can be exploited for transitivity reasoning and reasoning by composition of relations. (LT2), for example, tells us that we can validly perform transitivity reasoning. (LT5) and (LT6) tells us that we can validly compose information about (proper) parthood and location. Using (LT5) we can derive: patient x's heart is located in patient x's thoracic cavity from (i) patient x's heart is located in patient x's middle mediastinal space and (ii) patient x's middle mediastinal space is a proper part of patient x's thoracic cavity. Similarly from (i) my articular disc is located in my synovial cavity and (ii) my synovial cavity is (a proper) part of my TMJ we can derive that my articular disc is located in my TMJ. (This conclusion, of course also follows from the fact that the articular disc is a part of my TMJ by LT3). Using (LT6) we can derive for example that the left part of my articular disc is located in my synovial cavity from (i) the left part of my articular disc is part of my articular disc and (ii) articular disc is located in my synovial cavity.[1]

---

[1] In general, if relation S is transitive and relation R implies S then we can validly perform composition reasoning of the form (i) if $a\,R\,b$ and $b\,S\,c$ then derive $a\,S\,c$ and (ii) if $a\,S\,b$ and $b\,R\,c$ then derive $a\,S\,c$.

15.5 Connectedness Relations

A third primitive enables us to describe topological relations among individuals. On the intended interpretation, the connection relation C holds between individuals x and y if the distance between them is zero (where distance between extended individuals is here understood as the greatest lower bound of the distance between any point of the first individual and any point of the second individual). Intuitively, x is connected to y if and only if x and y overlap or x and y are in direct external contact. Two regions are connected at t if and only if they share at least a boundary point (they may share interior points). The left filled circle and the middle filled circle in Figure 15.2(a) are connected and so are the middle and the right filled circle in the figure. Moreover, the two filled circles in Figure 15.2(b) are (externally) connected.[2]

The following relations are defined using the connection relation: x and y are *externally connected* if and only if x and y are connected and x and y do not partially coincide (D_{EC}); x and y are *separated* if and only if x and y are not connected (D_{SP}). Symbolically:

$(D_{EC})\ ECxy \equiv Cxy \land \neg PCoin(x, y)$ (x and y are externally connected)
$(D_{SP})\ SPxy \equiv \neg Cxy$ (x and y are separated)

Consider Figure 15.1 (b). Every part of the TMJ in Figures 15.1 (a) and (c) is topologically equivalent to a filled circle[3] which is indicated by the labels of the nodes in the graph depicted in Figure 15.1(b). Thus, the nodes (the filled circles) in the graph represent proper parts of the TMJ. The edges of the graph represent external connectedness relations between parts of the TMJ depicted in Figures 15.1(a) and (c): the condyle (2), the articular disc (3), and the temporal bone (1) – all are separated from another – are externally connected to the posterior attachment (4) and to the lateral pterygoid muscle (5).

Axioms for the connection relation are as follows:

$(C1)\ Cxx$ (everything is connected to itself)
$(C2)\ Cxy \rightarrow Cyx$ (if x is connected to y, then y is connected to x)
$(C3)\ LocIn(x, y) \rightarrow (\forall z)(Czx \rightarrow Czy)$
         (if x is located in y, then everything connected to x is connected to y)

---

[2] Notice that we simplify matters here. Strictly speaking the relation that holds between material anatomical entities like the articular disc and the lateral pterygoid muscle is not the relation of external connectedness but the relation of adjacency. In this paper we ignore the distinction between these two relations. For a discussion of adjacency and qualitative distance relations see [2, 5].

[3] Two geometric figures are topologically equivalent if they can be made coincide by a transformation that involves change of shape (stretching, bending, ...) but no cutting, drilling holes, etc. For example a solid cube and a solid sphere are topologically equivalent but a sphere and a doughnut are not topologically equivalent since the doughnut has a hole and the solid sphere does not have a hole.

Thus C is reflexive and symmetric. It follows that EC and SP are irreflexive and symmetric. In addition, the following theorems can be derived:

$(CT1)$ $Cxy \leftrightarrow Cr(x)r(y)$ (x and y are connected if and only if their regions are connected)

$(CT2)$ $Pxy \rightarrow Cxy$ (if x is part of y, then x and y are connected)

$(CT3)$ $SPxy \rightarrow DSxy$ (if x and y are separated, then they are discrete)

$(CT4)$ $SPxy \rightarrow \neg PCoin(x, y)$ (if x and y are separated, then they are non-coincident)

$(CT5)$ $PCoin(x, y) \rightarrow Cxy$ (if x and y partially coincide, then x and y are connected)

$(CT6)$ $PCoin(x, y) \vee ECxy \vee SPxy$ (any two individuals either partially coincide, are externally connected, or are separated)

$(CT7)$ $Cxy \wedge LocIn(y, z) \rightarrow Cxz$

     (if x is connected to y and y is located in z then x is connected to z)

$(CT8)$ $LocIn(x, y) \wedge SPyz \rightarrow SPxz$ (if x is located in y and y is separated from z, then x is separated from z)

CT1-6 establish the interrelationships between connectedness (and relations defined in terms of connectedness) and other relations that are based on parthood and location. CT7 and CT8, again, are theorems that support reasoning by composition of relations. For example, by theorem (CT7) we validly can infer that my posterior attachment is connected to my synovial cavity from the facts that (i) my posterior attachment is connected to my articular disc and (ii) my articular disc is located in my synovial cavity.

15.6 Convexity

An anatomical ontology with only parthood relations, connectedness relations, and the region function alone is not powerful enough to sufficiently characterize the important properties of TMJs. Consider the graph in Figure 15.1 (b). It is a graph-theoretical representation of the mereotopological properties of the TMJs depicted in Figures 15.1 (a) and (c). Hence, in terms of mereotopological properties and relations we cannot distinguish the TMJs depicted in Figures 15.1 (a) and (c) because they are mereotopologically equivalent, i.e., indistinguishable.

Similarly, with only parthood relations, connectedness relations, and the region function we cannot introduce the kind of location relation that holds between my pleural space and my pleural membrane. The region of my pleural space does not overlap the region of my pleural membrane. The region of my pleural space lies within a region which is somewhat bigger than the region of my pleural membrane – convex hull the region of my pleural membrane.

A *convex* region is one which includes any straight line segment connecting any of its parts. For example, the region occupied by a solid ball is convex. Regions occupied by a drinking glass or my pleural membrane are not convex. The convex hull of an individual x is the smallest convex region of which x's region is part. For example, the convex hull of my pleural membrane extends over both the pleural membrane and the space inside the pleural membrane. See also [7].

We add a convex hull function (symbolized 'ch') which maps each individual to its convex hull. Important properties of the convex hull function are captured in the following axioms:

> $(CH1)$ $Pr(x)ch(x)$ (x's region is part of x's convex hull)
> $(CH2)$ $LocIn(x, y) \rightarrow Pch(x)ch(y)$
>   (if x is located in y, then x's convex hull is part of y's convex hull)
> $(CH3)$ $ch(ch(x)) = ch(x)$ (the convex hull of x's convex hull
>   is x's convex hull)

Using the convex hull function we can now generalize location relations between spatial individuals:

Individual x is *surrounded* by individual y if and only if x's region is part of y's convex hull and x's region does not overlap y's region (D_{SurrBy}); Individual x is *partly surrounded* by individual y if and only if x's region overlaps y's convex hull and x's region does not overlap y's region ($D_{PSurrBy}$); Symbolically:

$$D_{SurrBy} \quad SurrBy(x, y) \equiv Pr(x)ch(y) \land \neg Or(x)r(y)$$
$$D_{PSurrBy} \; PSurrBy(x, y) \equiv Or(x)ch(y) \land \neg Or(x)r(y)$$

Examples for surrounded-by are depicted in Figures 15.3(a) and (b). Other examples include: my pleural space is surrounded by my pleural membrane and the cavity of my stomach is surrounded by the wall of my stomach. The bolus of food in my stomach is surrounded wall of my stomach. An example for partly-surrounded by is depicted in Figure 15.3(c). Similarly all of the teeth in my mouth are partly surrounded by their sockets.

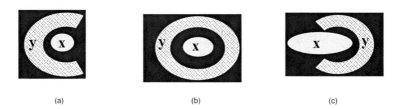

(a)   (b)   (c)

Fig. 15.3. (a) and (b): Two possibilities for x surrounded-by y. (c): x is partially surrounded by y. [12]

Consider Figure 15.1(a). The articular disc is surrounded by the fossa (the concave part of the temporal bone) and the head of the condyle is partly surrounded by the fossa. This is not the case in 15.1(c): neither the articular disc nor the head of the condyle are partially surrounded by the fossa. Hence taking into account the notion of convexity we are able to distinguish a TMJ in jaw open position from a TMJ in jaw closed position.

Consider the relation between my teeth and their sockets. Notice that this relation is different from the relation between x and y in Figure 15.3(c): my teeth are externally connected to their sockets while the individuals x and y in the figure are separated. Consequently, we could further refine the relations surrounded-by and partially-surrounded-by by taking the connectedness relations defined in Section 15.5 into account.

Neither *SurrBy* nor *PSurrBy* are transitive. For example, even though my tooth is partially surrounded by its socket, a filling or dental instrument may be partially surrounded by my tooth without also being partially surrounded by the tooth socket.

15.7 Ordering Relations between Extended Objects

Consider the TMJ depicted in Figure 15.4 [18, 5] and compare it with the TMJs depicted in Figures 15.1(a) and (c). Obviously it is critical to distinguish the TMJ in Figure 15.4 from the TMJs in Figures 15.1(a) and (c). It is the purpose of the disc in a TMJ to be *between* the condyle and temporal bone at all times. If we take the ordering relation of betweenness into account then the TMJs in Figures 15.1(a) and 15.1(c) can be distinguished from the clearly pathological TMJ in Figure 15.4 where the posterior attachment is between the condyle and the temporal bone and not the disc.

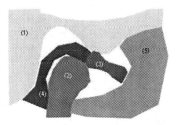

Fig. 15.4. TMJ with disc not positioned between condyle and temporal bone.

Ordering relations like betweenness describe situations where objects are placed in relative relation to each other. Besides betweenness, ordering relations include: left-of, right-of, in-front-of, above, below, behind, etc. The science of anatomy has developed a whole set of ordering relation terms to describe the arrangement of anatomical parts in the human body: superior, inferior, anterior, posterior, lateral, medial, dorsal, ventral, rostral, proximal, distal, etc. The FMA, for example, has an 'orientation network' in which these kinds of relations are represented [21].

Using ordering relations, frames of reference can be established. Consider the teeth as an example. Normally there is an upper set of teeth located in the skull, and a lower set located in the mandible. There are 16 upper teeth (8 to the left of the midline[4] and 8 to the right), and 16 lower teeth (8 to the left and 8 to the right of the midline). Each of the 32 teeth can be numbered. (At least two conventions are used.) Similarly, each tooth has five surfaces: mesial and distal; buccal and lingual; and occlusal. Mesial refers to the surface immediately adjacent to the tooth in front (toward the opening of the mouth) and distal refers to the surface immediately adjacent to the tooth in back (toward the throat). Buccal is the surface on the cheek side; lingual is on the tongue side. Occlusal refers to that surface of the which that faces its counterpart in the opposite (upper or lower) jaw.

Unfortunately, ordering relations between spatially extended objects are difficult to formalize. As [9] points out in her treatment of the relation of betweenness: 'The problem with trying to characterize the betweeness relation on extended objects is that we typically use the betweeness relation only on objects that have fairly uniform shapes and are nearly the same size. It is unclear whether or not the betweeness relation should hold in certain cases involving irregularly shaped objects and differently sized objects.' Similar problems face attempts to formalize qualitative direction relations between spatially extended objects, e.g., [19, 15, 15]. Similarly, it is very difficult to qualitatively describe distances between extended objects; particularly if they are of different size and shape, e.g., [31, 32].

15.8 Type-based Relations

In this chapter so far we have focussed on spatial relations between individuals. We used examples like: "*My* right ventricle is part of *my* heart" or "*My* brain is located in *my* cranial cavity".

Assertions of canonical anatomy, however, like:

the right ventricle is part of *the* heart
or
the brain is contained in *the* cranial cavity

are not limited to specific individuals but rather apply to all instances of the related anatomical types. On one interpretation, the first assertion tells us roughly that any right ventricle is part of a heart and any heart has a right ventricle as a part. The second assertion can be interpreted as saying roughly that any brain is contained in a cranial cavity and any cranial cavity contains a brain. Thus, these general statements

---

[4] The midline is a fiat boundary that bisects the body into right and left halves.

imply that certain spatial relations hold among very many specific individuals. The purpose of this section is to present a general procedure for extending a formal theory of spatial relations among individuals to also include relations among types corresponding to those made use of in the two assertions above. (For more details see also chapter 14 in this book.)

Let R be any binary relation on individuals – for example, the parthood relation (P), the overlap relation (O), the located in relation ($LocIn$), or any of the other relations introduced above. We can use R and the instantiation relation (see chapter 14) to define the following three relations among types. (See also [4, 14, 24, 28] where these distinctions are made for different versions of type-level parthood relations. [1] uses description logic for distinguishing versions of type-level parthood relations. [27] also define additional type-level relations like adjacent-to, participates-in, etc.)

(Type Relation Definition Schema 1)
$$R_1(A, B) \equiv (\forall x)(Inst(x, A) \rightarrow (\exists y)(Inst(y, B) \wedge Rxy))$$
(Type Relation Definition Schema 2)
$$R_2(A, B) \equiv (\forall y)(Inst(y, B) \rightarrow (\exists x)(Inst(x, A) \wedge Rxy))$$
(Type Relation Definition Schema 1-2)
$$R_{12}(A, B) \equiv R_1(A, B) \wedge R_2(A, B)$$

R_1 type-level relations place restrictions on all instances of the first argument. $R_1(A, B)$ tells us that something is true of all A's – each A stands in the R relation to some B. R_2 type-level relations place restrictions on all instances of the second argument. $R_2(A, B)$ tells us that something is true of all B's – for each B there is some A that stands in the R relation to it. R_{12} type-level relations place restrictions on all instances of both arguments. $R_{12}(A, B)$ tells us that something is true of all A's and something else is true of all B's– each A stands in the R relation to some B and for each B there is some A that stands in the R relation to it.

As an example, we consider how three such type-level relations are defined when R is the proper part relation (PP). PP_1 is the relation that holds between type A and type B if and only if every instance of A is a proper part of some instance of B. For example, every instance of *Human Female Reproductive System* is a proper part of some instance of *Human Being*. Thus, PP_1(*Human Female Reproductive System, Human Being*).

PP_2 is the relation that holds between type A and type B if and only if every instance of B has some instance of A as a proper part. For example, every instance of *Heart* has an instance of *Body Cavity* as a proper part. Thus, PP_2(*Body Cavity, Heart*). But notice that PP_2(*Human Female Reproductive System, Human Being*) does NOT hold, since not all human beings have female reproductive systems. Also notice that PP_1(*Body Cavity, Heart*) does NOT hold, since not all body cavities are part of a heart.

PP_{12} is the relation that holds between type A and type B if and only if: i) every instance of A is a proper part of some instance of B and ii) every instance of B has some instance of A as a proper part. For example, every instance of *Human Nervous System* is a proper part of some instance of *Human Being* and every instance of *Human Being* has some instance of *Human Nervous System* as a proper part. Thus, PP_{12}(*Human Nervous System, Human Being*). By contrast, neither PP_{12}(*Human Female Reproductive System, Human Being*) nor PP_{12}(*Cell, Heart*) hold.[5]

A few examples of assertions using other relations defined on types are the following:

O_{12}(*Bony Pelvis, Vertebral Column*)   (every bony pelvis overlaps some vertebral column and every vertebral column overlaps some bony pelvis)

O_1(*Male Genital System, Urinary System*)   (every male genital system overlaps some urinary system)

O_2(*Genital System, Male Urinary System*)   (every male urinary system overlaps some genital system)

$LocIn_{12}$(*Brain, Cranial Cavity*)   (every brain is located in some cranial cavity and some cranial cavity has a brain located in it)

$LocIn_2$(*PortionOfBlood, Cavity of the Right Ventricle*)   (some portion of blood is located in every cavity of a right ventricle)

$PCoin_{12}$(*Esophagus, Mediastinal Space*)   (every esophagus partially coincides with some mediastinal space and every mediastinal space partially coincides with some esophagus)

The examples given in this section should make clear that it is crucial for anatomical ontologists to explicitly distinguish individual-based relations and between the various kinds of type-level R_1, R_2, and R_{12} relations.

15.9 Reasoning about Type-level Relations

A detailed analysis of type-level relations and their logical properties can be found in [14]. In this section we briefly review some important logical properties of type-level relations and their consequences for automated reasoning.

Firstly, the logical properties of relations among individuals discussed in the first part of this chapter may not automatically transfer to the type-level relations that are defined in terms of them. This is one reason why it is important to always clearly distinguish the type-level relations from the individual-level relations. For example, overlap, O between individuals is a symmetric relation (PT5). The corresponding type-level relation, O_1, does not behave as a symmetric relation on anatomical types. For example it holds that every hand overlaps some nerve, O_1(*Hand, Nerve*), but it

---

[5] The relations P_1, P_2, and P_{12} correspond to the relations *part_of*, *is_part*, and *integral_part_of*, respectively, in chapter 14 of this book.

is not true that every nerve overlaps some hand, i.e., not $O_1(Nerve, Hand)$. Thus, although (PT5) is useful to draw inferences about relations between instances, there is no corresponding theorem that supports similar inferences about relations between types.

An overview of the correlation between some basic individual-level properties of a relation R and the logical properties of the corresponding type-level relations R_1, R_2, and R_{12} is given in Table 15.1. Fortunately, as the table shows, it does hold that if an individual-level relation R is transitive then so are the corresponding type-level relations R_1, R_2, and R_{12}. For example, since the proper parthood relation between individuals, PP, is transitive (PT3), it follows that the type-level relations PP_1, PP_2, and PP_{12} are transitive (CIT1-3)[6]. Similarly, since the located-in relation between individuals, $LocIn$, is transitive (LT2) it follows that the type-level relations $LocIn_1$, $LocIn_2$, and $LocIn_{12}$ are transitive (CIT4-6).

(CIT1-3) $PP_i(A, B) \wedge PP_i(B, C) \rightarrow PP_i(A, C)$ $i = 1, 2, 12$
(CIT4-6) $LocIn_i(A, B) \wedge LocIn_i(B, C) \rightarrow LocIn_i(A, C)$ $i = 1, 2, 12$

For example, it follows logically from $PP_2(Body\ Cavity, Heart)$ (every heart has some body cavity as a proper part) and $PP_2(Heart, Cardiovascular\ System)$ (every cardiovascular system has some heart as a proper part) that $PP_2(Body\ Cavity, Cardiovascular\ System)$ (every cardiovascular system has some body cavity as a proper part).

Thus transitivity reasoning between individuals generalizes to transitivity reasoning between types. Notice, however that we must not mix different kinds of type-level relations. For example we cannot derive anything from $PP_1(A, B)$ and $PP_2(B, C)$. PP_1 and PP_2 are distinct relations and transitivity reasoning cannot be applied to them.

There are limited ways of performing reasoning based on relation composition between type-level relations. We now show that the composition of type-level relations of the form:

> if $a\ R_{12}\ b$ and $b\ R_1\ c$ then derive $a\ R_1\ c$,
> if $a\ R_1\ b$ and $b\ R_{12}\ c$ then derive $a\ R_1\ c$,
> if $a\ R_{12}\ b$ and $b\ R_2\ c$ then derive $a\ R_2\ c$, and
> if $a\ R_2\ b$ and $b\ R_{12}\ c$ then derive $a\ R_2\ c$.

is valid. This kind of reasoning is valid because one important property of the R_{12} type-level relations is that they always imply the corresponding R_1 and R_2 type-

---

[6] To save pointless repetitions, we frequently condense into one line three distinct theorems which differ only in indexing of the type-level relations. Thus, for example, (CIT1-3) is a condensed representation of the following three theorems:

(CIT1) $PP_1(A, B) \wedge PP_1(B, C) \rightarrow PP_1(A, C)$
(CIT2) $PP_2(A, B) \wedge PP_2(B, C) \rightarrow PP_2(A, C)$
(CIT3) $PP_{12}(A, B) \wedge PP_{12}(B, C) \rightarrow PP_{12}(A, C)$.

Among Individuals	Among Types		
R is...	R_1 must also be...?	R_2 must also be...?	R_{12} must also be...?
Reflexive	Yes	Yes	Yes
Irreflexive	No	No	No
Symmetric	No	No	Yes
Asymmetric	No	No	No
Antisymmetric	No	No	No
Transitive	Yes	Yes	Yes

Table 15.1. Correlation between the logical properties of a relation R for individuals and the logical properties of the type-level relations R_1, R_2, and R_{12} [14].

level relations. More precisely, let R be any binary relation on individuals. Then the following two implications hold:

$$R_{12}(A, B) \rightarrow R_1(A, B) \quad R_{12}(A, B) \rightarrow R_2(A, B)$$

For example, we can prove:

$$\text{(ClT7-8)} \quad PP_{12}(A, B) \rightarrow PP_i(A, B) \qquad i = 1, 2$$
$$\text{(ClT9-10)} \; LocIn_{12}(A, B) \rightarrow LocIn_i(A, B) \; i = 1, 2$$

(ClT7)-(ClT10) allow us to substitute the stronger R_{12} relations for the weaker R_1 or R_2 relations in the antecedent of another implication. For example, in combination with the transitivity theorems (ClT1) - (ClT6) theorems (CLT7)-(CLT10) yield the following additional theorems:

$$\text{(ClT11-12)} \; PP_i(A, B) \wedge PP_{12}(B, C) \rightarrow PP_i(A, C) \qquad i = 1, 2$$
$$\text{(ClT13-14)} \; PP_{12}(A, B) \wedge PP_i(B, C) \rightarrow PP_i(A, C) \qquad i = 1, 2$$
$$\text{(ClT15-16)} \; LocIn_i(A, B) \wedge LocIn_{12}(B, C) \rightarrow LocIn_i(A, C) \; i = 1, 2$$
$$\text{(ClT17-18)} \; LocIn_{12}(A, B) \wedge LocIn_i(B, C) \rightarrow LocIn_i(A, C) \; i = 1, 2$$

Thus, from PP_1(*Uterus, Pelvis*) (every uterus is a proper part of some pelvis) and PP_{12}(*Pelvis, Trunk*) (every pelvis is a proper part of some trunk and every trunk has a pelvis as a proper part), it follows that PP_1(*Uterus, Trunk*) (every uterus is a proper part of some trunk). As another example, from PP_2(*Cartilage, Vertebra*) (every vertebra has some cartilage as a proper part) and PP_{12}(*Vertebra, Vertebral Column*) (every vertebra is a proper part of some vertebral column and every vertebral column has some vertebra as a proper part), it follows that PP_2(*Cartilage, Vertebral Column*) (every vertebral column has some cartilage as a proper part).

Another important point for reasoning about type-level relations is that the composition of relations between individuals generalizes to the composition of type-level relations in the following way. If implications of the form

$$Rxy \wedge Syz \rightarrow Rxz \quad \text{or} \quad Sxy \wedge Ryz \rightarrow Rxz$$

are theorems, then the three type-level counterparts of each of these formulae are theorems too, and thus permit the composition of relations at the type-level. For example, from theorems (LT5) and (LT6) about the composition of parthood and location at the levels of individuals, we can derive the following theorems at the type-level:

(CIT19-21) $LocIn_i(A, B) \land PP_i(B, C) \rightarrow LocIn_i(A, C)$ $i = 1, 2, 12$

(CIT22-24) $PP_i(A, B) \land LocIn_i(B, C) \rightarrow LocIn_i(A, C)$ $i = 1, 2, 12$

Thus, it follows from PP_{12}(*Middle Mediastinal Space, Thoracic Cavity*) and $LocIn_{12}$ (*Heart, Middle Mediastinal Space*), that $LocIn_{12}$(*Heart, Thoracic Cavity*).

Notice that, as in the case of transitivity, we are only permitted to combine relations with the same index, i.e., we cannot draw valid conclusions from $LocIn_1(A, B) \land PP_2(B, C)$. However, as discussed above, (CIT7)-(CIT10) allow us to substitute the stronger R_{12} relations for the weaker R_1 or R_2 relations in the antecedent of another implication. In combination with the composition theorems (CIT19) - (CIT24) theorems (CLT7)-(CLT10) yield additional theorems similar to (CLT11)-(CLT18).

All theorems discussed in this section are represented in compact form in Table 15.2.

	$PP_1(B,C)$	$PP_2(B,C)$	$PP_{12}(B,C)$	$L_1(B,C)$	$L_2(B,C)$	$L_{12}(B,C)$
$PP_1(A,B)$	$PP_1(A,C)$		$PP_1(A,C)$	$L_1(A,C)$		$L_1(A,C)$
$PP_2(A,B)$		$PP_2(A,C)$	$PP_2(A,C)$		$L_2(A,C)$	$L_2(A,C)$
$PP_{12}(A,B)$	$PP_1(A,C)$	$PP_2(A,C)$	$PP_{12}(A,C)$	$L_1(A,C)$	$L_2(A,C)$	$L_{12}(A,C)$
$L_1(A,B)$	$L_1(A,C)$		$L_1(A,C)$	$L_1(A,C)$		$L_1(A,C)$
$L_2(A,B)$		$L_2(A,C)$	$L_2(A,C)$		$L_2(A,C)$	$L_2(A,C)$
$L_{12}(A,B)$	$L_1(A,C)$	$L_2(A,C)$	$L_{12}(A,C)$	$L_1(A,C)$	$L_2(A,C)$	$L_{12}(A,C)$

Table 15.2. Composition table for type-level proper parthood and location relations. To save space we use L instead of *LocIn* as a symbol for located-in. [14]

Table 15.2 tells us which relation between type A and type C can be inferred from a given assertion about the relation between type A and type B (listed in row headings) in conjunction with an assertion about the relation between type B and type C (listed in the column headings). For example, given $PP_2(A, B)$ (row 2) and $LocIn_2(B, C)$ (column 5), it follows from the axioms given above that$LocIn_2(A, C)$ must also hold. (This is just theorem (CIT23).) A blank cell in the table tells us that, unless additional information is given, we cannot derive any assertion of the form $R_i(A, B)$ where R is an individual-level relation. For example, from $LocIn_1(A, B)$ (row 4) and $PP_2(B, C)$ we cannot in general make any inference about the relation between the types A and C. To see this, consider the following example. $LocIn_1$(*Prostate, Pelvic Cavity*) (every prostate is located in a pelvic cavity) and PP_2(*Pelvic Cavity, Female Pelvis*) (every female pelvis has a pelvic cavity as a proper part) both hold, but *Prostate* stands in none of the relations PP_1, PP_2, PP_{12}, $LocIn_1$, $LocIn_2$, or $LocIn_{12}$ to *Female Pelvis*.

15.10 Conclusions

The quality of any ontology depends heavily on an unambiguous use of its relation terms. However, ambiguity can be only avoided if we have a clear understanding of the relations that are denoted by these terms. The best way to achieve this goal is to embed the terms in a formal theory, because this approach enables us to analyze the connections between the relations and their logical properties. In this chapter we have presented, within the framework of a first-order logical theory, a formal account of those spatial relations which are critical for the representation of anatomical facts. At the level of relations between individuals we distinguished mereological, topological, and ordering relations, as well as relations that can be defined using the notion of convexity. We then discussed, how an individual-based ontology of spatial relations can be extended to take spatial relations between types into account. We showed that for every binary spatial relation at the level of individuals at least three type-level relations can be distinguished.

The content of this chapter is a compilation of material presented in a series of papers on spatial relations in biomedical ontologies: [5, 11, 12, 13, 14].

References

1. R. Beck and S. Schulz. Logic-based remodeling of the digital anatomist foundational model. In *AIMA Annual Symposium Proceedings*, pages 71–75, 2003.
2. T. Bittner and M. Donnelly. Logical properties of foundational mereotopological and adjacency relations in bio-ontologies. *Technical report, Department of Philosophy, State University of New York at Buffalo*, 2007.
3. T. Bittner and M. Donnelly. Logical properties of foundational relations in bio-ontologies. *Artificial Intelligence in Medicine*, 39:197–216, 2007.
4. T. Bittner, M. Donnelly, and B. Smith. Individuals, universals, collections: On the foundational relations of ontology. In A.C. Varzi and L. Vieu, editors, *Proceedings of the third International Conference on Formal Ontology in Information Systems, FOIS04*, volume 114 of *Frontiers in Artificial Intelligence and Applications*, pages 37–48. IOS Press, 2004.
5. T. Bittner and L. J. Goldberg. The qualitative and time-dependent character of spatial relations in biomedical ontologies. *Bioinformatics*, doi: 10.1093/bioinformatics/btm155, 2007.
6. R. Casati and A. C. Varzi. *Parts and Places*. Cambridge, MA: MIT Press, 1999.
7. A.G. Cohn, D.A. Randell, and Z. Cui. Taxonomies of logically defined qualitative spatial relations. *International Journal of Human Computer Studies*, 43:831–846, 1995.
8. I.M. Copi. *Symbolic Logic*. Prentice Hall, 1979.
9. M. Donnelly. *An Axiomatization of Common-Sense Geometry*. PhD thesis, University of Texas at Austin, 2001.

10. M. Donnelly. A formal theory for reasoning about parthood, connection, and location. *Journal of Artificial Intelligence*, 160:145–172, 2004.
11. M. Donnelly. On parts and holes: The spatial structure of the human body. In M. Fieschi, E. Coiera, and Y. J. Li, editors, *Proceedings of the 11th World Congress on Medical Informatics (MedInfo-04)*, pages 351–356, 2004.
12. M. Donnelly. Containment relations in anatomical ontologies. In *AIMA Annual Symposium Proceedings*, pages 206–210, 2005.
13. M. Donnelly and T. Bittner. Spatial relations between classes of individuals. In D. Mark and T. Cohn, editors, *Spatial Information Theory. Cognitive and Computational Foundations of Geographic Information Science. International Conference (COSIT 2005)*, number 3693 in Lecture Notes in Computer Science, pages 182 – 199. Springer Verlag, 2005.
14. M. Donnelly, T. Bittner, and C. Rosse. A formal theory for spatial representation and reasoning in bio-medical ontologies. *Artificial Intelligence in Medicine*, 36(1):1–27, 2006.
15. R. K. Goyal and M. J. Egenhofer. The direction-relation matrix: A representation for direction relations between extended spatial objects. In *UCGIS Annual Assembly and Summer Retreat*, Bar Harbor, ME, USA, 1997.
16. U. Hahn, S. Schulz, and M. Romacker. Partonomic reasoning as taxonomic reasoning in medicine. In *Proceedings of the 16th National Conference on Artificial Intelligence and 11th Innovative Applications of Artificial Intelligence Conference*, pages 271–276, 1998.
17. José L. V. Mejino Jr. and Cornelius Rosse. Symbolic modeling of structural relationships in the foundational model of anatomy. In *KR-MED*, pages 48–62, 2004.
18. D. M. Laskin, C. S. Greene, and W. L. Hylander, editors. *TMJs - An Evidence Based-Approach to Diagnosis and Treatment*. Quintessence Books, Chicago, 2006.
19. D. Papadias and T. Sellis. On the qualitative representation of spatial knowledge in 2d space. *VLDB Journal, Special Issue on Spatial Databases*, pages pp. 479–516, 1994.
20. J. Rogers and A. Rector. GALEN's model of parts and wholes: experience and comparisons. In *Proceedings of the AMIA Symp 2000*, pages 714–8, 2000.
21. C. Rosse and J. L. V. Mejino. A reference ontology for bioinformatics: The Foundational Model of Anatomy. *Journal of Biomedical Informatics*, 36:478–500, 2003.
22. S. Schulz and U. Hahn. Mereotopological reasoning about parts and (w)holes in bio-ontologies. In C. Welty and B. Smith, editors, *Formal Ontology in Information Systems. Collected Papers from the 2nd International Conference*, pages 210 – 221, 2001.
23. S. Schulz and U. Hahn. Parts, locations, and holes: Formal reasoning about anatomical structures. In S. Quaglini, P. Barahona, and S. Andreassen, editors, *Artificial Intelligence in Medicine. Proceedings of the 8th Conference on Artificial Intelligence in Medicine in Europe – AIME2001*, Lecture Notes in Artificial Intelligence, pages 293 – 303. Berlin: Springer, 2001.
24. S. Schulz and U. Hahn. Representing natural kinds by spatial inclusion and containment. In R. Lopez de Mantaras and L. Saitta, editors, *Proceedings of the 16th European Conference on Artificial Intelligence*, pages 283–287. IOS Press, Amsterdam, 2004.
25. S. Schulz, U. Hahn, and M. Romacker. Modeling anatomical spatial relations with description logics. In *AMIA 2000 – Proceedings of the Annual Symposium of the American Medical Informatics Association*, pages 779 – 783, 2000.
26. P. Simons. *Parts, A Study in Ontology*. Clarendon Press, Oxford, 1987.
27. B. Smith, W. Ceusters, B. Klagges, J. Köhler, A. Kumar, J. Lomax, C. Mungall, F. Neuhaus, A. Rector, and C. Rosse. Relations in biomedical ontologies. *Gnome Biology*, 6(5):r46, 2005.
28. B. Smith and C. Rosse. The role of foundational relations in the alignment of biomedical ontologies. In M. Fieschi, E. Coiera, and Y. J. Li, editors, *Proceedings of the 11th World Congress on Medical Informatics*, pages 444–448, 2004.

29. K.A. Spackman. Normal forms for description logic expressions of clinical concepts in SNOMED RT. *Journal of the American Medical Informatics Association*, pages 627–631, 2001. Symposium Supplement.

30. A. Varzi. Mereology. In Edward N. Zalta, editor, *Stanford Encyclopedia of Philosophy*. Stanford: CSLI (internet publication), 2003.

31. M. Worboys. Metrics and topologies for geographic space. In M.J. Kraak and M. Molenaar, editors, *Advances in Geographic Information Systems Research II: Proceedings of the International Symposium on Spatial Data Handling, Delft*, pages 7A.1–7A.11. International Geographical Union, 1996.

32. M. F. Worboys. Nearness relations in environmental space. *International Journal of Geographical Information Science*, 15(7):633–651, 2001.

CARO — The Common Anatomy Reference Ontology

Melissa A. Haendel, Fabian Neuhaus, David Osumi-Sutherland, Paula M. Mabee, José L.V. Mejino Jr., Chris J. Mungall, and Barry Smith*

Summary. The Common Anatomy Reference Ontology (CARO) is being developed to facilitate interoperability between existing anatomy ontologies for different species, and will provide a template for building new anatomy ontologies. CARO has a structural axis of classification based on the top-level nodes of the Foundational Model of Anatomy. CARO will complement the developmental process sub-ontology of the GO Biological Process ontology, using the latter to ensure the coherent treatment of developmental stages, and to provide a common framework for the model organism communities to classify developmental structures. Definitions for the types and relationships are being generated by a consortium of investigators from diverse backgrounds to ensure applicability to all organisms. CARO will support the coordination of cross-species ontologies at all levels of anatomical granularity by cross-referencing types within the cell type ontology (CL) and the Gene Ontology (GO) Cellular Component ontology. A complete cross-species CARO could be utilized by other ontologies for cross-product generation.

16.1 Necessity of a Common Anatomy Reference Ontology

Genomes are modified over evolutionary time to produce a diversity of anatomical forms. Understanding the relationship between a genome and its phenotypic outcome requires an integrative approach that synthesizes knowledge derived from the study of biological entities at various levels of granularity, encompassing gene structure and function, development, phylogenetic relationships, and ecology.

Many model organism databases collect large amounts of data on the relationship between genetic/genomic variation and morphological phenotypes in databases. Model organism databases standardize the description of morphological phenotypes and gene expression patterns by using types from anatomy ontologies that are specific to their focus species of interest. These ontologies have allowed the model organism databases to group phenotypic and gene expression data pertaining to partic-

---

* Melissa Haendel, Fabian Neuhaus, and David Osumi-Sutherland contributed equally to this chapter.

ular anatomical types.[2] Methods of phenotype curation are being extended and standardized as part of the work of the National Center for Biomedical Ontology, which aims to provide data-mining tools that can be applied across all species. In particular these tools will facilitate queries relating to anatomical structures and associated genes. Currently, however, there is no system for standardizing the representation of anatomy in ontologies.

Cross-species standardization among anatomy ontologies would bring a number of benefits. First, it would allow the development of standardized tools for grouping and querying anatomy-linked data. Second, it is a prerequisite for inference of anatomically based phenotypic and gene expression data within and across species. Third, if anatomy ontologies were standardized, then a method for representing homology between anatomical types in different anatomy ontologies could be devised. Fourth, standardization would allow better interoperability between anatomy ontologies and other ontologies.

In this chapter, we propose a common anatomy reference ontology (CARO), which is designed to serve as a standardized, generic structural classification system for anatomical entities. We also propose a standardized set of relations for use in building anatomy ontologies, extending the set of relations already defined as part of the OBO Relations Ontology [17]. By necessity, this proposal also begins to address the key issue of representation of homology between anatomical types in the context of anatomy ontologies.

This chapter summarizes progress on creating CARO, drawing on conclusions reached during an anatomy ontology workshop held in Seattle, WA, in September of 2006 sponsored by the National Center for Biomedical Ontology.[3]

16.2 What is CARO?

CARO is an ontology of common anatomy. At its core is a single, structural classification scheme based on that developed by the Foundational Model of Anatomy (FMA), a well established ontology of human anatomy [11] – see also Rosse and Mejino, in this volume – which adheres to the principles laid out by the OBO Foundry.[4] CARO has also adopted the FMA policy of single inheritance. This policy is based principally on the empirical observation that ontologies that allow multiple

---

[2] In keeping with the nomenclature of Smith et al. [18], we prefer the term 'type' to 'class'. Ontologies contain terms that refer to types of things in the real world. A type should not be confused with its instances. For example, a human anatomy ontology might contain the term 'foot'. This refers to the type *human foot*, of which your left foot is an instance. The collection of all such instances is the extension of the corresponding type.

[3] http://bioontology.org/wiki/index.php/Anatomy_Ontology_Workshop

[4] http://www.obofoundry.org/

inheritance, while easier to build, are marked by characteristic errors, which generally result from the use of multiple classification schemes within a single ontology, leading to what has been called '*is_a* overloading'. This can be avoided by utilizing *genus-differentia* definitions of the terms in ontologies, in which each type is specified as a refinement (via some *differentia*) of an existing more general type (the *genus*, i.e. the corresponding parent type, in the *is_a* hierarchy). Definitions following this form are typically written along the lines of 'An S *is_a* G which D'. This provides unambiguous definitions that can be applied consistently and leads, if done properly, to clean classification hierarchies in which all types have a single (*is_a*) parent and all children of a given type are disjoint (so that nothing can be an instance of both a type and its sibling).

CARO provides relations and the definitions for high-level anatomical types for canonical anatomies. A canonical anatomy gives an account of the 'prototypical' composition of the members of a given species.[5] This simplifies the task of constructing anatomy ontologies, because information captured in them, for example pertaining to part and location relationships, can differ radically in non-canonical types. Scientific communities have different perspectives on what constitutes canonical anatomy. Biologists working on model organisms generally have a standard strain or strains that are considered 'wild-type' for their chosen species. Within medicine, canonical anatomy is a generalization deduced from qualitative observations that are implicitly sanctioned by their accepted usage by anatomists [12, 18]. Defining canonical anatomy is even more problematic in the context of evolutionary biology, where natural variation within a species is often the object of study. Taxonomists therefore utilize voucher or 'type' specimens to define what is representative for a given species.[6] Extensions of CARO to enable integration with the disease ontology (DO) or other ontologies representing pathology or non-canonical anatomy can be accomplished in due course; but such integration will be unfeasible except on the basis of a foundation of canonical anatomy in relation to which relevant deviations can be defined.

CARO includes structural definitions of many generic anatomical types such as cell, portion of tissue, complex organ, anatomical system, and multicellular organism (see appendix for a complete list), organized in an *is_a* hierarchy. *Part_of* and other relations between these types will also be represented. CARO thereby provides a standardized reference ontology on which to build single or multi-species anatomy ontologies or from which to reorganize existing ontologies. This can be achieved by using a clone of CARO to create upper-level types for a single or multi-species ontology. As part of a single or multi-species ontology, the cloned types will refer to anatomical types in the species or taxon in question. Each of these types cloned

---

[5] For a more detailed analysis see Chapter 14.

[6] International Commission on Zoological Nomenclature, International Code of Zoological Nomenclature online, chapter 13: The type concept in nomenclature, Article 61. Principles of Typification. http://www.iczn.org/iczn/index.jsp

from CARO will have an *is_a* relationship to the corresponding CARO type, and will inherit from the latter its definition.

The CARO types *cell* and *cellular component* are potential root nodes for two existing non-species-specific anatomy ontologies: GO *cell component* and OBO *cell type*. Work is already under way to coordinate definitions and type names that are common to CARO and the latter ontologies, and definitions in all three ontologies will cross-reference each other.

A structural classification alone is not sufficient for the complete representation of anatomy. Other classification systems required for this task include an ontology of functions applicable to anatomical structures and an ontology of phenotypic qualities such as shape (see Figure 16.1). Types from ontologies of function and quality can be used in conjunction with CARO types to build combined anatomy ontologies for single species with multiple inheritance 'views'. For example, components of the immune system are grouped based on the function 'body defense'; they are not part of some single structure or group that can be structurally defined in CARO. Some suitable ontologies of functions are already in existence or are planned (GO Molecular Function [5]; FMP [3]). However, it may be necessary to supplement these ontologies with others still to be created.

Anatomical types classified under CARO can also be linked to types representing biological processes in which they participate, such as those found in the Biological Process Ontology (GO) or in developmental stage ontologies (see Section 16.6). The formalism for combining definitions of types from different parent ontologies in a definition follows the genus and differentia methodology described earlier.

CARO is an ontology of independent anatomical continuants. Continuants have a continuous existence through time. Dependent continuant entities are things that inhere in independent continuant entities such as qualities and functions. Occurrents (processes) have temporal parts which unfold in time (every occurrent depends on one or more independent continuants as its participant or bearer). The prefixes shown in parentheses in Figure 16.1 refer to ontologies that are either under development (FMP, RnaO, PrO) or are available at OBO web site.[7]

16.3 CARO Structure and Definitions

At time of writing, the first version of CARO is under active development. A CARO listserve and wiki track discussion of the ontology and related subjects. CARO can be downloaded in obo and owl formats.[8]

The CARO types and definitions are based on the topmost nodes of the FMA (see [11]; and also Rosse and Mejino, elsewhere in this volume). The top levels of

---

[7] http://obo.sourceforge.net/browse.html
[8] http://obo.sourceforge.net/cgi-bin/detail.cgi?caro

RELATION TO TIME GRANULARITY	CONTINUANT				OCCURRENT
	INDEPENDENT		DEPENDENT		
ORGAN AND ORGANISM	Organism (NCBI Taxonomy)	Anatomical Entity (CARO)	Organ Function (FMP)	Phenotypic Quality (PaTO)	Biological Process (GO)
CELL AND CELLULAR COMPONENT	Cell (CL)	Cellular Component (GO)	Cellular Function (GO)		
MOLECULE	Molecule (ChEBI, SO, RnaO, PrO)		Molecular Function (GO)		Molecular Process (GO)

Fig. 16.1. Coverage of species-independent ontologies relevant to biology

the FMA provide a rich set of abstract structural classifications that take into account qualities such as dimensions and contiguity and cover many levels of granularity from whole organism down to cell parts. All of these characteristics have made the FMA an ideal starting point for CARO. However, many of the FMA type definitions are not applicable to all species; some are mammal-specific, some are human-specific, and some are specific to only adult humans. The definitions of these types have been generalized in CARO to be inclusive of more species. Organismal domain specialists will be required to validate the CARO types, in much the same way that human anatomists were required to build and validate the FMA. In addition, the FMA is incomplete in its treatment of developmental structures and developmental relations. Because the representation of developmental anatomy in ontologies is central to the functioning of multiple model organism databases, we have begun to extend the CARO classification scheme to fill this gap. Figure 16.2 shows the taxonomy of the types in CARO. At the end of this chapter we have appended a table that lists all types of CARO including their definitions. Definitions which have been modified from those used by the FMA for use in CARO are discussed below.

16.3.1 Representing Granularity

In order to represent different levels of granularity in CARO, the appropriate types must be specified in such a way as to be applicable across all taxa. The FMA has a well developed system for classifying structural types according to a hierarchy of granularity. Each level of the hierarchy defines the basic building blocks for the level above; for example, portions of tissues are defined as aggregates of cells. However, because the FMA applies only to human anatomy, the FMA developers have used

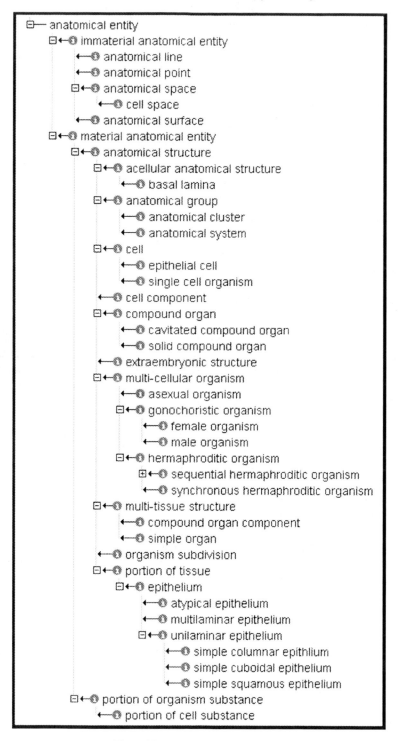

Fig. 16.2. The taxonomy of CARO.

both this bottom up definition of structural types along with a human-specific top down naming system: a cardinal organ part is made up of multiple portions of tissues and an organ is made up of multiple cardinal organ parts. The term 'organ' in the FMA scheme is therefore restricted to structures with a high level of granularity. We have retained this scheme, but have renamed 'cardinal organ part' as *multi-tissue structure* and redrafted the definition so that it also applies to aggregates of portions of tissue that are not themselves part of compound organs. This results in two subtypes of multi-tissue structure. The first, *simple organ*, is representative of many structural units in anatomically simpler organisms and during the development of more anatomically complex organisms. The second, *compound organ component*, refers to discrete multi-tissue structures found within compound organs.

In order to accommodate anatomical structures which are comprised of other anatomical structures of varying levels of granularity, we propose the type *anatomical group*. The subtypes of anatomical group are *anatomical cluster* and *anatomical system*, which permit classification of structures connected either directly or distally. In contrast to an anatomical cluster, the major elements of an anatomical system are discrete, localized anatomical structures of any granularity, or anatomical clusters of varying granularity, distributed across an organism. It has components that while connected, are not adjacent to each other and are separated by intervening structures that are not part of the system. Particularly illustrative examples are the nervous system, the vascular system of vertebrates and the tracheal tree of arthropods. In these examples, the system is in the form of trees or networks that are woven into the fabric of other tissues and organs. The type *anatomical group* and its children allow representation of systems or clusters of anatomical structures for all organisms, where the component parts may vary in their degree of granularity.

Portion of tissue: The term 'tissue' is used sometimes as a mass noun (compare: 'luggage', 'sugar') in such a way as to refer ambiguously to indeterminate amounts of cellular material. We prefer *portion of tissue* (a count noun analogous to 'suitcase' or 'sugar-lump') to make it clear that the term refers unambiguously to a single discrete structure. In addition, we have altered the definition to make 'cells of one or more types spatially arranged in a characteristic pattern' one of the defining features of tissue, rather than 'similarly specialized cells' as we believe this to be more inclusive of different taxa and of developing structures. 'Characteristic' is used to signify that each type of portion of tissue is marked by a distinctive pattern of organization of cells of distinctive types.

16.3.2 Defining Organism Subdivisions

Definitions based on the level of granularity are not sufficient to define all types of anatomical structure. Some types need to be defined as divisions of a whole organism. The segmental organization of the anterior-posterior body axis in arthropods and annelids provides a particularly clear example. Segments are not defined by their level of granularity (e.g. portions of tissue, multi-tissue structures, etc.), but by morphological boundaries distributed along the anterior-posterior axis of the animal. However, within a particular taxonomic group, it may be possible to develop specific definitions of divisions of the whole organism that specify the granularity of these regions as well as defining them in relation to other such divisions. For example, the FMA's definition of cardinal body part subtypes (head, neck, trunk and limbs) is defined relative to the skeletal system. Because the particular ways that organisms are divided up differs between taxonomic groupings, we have added a generic node in place of 'cardinal body part', *organism subdivision*. This can be used as a parent term for more detailed definitions, including specification of granularity if appropriate, in more taxonomically restricted anatomy ontologies.

16.3.3 Cross-ontology Coordination of CARO Types

A number of types in CARO are present in other ontologies, such as the Gene Ontology Cellular Component (GO CC), and the Cell Type ontology (CL) (see Table 16.1). Specifically, these types represent integration of different levels of anatomical granularity. Coordination of definitions between the GO CC, the CL, and CARO ontologies has begun, and these types will be linked via cross-references.

Table 16.1. CARO types and their corresponding types in other OBO ontologies

CARO	other OBO ontologies
acellular anatomical structure	GO:0044421 *extracellular region part*
cell	GO:0005623 *cell* and CL:0000000 *cell*
epithelial cell	CL:0000066 *epithelial cell*
cell component	GO:0044464 *cell part*
basal lamina	GO:0005605 *basal lamina*

16.3.4 The Organism Types

We include the whole organism as an anatomical structure to allow the formulation of part relations of sexually dimorphic anatomical structures. For example, humans have as parts gonads, but only male humans have testes. Different life strategies for reproduction have different corresponding anatomical structures, requiring that these organism types be defined in CARO.

16.4 Developing Structure Types

Prior to extensive morphogenesis and differentiation, most developing structures are sufficiently simple that they can be defined as a subtype of the CARO type *portion of tissue*. In some cases, types originally defined for adult structures are clearly applicable to developing structures. For example, the regions of the imaginal discs of

Drosophila that will develop into adult appendages have a structure consistent with our definition of *columnar epithelium*. However, other developing tissues share many but not all of the qualities of mature tissues. For example, many tissues of the early *Drosophila* embryo fit the definition of *epithelium* except that they lack a basal lamina. For this reason, the number of generic structural types will be expanded in future versions of CARO to ensure applicability to developing tissues.

Our system also allows the gradual increases in granularity that occurs during development to be captured in a consistent fashion. As development proceeds, developing structures of different granularity levels are formed. As they do, such structures can be reclassified from *portion of tissue* to *multi-tissue structure*, etc. Use of structurally classified developmental types to curate gene expression and phenotypic data will make it possible to look for genes common to the development and maintenance of particular structural types and to the transitions from one structural type to another.

These generic structural types will provide a basic structural classification of developing structures. However, many important details of structural types specific to a single species or taxonomic group will need to be captured in the relevant leaf nodes (the lowest nodes) of species-specific anatomy ontologies. These details can be formalized by referencing structural qualities specified in the Phenotype Attribute and Trait Ontology.[9]

Structural classification is limited in its ability to capture some of the dynamic structural changes which are important to developmental biologists. Specifically, they are interested in defining and classifying portions of developing tissue. CARO cannot provide terms that refer to specific regions of portions of tissue that do not have a structural differentia, but we think it important to specify how this might best be achieved in species-specific or multi-species anatomy ontologies built using CARO as a template. In the following we will first discuss an example and afterwards present the template that allows us to define structures by shared cell fate.

Developmental biologists traditionally define and name portions of tissue, at least in part, on the basis of some shared fate: lens placode, limb field, limb bud, fat-body primordium, and so on. The boundaries of these regions delimit groups of cells that are precursors of some specific type or types of anatomical structure. For example, each of the pair of heart primordia in a zebrafish embryo consists of all the members of a connected group of heart precursor cells, and the *Drosophila* wing pouch consists of all members of a connected group of cells that give rise to the wing. This can be made explicit by the following definition:

d 13 *x is a wing pouch if and only if:*

---

[9] http://www.bioontology.org/wiki/index.php/PATO:Main_Page

1. *x is a portion of columnar epithelium such that some cells that are part of x are ancestors of some cells that are part of some instance of the type wing; and*
2. *for all y, z: if y is a cell that is part of x and y is the ancestor of the cell z, then there is some type C and some instance c such that c is an instance of C, z is part of c and (either C is identical with the type wing or wing develops_from C).*

The underlying template of this definition is:

d 14 *x is a P if and only if:*

1. *x is an instance of Q such that some cells that are part of x are ancestors of some cells that are part of some instance of the type D; and*
2. *for all y, z: if y is a cell that is part of x and y is the ancestor of the cell z, then there is some type C and some instance c such that c is an instance of C, z is part of c and (either C is identical with the type D or D develops_from C).*

In our example P is the developing type *wing pouch*, Q is the structurally defined supertype *columnar epithelium*, and D is the 'mature' type *wing*. The details of this formalization ensure that it is compatible with the apoptosis of cells that are part of precursor structures during development and can apply to precursor anatomical structures where cell division has ceased but which have yet to differentiate.

In order to apply this approach to structures that are the precursors of multiple later types we need to generalize the definition. Let P again be the developing type, Q the structurally defined supertype, and let S be a set of types of compound organs, multi-species structures, and (maximal) portions of tissue. (S is the set of types of entities that the instances of P develop into.) We now define:[10]

d 15 *x is a P if and only if:*

1. *x is an instance of Q such that for every element D of S the following holds: some cells that are part of x are ancestors of some cells that are part of some instance of D; and*
2. *for all y, z: if y is a cell that is part of x and y is the ancestor of the cell z, then there is some type C and some instance c such that c is an instance of C, z is part of c and (either C is an element of S or there is some element D of S such that D develops_from C).*

Note that the differentia of this definition schema distinguishes precursor tissues from other portions of developing tissues that do not consist of a group of cells sharing some fate. Hensen's node in the chicken embryo, for example, contains different precursors at different stages of gastrulation, and does not delimit a connected group of cells sharing some particular fate [14].

The definition schema 15 provides a template for definitions of types of precursor tissues, which can be used in species specific ontologies. As mentioned above,

---

[10] Definition schema 15 is a generalization of schema 14, since schema 14 is the consequence of schema 15 if we assume that $S = \{D\}$.

this approach is especially useful in cases where developing types cannot be defined on a purely structural bases, because the precursor tissues are not yet mopholog-ically distinct from their surroundings, but have been experimentally defined. The approach also provides a way to define germ-layers, mesoderm, ectoderm and endo-derm according to the classes of mature structure whose precursor cells they contain. Finally, as mature structures are named in these definitions, it is possible to use this information to group developing structures according to what they will develop into.

16.5 Relations in CARO

An ontology is a controlled vocabulary that encapsulates the meanings of its terms in a computer parsable form. An anatomy ontology consists of statements composed of two kind of terms, denoting types and relations, respectively. Typically such state-ments involve two type terms A and B, so that they are of the form: A *rel* B. Relations commonly encountered in anatomical ontologies include the *is_a* relation, indicating that one type is a subtype of another, and the *part_of* relation, indicating that every instance of the first type is, on the instance level, a part of some instance of the sec-ond type. Examples of use include *pancreas is_a lobular organ* in the FMA and *cell nucleus part_of cell* in the GO Cellular Component ontology. However, anatomical ontologies are by no means limited to these two relations; the FMA employs a large number of spatial relations [11][11] and ontologies that encompass entities at various developmental stages typically link types using relations such as *develops_from*, as in the OBO Cell Type ontology (CL) and in anatomical ontologies for model organisms such as fly and zebrafish.

Relations play an essential role in ontologies, since they are the primary bearer of semantic content (see Chapter 14). To ensure a consistent use of terms that de-note relationships within and across ontologies, it is important to agree on shared, unambiguous definitions of these terms. These definitions utilize the dependence of relationships between types (e.g. *cell nucleus* and *cell*) on the relationships between instances of these types (e.g. concrete cell nuclei and the cells which contain them), as is discussed in detail in the Chapters 14 and 15 of this book. In this section, we will discuss the extension of the OBO Relations Ontology [17] to provide relations that are necessary for CARO and species-specific anatomies. This extension comes in different flavors: (a) in some cases, we need to add new relations to capture impor-tant aspects of anatomical entities, (b) in other cases, we need to add new relations that further specify existing ones in order to better represent the dynamic changes within developing organisms, and (c) we need to consider relations that link anatomy ontologies to other ontologies.

---

[11] Also see 'spatial association relationship' at:
http://fme.biostr.washington.edu:8089/FME/index.html

16.5.1 Defining *develops from*

The OBO Relations Ontology covers the most important relationships for anatomy ontologies, but lacks explicit definitions of many spatial relations that it would be desirable to include. Some of these are discussed in chapter 15 of this book. Further, for CARO to provide a representation of developmental anatomy, we need to define a relationship that represents the various ways that anatomical structures change through development. We lack a single, transitive relationship that can represent the transformation, fission and fusion of developing structures over time. Here we outline the relationship *develops from*, which fulfills these criteria. In order to define *develops from* we need to distinguish two cases. In the first case, some entity changes its properties but remains numerically identical; for example, if an adult develops from a child, then the adult will have different properties (e.g. a different weight and height) but it will be still the same individual. In contrast, if a zygote develops from a sperm cell and an ovum, then the zygote is not identical with either; but the zygote arises from the sperm cell and the ovum. These two relations are used to define the type level relationships *transformation of* and *derives from*[12] in the OBO Relations Ontology. Since it is often unknown during development whether one structure arising during development is a transformation of another or whether some portion of a structure arises from another one, we need a *develops from* relation which covers both cases.

More formally, the *develops from* relationship is defined as follows:[13]

d 16 *C develops from D if and only if, for any x and any time t, the following holds: if x instantiates C at time t, then*

1. EITHER *for some time t_1, x instantiates D at t_1 and t_1* **precedes** *t, and there is no time interval t_2 such that x instantiates C at t_2 and x instantiates D at t_2;*
2. OR *for some time t_1, there is some y such that y instantiates D at t_1 and x* **arises from** *y.*

The relation **succeeds** is defined with the help of the relations **buds from** and **arises from**. Note while *develops from* is a relationship between types, **precedes**, **buds from**, **succeeds**, and **arises from** hold between instances.

d 17 *x* **arises from** *y is defined recursively in the following way:*

1. *if x* **succeeds** *y, then x* **arises from** *y;*
2. *if x* **buds from** *y, then x* **arises from** *y;*
3. *if x* **arises from** *y and y* **succeeds** *z, then x* **arises from** *z;*

---

[12] To avoid confusion with the very different meaning of 'derives from' in an evolutionary context, we plan to rename this type level relationship '*arises from*'. *The corresponding instance level relationship is referred to as* '**arises from**' *in the following text.*

[13] These definitions, and the definitions below, are provided for the sake of technical completeness. They will not play any role in the actual use of CARO in day-to-day annotation and information retrieval purposes.

4. *if x* **arises_from** *y and y* **buds_from** *z, then x* **arises_from** *z;*
5. *x* **arises_from** *y holds only because of (1)-(4).*

With other words **arises_from** is the transitive closure of **buds_from** and **succeeds**. The relations **succeeds** and **buds_from** are defined in the following way.[14]

d 18 *x* **succeeds** *y if and only if*

1. *x and y are instances of the type anatomical entity; and*
2. *x begins to exist at the same instant of time at which y ceases to exist; and*
3. *there is some anatomical structure z such that z is* **part_of** *y when y ceases to exist and z is* **part_of** *x when x begins to exist.*

d 19 *x* **buds_from** *y if and only if*

1. *x and y are anatomical entities; and*
2. *at no time t, x is* **part_of** *y at t; and*
3. *there is some anatomical structure z such that z is* **part_of** *y immediately before x begins to exist, and x* **succeeds** *z; and*
4. *x continues to exist for some interval of time from the point when y begins to exist.*

16.5.2 Defining Time-Restricted Part Relationships

The parthood relations as defined in the OBO Relations Ontology [17] do not adequately represent some dynamic aspects of developmental anatomy. In particular, the relationships *has_part* and *part_of*, both apply at all stages: *C has_part D* means that every *C*, regardless of stage, has some *D* as instance-level part. The *Drosophila* anatomy ontology, however, contains types of neuroblasts that are part of the ventral nerve cord primordium (VNC). As these neuroblasts divide, more types become identifiable – at stage 9 there are 10 types but by stage 11 there are 34 [1]. We cannot capture the part relationship between these cell types and the VNC primordium using the *has_part* relation, because this would imply that all instances of the VNC have instances of each of these neuroblast types as a part at all stages. Similarly, the relation *part_of* also applies irrespective of stage. We can solve this dilemma by defining versions of *part_of* and *has_part* which are applicable only during the stages in which both partners in the relationship exist. The formal definitions of these relationships are:

d 20 *C* **time_restricted_part_of** *D if and only if the following holds for any x and any time t: if x instantiates C at time t, then there is a y such that*

1. *for some time t_1, y instantiates D at t_1 and x* **part_of** *y at t_1; and*
2. *for all times t_2: if x* **exists_at** *t_2 and y* **exists_at** *t_2, then x is* **part_of** *y at t_2.*

---

[14] The observant reader will notice that these definitions are less rigorous than the previous ones. For a full logical analysis of '**buds_from**' and '**succeeds**' we would need to spell out the underlying temporal theory; which is beyond the scope of this chapter.

d 21 *C time_restricted_has_part D if and only if the following holds for any x and any time t: if x instantiates C at time t, then there is a y such that*

1. *for some time t_1, y instantiates D at t_1 and y* **part_of** *x at t_1; and*
2. *for all times t_2: if x* **exists_at** t_2 *and y* **exists_at** t_2, *then y is* **part_of** *x at t_2.*

16.5.3 Relationships Linking Separate Ontologies

As mentioned above, the structural classification of anatomical entities in CARO is separate from the treatment of functional classification and of homology between anatomical entities across different species. In order to record function and homology information, the anatomical types within a species-specific anatomy ontology need to be linked to types in other ontologies, and the necessary relations – including *has_function* and *homologous_to* – will be added to the OBO Relations Ontology in due course. We discuss relations between developmental stage and anatomical types in the following section. Note that the spatial relations and the *develops_from* relation mentioned above are relations that are used within a given anatomical ontology. In contrast relations such as *has_function*, *homologous_to*, *starts_during* and *ends_during* are relationships that link types across different ontologies. Similarly, *is_a*, too, can link types across different ontologies, as for instance when we make the assertion that *mouse compound organ is_a CARO:compound organ.*

16.6 Representing Stages

Development can be considered a process that *has_participant* [17] whole organism. For any single species, events during development occur in a predictable order. However, the precise timing of these events is dependent on environmental conditions. Developmental biologists traditionally measure progress through (the process of) development relative to the occurrence of some standard series of events which can be easily and reliably scored [2, 10]. A standard table of development divides the process of development into stages, each delimited by a pair of events, and it describes key events occurring within each stage.

For some organisms, not only is the order of events consistent, but under standard laboratory conditions their timing relative to a reference event (e.g. fertilization) shows little variation. In these cases it is possible to define stages in terms of the period of time that elapsed since the reference event. This method of defining stages is particularly useful if no easily score-able morphological stage criteria are available. For example, in the zebrafish, early stages are often referred to either by morphological criteria or by time since fertilization, while the later stages are referred to exclusively by time since fertilization [8].

As stage series are necessarily species-specific, ontologies representing individual stage series have to be constructed for each species. Minimally, a stage ontology will contain types for the stages that make up a standard table of development. The relative timing of these stages can be recorded using the relation *preceded_by* [17]. Stages can be grouped together into super-stages, or divided into sub-stages, with the latter having a *part_of* relationship to the stages themselves, which are in turn *part_of* super-stages. While stage series are species-specific, many of the developmental processes described in standard tables of development are not. Information about the relative timing of developmental processes described in each standard table of development can be captured within species-specific stage ontologies. The relative timing of these processes to each other and to stage boundaries can be recorded using the relations *part_of*, *preceded_by* and an additional relationship *simultaneous_with*[15]. Linking these to relevant GO types such as cellularization (see Figure 16.3) will facilitate reasoning between species-specific stage ontologies.

We propose that these species-specific stage ontologies be used to record the periods of development during which anatomical entities exist by using the relationships *starts_during* and *ends_during* (a formalized version of the strategy used by ZFIN). These relationships link anatomy ontology types to appropriate types in the stage ontology. This will give a crude resolution to records of timing: the existence of X begins some time during stage N and ends some time during stage N'. The temporal resolution of these links could be improved, as data allows, in two ways. Where some standard system of substages has been defined, we can simply make *starts_during* and *ends_during* links to these substages. Alternatively, we can refine our record of the timing of the beginning or end of existence of an anatomical entity by instantiating these as events within the stage ontology and using *preceded_by* relations to processes beginning or ending within a stage (see Figure 16.3).

16.7 CARO Depth and Application

The question of CARO depth is closely related to its utility in building new anatomy ontologies. The top-level types in CARO together with the relationships defined above can be used to structure application anatomy ontologies. However, the types in CARO are very generic relative to the types commonly defined within a species-specific anatomy ontology. This is because it is very difficult to further subtype CARO and remain within the bounds of disjoint structural definitions. For example, the compound eye of a *Drosophila* and the camera-lens eye of a human have little in common structurally, making it unlikely that the type *eye* would be included in CARO (though these types might be grouped, outside of CARO, using the function 'to see'). However, it may be possible to achieve a disjoint set of structural definitions for particular monophyletic groups within multi-species anatomy ontologies.

---

[15] To be defined in a future publication.

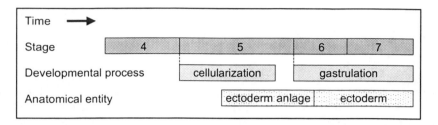

Fig. 16.3. Relationship between anatomical entities, stage, and process. For each species, an ontology will be constructed containing types for stage and developmental process in a single ontology of occurents. Anatomical entities are contained in a separate ontology of continuants. The ends of each bar represent events for which relative timing can be recorded using the relations *preceded_by* and *simultaneous_with*. These ordering relations will be used in conjunction with *starts_during* and *ends_during* to define the period during which an anatomical entity exists. This example illustrates ectoderm development in the *Drosophila* embryo, wherein the ectoderm anlage *starts_during* stage 5, the ectoderm anlage *ends_during* stage 6, the ectoderm *starts_during* stage 6, the process gastrulation *preceded_by* cellularization, and gastrulation *simultaneous_with* stage 6 and stage 7.

A number of projects aim to generate anatomy ontologies of multiple taxa. In particular, the Cypriniformes Tree of Life (CToL)[16], the plant ontology[17], as well as the amphibian[18], and Hymenoptera[19] anatomy ontologies. As in the case of species-specific anatomy ontologies, multi-species anatomy ontologies can also clone the CARO types for use as their topmost nodes. Within a multi-species anatomy ontology, a type that satisfies the definition of a CARO type will have an *is_a* relation to the CARO type with the *differentia* of a taxon rather than a species. For example, for the cypriniform fish anatomy ontology, the cypriniform type *compound organ is_a* CARO:*compound organ*, with the differentia being that it is a compound organ of a type found in Cypriniformes. CARO can in this way be used as a template for multi-species anatomy ontologies as well as for species-specific ones.

Currently, many ontology developers use an existing ontology when building a new one (as CARO itself is modeled on the FMA). For example, the zebrafish anatomy ontology has been used as a template for both fish and amphibian multi-species ontologies. This is because the zebrafish anatomy ontology refers to anatomical structures that evolved within chordates – a post-anal tail evolved at the level of Chordata, the lateral line system evolved at the level of Craniata, jaws evolved at the level of Gnathostomata, and bone at the level of Vertebrata (Figure 16.4).

Within multi-species anatomy ontologies it is necessary to specify in which organisms the anatomical entities are applicable. This can be accomplished with

[16] http://www.nescent.org/wg_fishevolution
[17] http://www.plantontology.org
[18] http://www.morphologynet.org
[19] http://ceb.scs.fsu.edu/ronquistlab/ontology/wiki/index.php/Main_Page

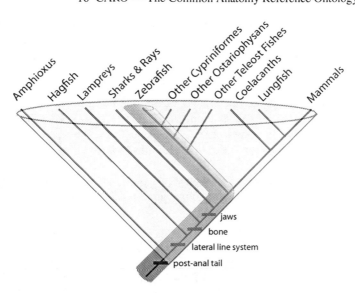

Fig. 16.4. Species-specific anatomy ontologies contain types applicable to more diverse taxa. The zebrafish anatomy ontology (inner lighter cylinder) includes terms referring to features that evolved at various times in the chordate lineage. This ontology could be expanded to include anatomical structures found in all vertebrates (entire cone).

the relation, *part_of_organism*, proposed by the CToL-ZFIN working group to link anatomical entities to taxa within a taxonomy ontology. Similarly, the types in CARO are not applicable to all organisms. For example, diploblastic animals such as cnidarians (a phylum that includes jellyfish and sea anemones) lack compound organs (a proposed CARO term) while sponges may have no distinct multi-tissue structures at all [6]. CARO classes could also be linked to a taxonomy ontology to indicate which classes are applicable at various taxonomic levels. The purpose of cross-referencing multi-species anatomy ontologies and CARO to a taxonomic ontology would be to provide a user with choice of appropriate types. A similar method has been proposed to limit classes to specific taxa in other species-independent ontologies such as the GO or the CL (Waclaw Kusnierczyk, personal communication). It is important to note that cross-referencing anatomy and taxonomy ontologies in this manner does not specify homology.

16.8 Representing Homology

Methods for recording homology between types in anatomy ontologies are extremely important both to provide resources for evolutionary biologists and for the development of tools for inter-species inference regarding the molecular basis of morphological phenotypes or traits. Structures (including genes) are homologous if they evolved from some structure in a common ancestor, and homology implies genealogical descent as the vehicle of transfer of information. Homology must be addressed within the context of multi-species anatomy ontologies because of the very nature of how anatomical structures evolve. The reason anatomical types are structurally or functionally similar, and therefore classified together in some ontology, may be because they are evolutionarily related. However, many well documented counter examples exist. For example, both zebrafish and humans have a skull bone named the parietal bone, and another named the frontal bone. These could be grouped in an ontology on the basis of position within the skull and name; but there is good evidence that the parietal bone in humans is homologous to the frontal bone in zebrafish [7, 13]. Thus, one cannot assume homology based on structural similarity or name.

We propose that homology information be captured independently of both structure and function information. Specifically, statements of homology are hypotheses and require evidence (codes) and attribution. This is particularly important to evolutionary biologists creating phylogenies, where different evidence is often used to generate different phylogenetic views. In light of this need to capture homology, a new relationship, *homologous_to*, is proposed to be included in the OBO Relations Ontology, but its definition is still under discussion. The ontological implications for this new relationship are as yet untested. For instance, if two structures are deemed homologous, is this information transitive down *is_a* chains? Can two structures be homologous if none of their parts are homologous? Erwin and Davidson [4] have suggested that the regulatory processes that underlie development may be homologous, whereas the creation of gross anatomical structures is specific to phyla or classes (and may not be homologous). In this respect, it is the processes or functions that are homologous whereas the structures are not.

To establish a homology relation between sister anatomical entities may require the determination of an evolutionary precursor in order to create sister subtypes within a multi-species anatomy ontology. It may prove difficult in some cases to define an evolutionary precursor purely on a structural basis and will require domain experts whose expertise spans large branches of the tree of life. However, it is possible that a function ontology used in combination with homology statements could overcome this difficulty. Multi-species anatomy ontologies will have to reconcile these homology issues with maintenance of disjoint definitions based on structure. It is important to note that even though one intended use for CARO is as a template for building multi-species anatomy ontologies, no homology between types is implied by common treatment within CARO, since CARO types are classified purely on the basis of structural criteria and not on evolutionary history.

16.9 Long Term CARO Goals

One of the long-term goals of CARO is to provide the source of standardized representations of anatomical types used in creating composite types of the kind found in ontologies such as the GO's Biological Process ontology. Like CARO, GO is cross-species, describing types of biological process that occur across a wide variety of species, encompassing types such as *heart development* and *neural tube closure*. Like CARO, GO is also canonical – it describes the features of typical, wild-type instances. At the present time, GO does not contain explicit references to types from an anatomical entity ontology. Instead, rough definitions of types such as *heart* and *neural tube* are 'embedded' inside the definitions of the corresponding GO types. This leads to redundancy, duplication of effort, inconsistency and a poor basis for cross-domain inference.

Once CARO is in use as a template for species-specific or multi-species anatomy ontologies, types from these ontologies along with their taxonomic reference can be referenced by the GO. GO will retain types such as *neural tube closure*, but the corresponding definitions can refer to definitions taken from CARO or from one of the multi-species or single-species anatomy ontologies created in a way which will allow the ontologies to be kept synchronized [9].

While the primary axis of classification in CARO is structural, not functional, this does not mean that CARO ignores function. Rather, CARO insists that function be treated as a separate *orthogonal* ontology. Instead of stating that *verterbrate eye is_a sense organ* as we may do in a mixed classification, we instead state that *vertebrate eye has_function visual perception*, with the *is_a* parent of verterbrate eye being the appropriate structural supertype (i.e. cavitated compound organ). Separating structure from function in this way leads to cleaner ontology design, with each type having a single *is_a* parent. At the same time, this methodology still allows for cross-ontology queries, such as 'find all genes active in seeing structures'. The organismal function ontology that will be used in conjunction with CARO or other anatomy ontologies is yet to be developed. Like CARO, this ontology will adhere to OBO Foundry principles and be itself placed in the OBO Foundry.[20] Many of these functions will be realized in biological processes of the kind found in the GO, so this ontology will be developed in coordination with the Gene Ontology Consortium.

One final consideration is that CARO compliance can be exploited to help build phylogenetic views of a given set of taxa. Since all species-specific and multi-species anatomy ontologies will have *is_a* links to CARO nodes, it will be possible to view an assembly of anatomical structures by limiting the taxonomic level. In combination with a set of homology statements, one could build different phylogenies based on different evidence. This is not unlike the current method of creating phylogenies, except that the anatomical structures are named and assigned to taxa in a standardized manner thereby providing links to other relevant data. For example, the development and function of homologous structures in two different species are likely to retain at least some of the molecular mechanisms present in the ancestral structure in their most recent common ancestor. CARO should in this way prove a useful organizational tool to facilitate the inference of molecular mechanisms underlying morphology.

---

[20] http://obofoundry.org/

Acknowledgements

This work was funded in part by the National Institutes of Health through the NIH Roadmap for Medical Research, Grant 1 U 54 HG004028. Information on the National Centers for Biomedical Computing can be found at `http://nihroadmap. nih.gov/bioinformatics`. The CToL-ZFIN working group has been supported by the National Evolutionary Synthesis Center. Additional funding came from a grant to FlyBase, `http://flybase.org`, from the National Human Genome Research Institute (USA), Grant 5 P41 HG 000 739. Thanks go to Michael Ashburner, Rachel Drysdale, Monte Westerfield, and Cornelius Rosse for their valuable comments.

References

1. C. Berger, J. Urban, and G.M. Technau. Stage-specific inductive signals in the Drosophila neuroectoderm control the temporal sequence of neuroblast specification. *Development*, 128: 3243-3251, 2001.
2. J.A. Campos-Ortega and V. Hartenstein. *The Embryonic Development of Drosophila melanogaster*. (Second Edition), Springer-Verlag, Berlin, 1999.
3. D.L. Cook, J.L.V. Mejino, and C. Rosse. Evolution of a foundational model of physiology: symbolic representation for functional bioinformatics. *Medinfo*, 11(1): 336-340, 2004.
4. D.H. Erwin and E.H. Davidson. The last common bilaterian ancestor. *Development*, 129: 3021-3032, 2002.
5. Gene Ontology Consortium. The Gene Ontology (GO) project in 2006. *Nucleic Acids Res.*, 34: D322-6, 2006.
6. P.W.H. Holland. Major transitions in animal evolution: a developmental genetic perspective. *American Zoologist*, 38(6): 829-842,1998.
7. M. Jollie. *Chordate Morphology*. New York, Reinhold Books, 1962.
8. C.B. Kimmel, W.W. Ballard, S.R. Kimmel, B. Ullmann, and T.F. Schilling. Stages of embryonic development of the zebrafish. *Dev. Dyn.*, 203: 253-310, 1995.
9. C. Mungall. Obol: integrating language and meaning in bio-ontologies. *Comparative and Functional Genomics*, 5 (6-7), 509-520, 2004.
10. P.D. Nieuwkoop and J. Faber. *Normal Table of Xenopus laevis*. 3rd Ed, 1994.
11. C. Rosse and J.L.V. Mejino. A reference ontology for bioinformatics: The Foundational Model of Anatomy. *Journal of Biomedical Informatics*, 36: 478-500, 2003.
12. C. Rosse, J.L.V. Mejino, B.R. Modayur, R. Jakobovits, K.P. Hinshaw, J.F. Brinkley. Motivation and organizational principles for anatomical knowledge representation: the Digital Anatomist symbolic knowledge base. *Journal of the American Medical Informatics Association*, 5(1): 17-40, 1998.
13. H. P. Schultze and M. Arsenault. The panderichthyid fish Elpistostege: A close relative of tetrapods? *Paleontology*, 28: 293-309, 1985.
14. M.A. Selleck and C.D. Stern. Fate mapping and cell lineage analysis of Hensen's node in the chick embryo. *Development*, 112(2): 615-626, 1991.
15. J. Slack. *From egg to Embryo* (2nd Ed). Cambridge University Press, 1991.
16. B. Smith. Fiat objects. *Topoi*, 20(2): 131-148, 2001.

17. B. Smith, W. Ceusters, B. Klagges, J. Köhler, A. Kumar, J. Lomax, C. Mungall, F. Neuhaus, A. Rector, and C. Rosse. Relations in biomedical ontologies. *Genome Biology*, 6(5):r46, 2005.
18. B. Smith, W. Ceusters, and C. Rosse. On carcinomas and other pathological entities. *Comparative and Functional Genomics*, vol 6, 7-8, 379-387, 2005.

Appendix

The following table contains the types of CARO and their definitions in the order they appear in Figure 16.2.

CARO Definitions

anatomical entity	Biological entity that is either an individual member of a biological species or constitutes the structural organization of an individual member of a biological species.
immaterial anatomical entity	Anatomical entity that has no mass.
anatomical line	Non-material anatomical entity of one dimension, which forms a boundary of an anatomical surface or is a modulation of an anatomical surface.
anatomical point	Non-material anatomical entity of zero dimension, which forms a boundary of an anatomical line or surface.
anatomical space	Non-material anatomical entity of three dimensions, that is generated by morphogenetic or other physiologic processes; is surrounded by one or more anatomical structures; contains one or more organism substances or anatomical structures.
cell space	Anatomical space that is part of a cell.
anatomical surface	Non-material anatomical entity of two dimensions, that is demarcated by anatomical lines or points on the external or internal surfaces of anatomical structures.
material anatomical entity	Anatomical entity that has mass.
anatomical structure	Material anatomical entity that has inherent 3D shape and is generated by coordinated expression of the organism's own genome.
acellular anatomical structure	Anatomical structure that consists of cell parts and cell substances and together does not constitute a cell or a tissue.
basal lamina	Acellular anatomical structure that consists of a thin sheet of fibrous proteins that underlie and support the cells of an epithelium. It separates the cells of an epithelium from any underlying tissue.
anatomical group	Anatomical structure consisting of at least two non-overlapping organs, multi-tissue aggregates or portion of tissues or cells of different types that does not constitute an organism, organ, multi-tissue aggregate, or portion of tissue.

CARO Definitions

anatomical cluster	Anatomical group that has its parts adjacent to one another.
anatomical system	Anatomical group that is has as its parts distinct anatomical structures interconnected by anatomical structures at a lower level of granularity.
cell	Anatomical structure that has as its parts a maximally connected cell compartment surrounded by a plasma membrane.
epithelial cell	Cell which has as its part a cytoskeleton that allows for tight cell to cell contact and which has apical-basal cell polarity.
single cell organism	Cell that is an individual member of a species.
cell component	Anatomical structure that is a direct part of the cell.
compound organ	Anatomical structure that has as its parts two or more multi-tissue structures of at least two different types and which through specific morphogenetic processes forms a single distinct structural unit demarcated by bona fide boundaries from other distinct anatomical structures of different types.
cavitated compound organ	Compound organ that contains one or more macroscopic anatomical spaces.
solid compound organ	Compound organ that does not contain macroscopic anatomical spaces.
extraembryonic structure	Anatomical structure that is contiguous with the embryo and is comprised of portions of tissue or cells that will not contribute to the embryo.
multi-cellular organism	Anatomical structure that is an individual member of a species and consists of more than one cell.
asexual organism	Multi-cellular organism that does not produce gametes.
gonochoristic organism	Multi-cellular organism that has male and female sexes.
female organism	Gonochoristic organism that can produce female gametes.
male organism	Gonochoristic organism that can produce male gametes.
hermaphroditic organism	Multi-cellular organism that can produce both male and female gametes.
sequential hermaphroditic organism	Hermaphroditic organism that produces gametes first of one sex, and then later of the other sex.
synchronous hermaphroditic organism	Hermaphroditic organism that produces both male and female gametes at the same time.
multi-tissue structure	Anatomical structure that has as its parts two or more portions of tissue of at least two different types and which through specific morphogenetic processes forms a single distinct structural unit demarcated by bona-fide boundaries from other distinct structural units of different types.
compound organ component	Multi-tissue structure that is part of a compound organ.
simple organ	Multi-tissue structure that is not part of a compound organ.

CARO Definitions

organism subdivision	Anatomical structure which is a primary subdivision of whole organism. The mereological sum of these is the whole organism.
portion of tissue	Anatomical structure, that consists of similar cells and intercellular matrix, aggregated according to genetically determined spatial relationships.
epithelium	Portion of tissue, that consists of one or more layers of epithelial cells connected to each other by cell junctions and which is underlain by a basal lamina.
atypical epithelium	Epithelium that consists of epithelial cells not arranged in one ore more layers.
multilaminar epithelium	Epithelium that consists of more than one layer of epithelial cells.
unilaminar epithelium	Epithelium that consists of a single layer of epithelial cells.
simple columnar epithlium	Unilaminar epithelium that consists of a single layer of columnar cells.
simple cuboidal epithelium	Unilaminar epithelium that consists of a single layer of cuboidal cells.
simple squamous epithelium	Unilaminar epithelium that consists of a single layer of squamous cells.
portion of organism substance	Material anatomical entity in a gaseous, liquid, semisolid or solid state; produced by anatomical structures or derived from inhaled and ingested substances that have been modified by anatomical structures as they pass through the body.
portion of cell substance	Portion of organism substance located within a cell.

Index